Beverly McMillan

the human body

A VISUAL GUIDE

Published by The Reader's Digest Association Limited
London • New York • Sydney • Montreal

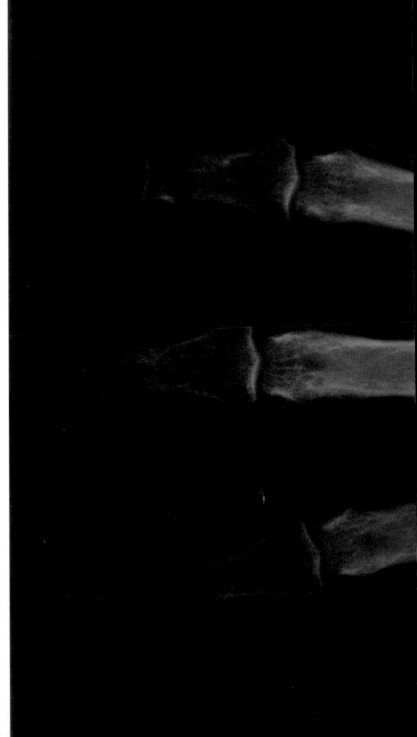

A READER'S DIGEST BOOK
Published by The Reader's Digest Association Limited
11 Westferry Circus
Canary Wharf
London E14 4HE
www.readersdigest.co.uk

We are committed to both the quality of our products and the service we
provide to our customers. We value your comments so please feel free to
call us on 08705 113366, or via our website at www.readersdigest.co.uk.
If you have any comments about the contents of any of our books, you
can contact us at gbeditorial@readersdigest.co.uk.

Conceived and produced by Weldon Owen Pty Ltd
61 Victoria Street, McMahons Point
Sydney, NSW 2060, Australia
Copyright © 2006 Weldon Owen Inc

Chief Executive Officer John Owen
President Terry Newell
Publisher Sheena Coupe
Creative Director Sue Burk
Editorial Coordinator Helen Flint
Production Director Chris Hemesath
Production Coordinator Charles Mathews
Vice President International Sales Stuart Laurence
Administrator, International Sales Kristine Ravn

Project Editor Carol Natsis
Consultant Robin Arnold
Designer Mark Thacker/Big Cat Design
Design Assistant Alex Stafford
Cover Design Stephen Smedley
Picture Researcher Joanna Collard
Illustrators Peter Bull Art Studio, Leonello Calvetti
Information Graphics Andrew Davies/Creative Communication

ISBN (13) 978-0-276-44090-8
ISBN (10)   0-276-44090-0

Color reproduction by Chroma Graphics (Overseas) Pte Ltd
Printed by SNP Leefung Printers Ltd
Printed in China

A WELDON OWEN PRODUCTION

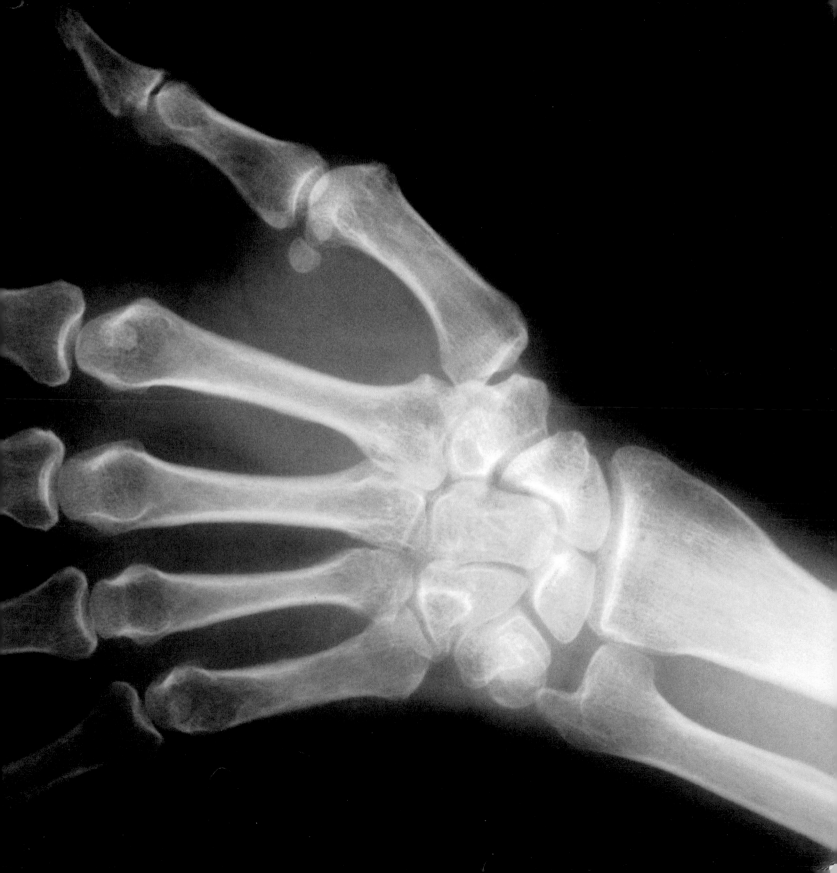

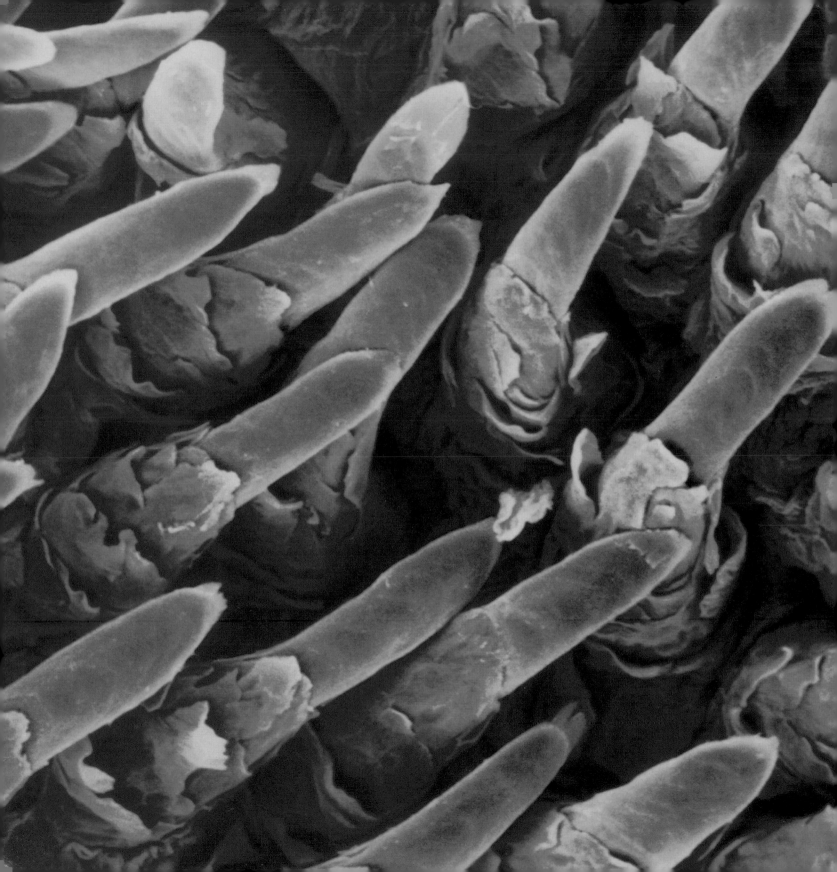

Contents

13 Introduction

14 From the outside in

100 Sensing the world

58 Control and communication

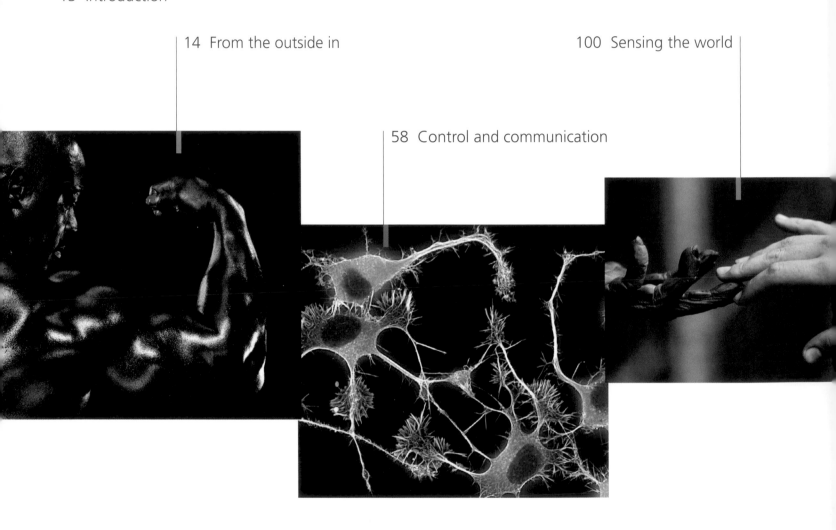

134 Breath, blood and defense

190 Nourishment and maintenance

232 Continuity

280 Factfile

292 Glossary

296 Index

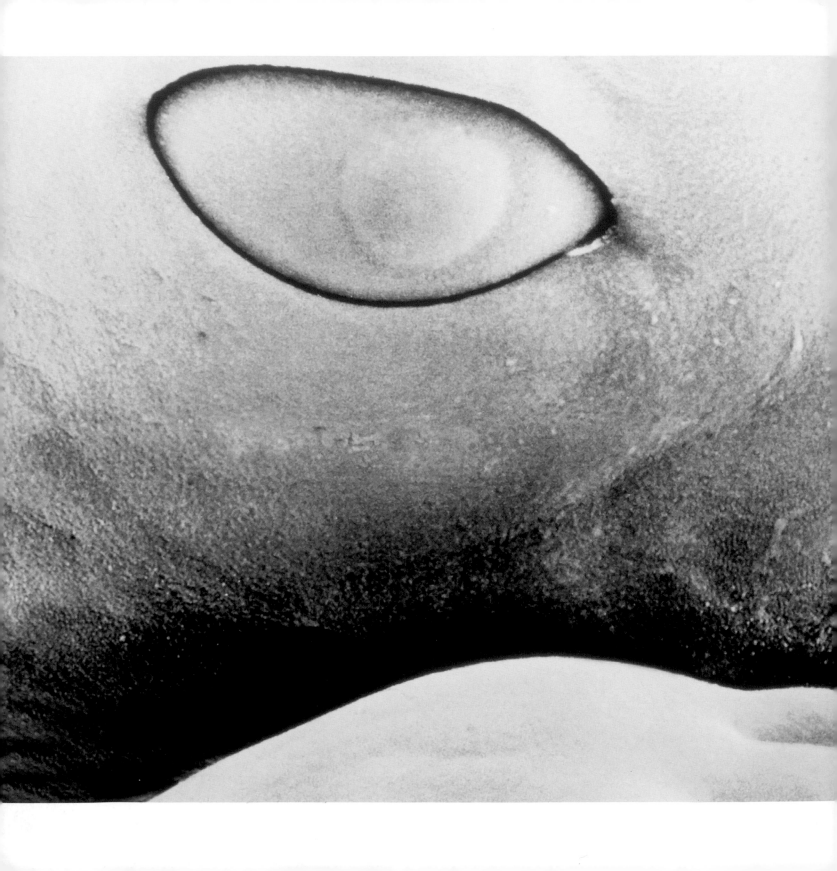

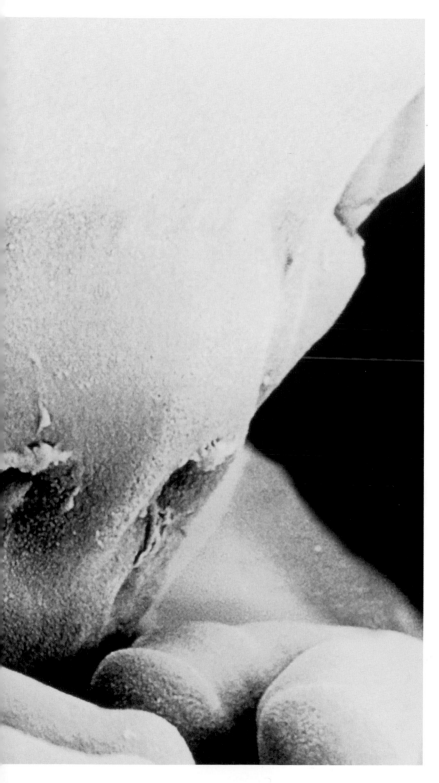

Introduction

The human body is a natural wonder. Its unique blend of bones and muscles, nerves and sense organs, skin and linings, and many other components make up a whole that far exceeds the sum of its parts. These parts are integrated and interconnected, with individual cells organized into tissues, tissues into organs, and organs into organ systems, such as the heart and blood vessels or the lungs and airways. In all, it takes the coordinated functioning of eleven organ systems simply to keep each of us alive. Much of the responsibility for managing these activities falls to the human brain, dazzling in its complexity, as it also bestows capacities such as complex reasoning, memory, emotions and self-expression through language and creativity. Exactly how the brain accomplishes these feats continues to be one of the enduring mysteries of human existence.

Another marvel is deoxyribonucleic acid (DNA), which provides in chemical code the instructions needed to build and operate each individual. The stuff of our genes, DNA is the foundation both for day-to-day functioning of tissues and organs and for the inner and outer features that make each person biologically unique.

The ebb and flow of life-sustaining processes—breathing, the heart's rhythmic pumping, digesting food and eliminating wastes, fighting off microbes—take place in a state of dynamic balance called homeostasis. Every sniffle, ache, or more serious illness or disorder reflects a short- or long-term shift away from some aspect of this balance.

Balance, astonishing complexity, intricately choreographed activities of its parts—this is the remarkable universe of the human body.

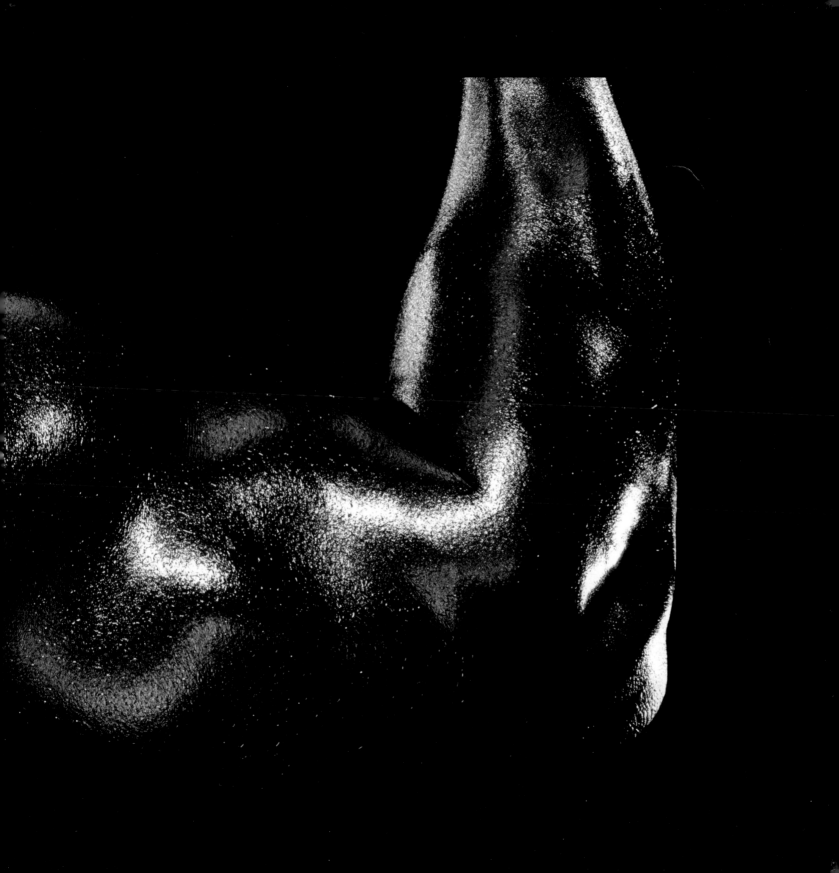

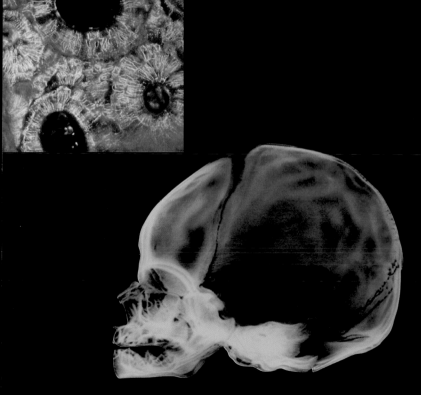

From the outside in

Humans consist of trillions of cells that are organized into tissues, which form organs and organ systems. Three components, the skin, skeleton and skeletal muscles, serve as a protective covering for the body, provide its bony framework and generate movement of the body and its parts.

Building blocks: cells and tissues	18
The skin we're in	20
Below the skin's surface	22
Derivatives of the skin	24
The skin under siege	26
Skin as self	28
The three kinds of muscle	30
Skeletal muscles: 600 strong	32
Tendons and ligaments	34
How muscles move bones	36
Whole muscles at work	38
Strength, tone and fatigue	40
Making the most of muscles	42
The nature of bone	44
The dynamic life of a bone	46
The skeleton	48
Limbs: the "hanging" skeleton	50
Linking bone to bone	52
The skeleton under siege	54
Body cavities and linings	56

Building blocks: cells and tissues

The cell is the smallest unit of life. Some life forms, such as bacteria, consist of only a single cell. By contrast, humans each are built of an estimated 60 trillion cells, all engaged in the biochemical commerce that keeps us alive. Those cells come in an array of shapes—long, wispy nerve cells, red blood cells shaped like flattened doughnuts and plump, round fat cells. In size human cells range from minuscule—1/12,000 of an inch (2 micrometers)—to nerve cells that stretch from the base of the spine to the tips of the toes. Each of these cells has the defining features of life—among others, operating instructions carried in the DNA of genes, the ability to harness and use energy, and the capacity to reproduce. The more than 200 different types of human cells are organized into four basic kinds of tissues: epithelial tissue such as skin, various kinds of muscle tissue, connective tissues such as cartilage and bone, and nervous tissue composed of cells that detect, process and coordinate information. Tissues in turn make up organs, and organs make up organ systems—the cardiovascular system, nervous system, muscular system and so forth. The sum of these parts is the marvelous human body.

→ **The single-cell *Pseudomonas* bacterium** and its kin may cause serious food poisoning and lung infections. Their long flagella propel the tiny microbes like the churning propellers of a torpedo. Bacteria reproduce simply by dividing in two, a process that is under way in the lower cell in this image.

STRUCTURE OF A HUMAN CELL

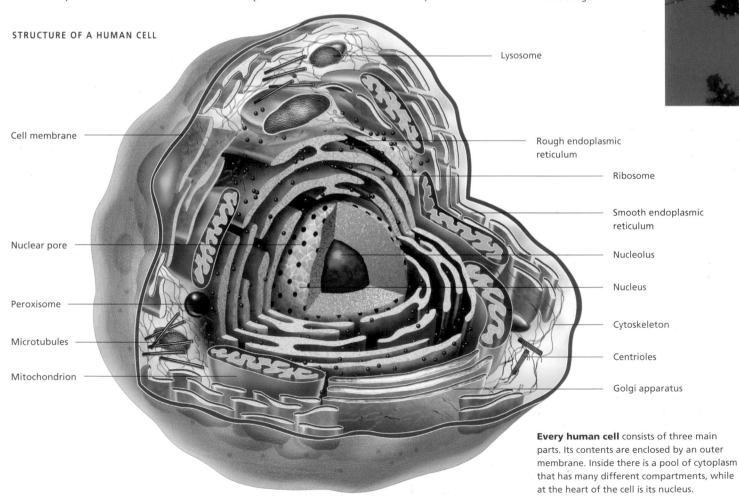

Cell membrane

Nuclear pore

Peroxisome

Microtubules

Mitochondrion

Lysosome

Rough endoplasmic reticulum

Ribosome

Smooth endoplasmic reticulum

Nucleolus

Nucleus

Cytoskeleton

Centrioles

Golgi apparatus

Every human cell consists of three main parts. Its contents are enclosed by an outer membrane. Inside there is a pool of cytoplasm that has many different compartments, while at the heart of the cell is its nucleus.

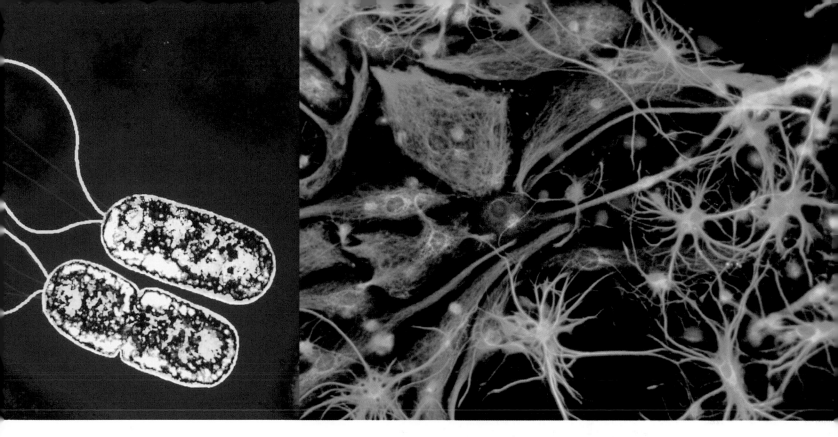

HOW CELLS FUNCTION

Cells are dynamic, always interacting with their surroundings and making adjustments as conditions vary. The outer membrane helps control what can enter and leave the cell. The jelly-like inner cytoplasm contains organelles, compartments that are specialized for different operations. The nucleus is the organelle that houses the cell's genes   DNA instructions for building cell parts and for the myriad tasks those parts must perform. The Golgi apparatus and endoplasmic reticulum, an interconnected labyrinth of sacs and membranes, are a cell's assembly line and packaging center for proteins and other molecules. Lysosomes dismantle materials to smaller components that can be reused or eliminated, and peroxisomes detoxify harmful substances. Mitochondria reconfigure nutrients from food into ATP, a chemical the cell uses as fuel.

The cell's outer membrane and its internal scaffolding, called the cytoskeleton, play key roles. Proteins embedded in the membrane receive incoming chemical signals or serve as identity tags; others provide passageways for substances moving into and out of cells, and still others weld together neighboring cells into tissues. The cytoskeleton is a blend of protein fibers that provides physical support to different cell parts, helps move them from place to place or plays a role in cell division.

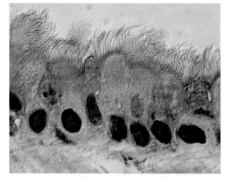

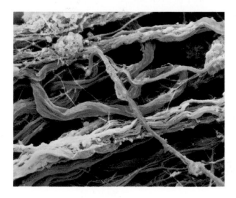

↑ **Star-shaped astrocytes** (bottom right) occur only in the brain, and make up more than half the volume of the nervous system. They are one of several types of neuroglia, or support cells, which brace and help to nourish nerve cells. Other types defend against microbes, clean up debris or perform other functions.

← **Ciliated cells** are the hallmark of the epithelium that lines the bronchi, the airways leading to the lungs. Nearby glands in the tissue produce mucus that traps inhaled dirt, debris and microbes. As the hair-like cilia move, they sweep the mucus up and out, so it can be expelled from the body.

← **Connective tissue** physically supports, anchors and binds other body parts together. A key ingredient is the sturdy protein collagen (colored orange), which in tendons helps prevent tearing and in cartilage adds strength. Elastin (blue), another connective tissue protein, has more give, making it suitable for tissues that must stretch, such as in the lungs.

The skin we're in

Every animal has a body covering that stands between its internal parts and the outside world. For some, like a clam or a beetle, this barrier is a hard shell that protects the soft tissues within and does double duty as a skeleton. Large animals that have a hard internal skeleton—like humans—have a different sort of external protection, which is known as an integument, after the Latin *integere*, meaning to cover. In humans, the integument that protects us from the environment around us is our skin, and a remarkable cover it is.

Skin is the body's largest organ, covering roughly 27 square feet (2.5 square meters) and weighing in at about 9 pounds (4 kg) in an adult. It also does much more than simply keep our insides under wraps. Human skin is a lifelong bastion of defense against all kinds of minor wounds and abrasions, against bacteria patrolling to find a portal into the body, against exposure to harmful chemicals, solar radiation and the drying effects of air. It is washable, mostly waterproof and airtight. Skin can be scented and decorated, serving as a palette for each one of us to express our individuality. It can also convey the caring touch of a lover, a friend or a child. Perhaps most amazing of all, these services are provided by an organ that in most places is no thicker than the page you are reading.

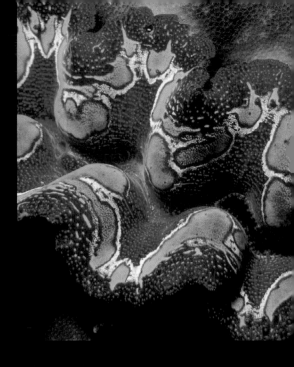

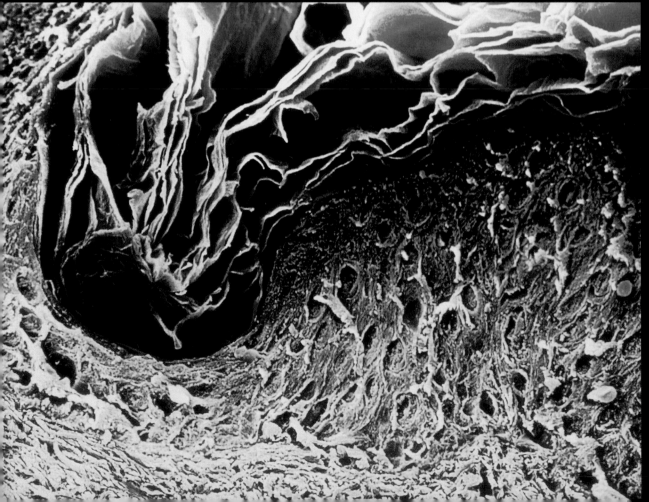

↑ **Nature provides body protection** suitable to a species' lifestyle. The colorful mantle of the **giant clam** (above) sits within its massive hinged shell. Cells filled with sacs of pigment allow the **banded octopus** (right) to change its color when it is stalking prey or confronted by a predator, while the sculpted scales of a **shark** (far right) reduce both friction and noise as the shark swims.

← **The top layer of human skin** is the epidermis (colored brown), a thin but tough tissue covered with flaky-looking dead cells, which protect the living cells underneath. Below is the more flexible dermis (pink), which contains glands, hair follicles, blood vessels and nerves.

A PROTECTIVE BARRIER

Body coverings are designed for two basic functions: to separate internal tissues from the outside environment, and to protect them from the often harsh conditions there. In the animal world, each species' way of life determines which environmental threats it faces. Nearly all mammals, the animal group that includes humans, are active land dwellers that maintain a warm body temperature. They need a body covering that is relatively flexible and lightweight, that has mechanisms for managing gains and losses of body heat and water, and that helps protect against damage from abrasion, harsh chemicals, punctures and the like. To address these needs, mammals have a thick but pliable skin, and in most species it is fortified with fur or hair—additions derived from the skin itself and that blend protection with lightweight insulation. Human skin also includes structures, such as two million sweat glands, that help cool the body when it is too hot. When cold is the problem, blood vessels in the skin can constrict so that less heat is lost. Human ancestors were much hairier than we are today; we compensate by wearing clothing, heating our homes and sheltering ourselves from the elements in other ways.

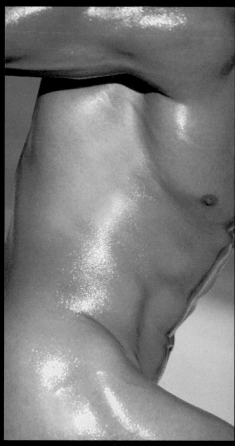

→ **The outer surface of skin** bars many microbes and other substances from easy entry to the body. Yet even the most pristine-looking skin still harbors an invisible universe of microorganisms, including bacteria, fungi and mites—tiny relatives of spiders that often live in hair follicles.

↑ **Epidermis grown in the laboratory** can be used to treat burns and other skin damage. When possible it is cultured from a patient's own healthy skin, avoiding the chance that it will be rejected when grafted at the damaged site.

Below the skin's surface

Strength, pliability, durability—look below the surface of skin and you begin to grasp why the human body's covering is so versatile. Its layered regions, the epidermis and dermis, divide the labor of skin's functions. The scale-like layer of dead, keratin-rich cells at the surface of the epidermis helps counter the abuse of constant abrasion and exposure to the elements. The living cells beneath are tightly bound together into a continuous sheet, a second line of defense against punctures and microbes. But the sturdiest skin layer is the dermis below. Built of collagen and elastin, dermis resists tears and remains resilient for decades. At the same time, it supports and nourishes the epidermis. According to some estimates every square inch (6.4 sq. cm) may be laced with as much as 230 feet (70 m) of nerves and 16 feet (about 5 m) of thread-like blood vessels. Under the dermis, a non-skin layer called the hypodermis anchors the skin to muscles and bones.

HOW SKIN REGENERATES

The skin's surface is constantly renewed. Like an epidermal escalator, its cells arise deep in the epidermis and are pushed toward the surface as new ones are produced beneath them. Most cells of the epidermis are keratinocytes, which make the water-resistant protein keratin. They die during their upward journey, and by the time they reach the surface, all that remains are flattened bags of keratin that blanket the skin with a protective, waterproof coating. As the dead cells flake or rub off, new ones take their place. Over the course of 4 to 6 weeks, almost the entire epidermis is replaced. Deep in the epidermis other cells, called melanocytes, produce the pigment melanin. Melanin can range from yellow-orange to red-brown to black and gives skin its color; when a pocket of melanin accumulates in a limited area, a freckle or mole is the result.

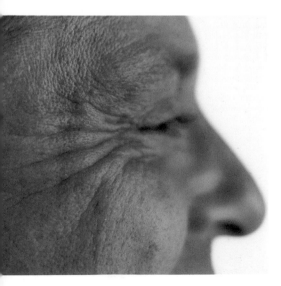

← **As we age** some subcutaneous fat and collagen is lost. The remainding collagen stiffens and elastin fibers degenerate. The skin gradually sags and wrinkles develop, especially in areas that have been frequently creased by smiling or squinting.

→ **The color and the quantity of melanin** in skin are determined by genes and exposure to sunlight. Differences in skin color are a hallmark of human diversity.

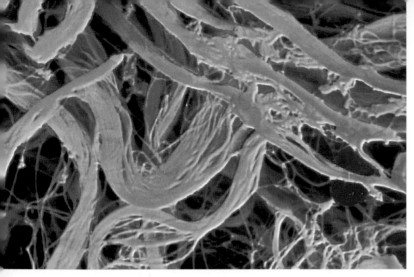

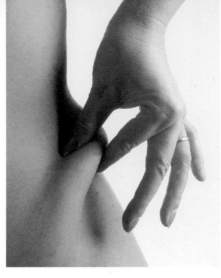

← **Most of our body fat** is contained in the hypodermis, the tissue layer under the dermis.

⟵ **Collagen fibers** in the dermis render it tough and durable.

The architecture of the skin is highly complex. In the dermis, blood vessels and nerves weave around hair follicles. Hairs, the channels of oil glands, and coiled sweat glands thread upward to the surface. Visible skin pores are the openings of hair follicles. Below the surface, muscle cells attach to follicles. When these muscle cells contract in response to fear or cold, hair follicles are pulled upright and a person gets "goosebumps."

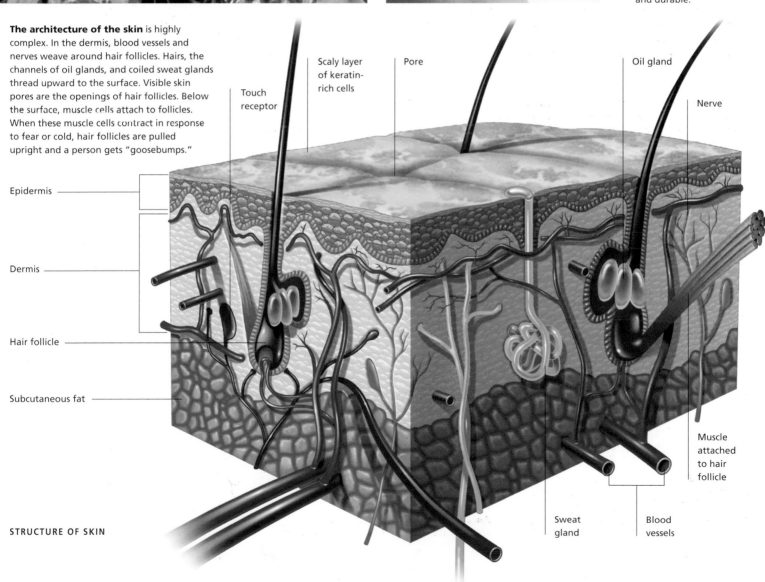

Touch receptor

Scaly layer of keratin-rich cells

Pore

Oil gland

Nerve

Epidermis

Dermis

Hair follicle

Subcutaneous fat

Muscle attached to hair follicle

Sweat gland

Blood vessels

STRUCTURE OF SKIN

Derivatives of the skin

Skin not only covers our bodies, it also gives rise to structures that bolster its protection, add insulation or extend other benefits, including giving us the means to scratch an itch. Hair, fingernails and toenails all develop from the skin's outer layer, the epidermis, and each is built from keratin. Dense and fibrous, this exceedingly tough protein resists tears and breaks—a property that makes it an ideal building block for body parts that are subject to friction and other mechanical stresses. Hair, which is mostly keratin, is weight for weight the strongest material in the body. Only the lips, the palms of the hands, the soles of the feet and parts of the genitals are completely hair-free. Nails protect the delicate tissues at the tips of our fingers and toes. Each nail grows out from its root, which is embedded in the skin below it. The part of the nail that protrudes is actually dead, making it able to withstand everyday wear and tear.

KERATIN IN THE ANIMAL KINGDOM
Keratin's tensile strength and durability are utilized by many species of animals in different forms—in fur and feathers, in spines and whiskers, in claws and hooves, and in shells, scales and horns. Like hair and fur, feathers can be both strong and lightweight because they are made of keratin. Many of the colors of animal fur and feathers are derived from melanin or other pigments, but in some birds, such as peacocks, it is the structure of the feathers that scatters light in a way that produces iridescent blues and greens.

→ **Fingernails and toenails** are human versions of animal claws and hooves. Healthy nails typically have a pinkish-tan color, which is due to the presence of blood vessels in the dermis that lies underneath the nail bed.

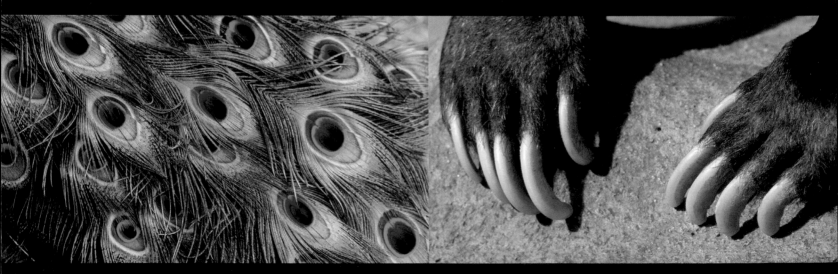

A HEAD OF HAIR

The average human head is endowed with about 100,000 hairs, each growing out of a follicle embedded in the scalp. A follicle is active for 3 to 5 years, and then rests for a few months before a new hair begins to develop. As a result, roughly 100 hairs are shed each day. Hormones, nutrition, aging and other factors affect hair growth. Testosterone promotes the growth of body hair, while a poor diet slows it. As we age the slower activity of hair follicles may cause thinning.

↑ **The shape of hair follicles** determines what kind of hair we have. Rounded follicles produce straight hair (left), oblong ones result in wavy hair (right) and curving ones are responsible for curly hair (center).

→ **A hair is a flexible, multi-layered strand** of cells packed with keratin. The outer layer consists of flattened, overlapping cells that form a durable cuticle. Unlike the keratinized cells at the skin's surface, cells in a hair's cuticle contain a particularly dense form of keratin, and so do not flake off. The strand's shingle-like surface helps keep adjacent hairs from clumping together.

↓ **Strong but lightweight, keratin** gives the peacock its fine feathers and the Malayan sun bear its claws (opposite), while the shell fused to the skeleton of a green sea turtle (below) and the horns of a mountain sheep (below right) both have an outer, horny layer of keratin blanketing a core of bone.

The skin under siege

Hundreds of ailments can afflict the skin, yet it has remarkable recuperative powers, including the capacity to heal a lifetime of nicks, cuts, scrapes and minor burns. A wound that penetrates the epidermis triggers a sequence of responses that stop the flow of blood, make initial repairs and ultimately restore the damaged area to health. Without these resources, the body would be open to the millions of bacteria and other microbes on its surface. Other threats include ultraviolet radiation, which can cause mutations in DNA, and the toxins in tobacco smoke. The pigment melanin helps shield skin cells; when a person tans the skin becomes browner as melanocytes boost their melanin production. In the longer term, however, tanning breaks down the elastin that makes skin resilient, frequently leading to deep facial wrinkles. Ultraviolet radiation also suppresses the immune system, and a day in the sun can cause chronic viral infections, such as herpes simplex, to flare up. Over time, excessive sun exposure can also activate proto-oncogenes—genes that trigger the first step toward cancer.

↓ **Exposing the skin to sunlight** makes it vulnerable to a variety of harmful effects over time. In fair-skinned people, the first line of biological defense against solar radiation is tanning. Too much time in the sun, however, can lead to sunburn, which kills living cells in the top skin layer, the epidermis. In a day or two the dead cells slough away or peel off in sheets. Frequent or severe episodes of sunburn during childhood may make a person more susceptible to developing skin cancer later in life.

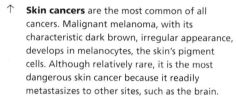

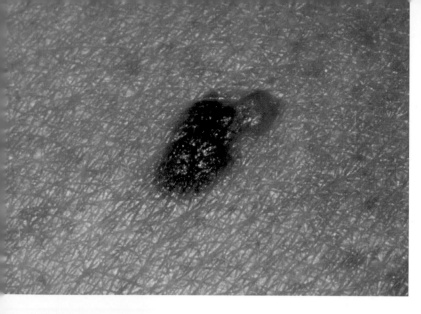

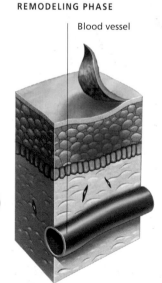

← **Extreme wrinkling,** known as "smoker's face," results from prolonged, heavy tobacco smoking, which breaks down the collagen and elastin in the skin.

↑ **Skin cancers** are the most common of all cancers. Malignant melanoma, with its characteristic dark brown, irregular appearance, develops in melanocytes, the skin's pigment cells. Although relatively rare, it is the most dangerous skin cancer because it readily metastasizes to other sites, such as the brain.

↓ **A minor skin wound gradually heals** over a few days in four main stages.

THE HEALING PROCESS

Minor skin wounds are repaired in a sequence of steps. At the outset, a blood clot forms, sealing off torn or severed small blood vessels and binding together the edges of the break. The surface of the clot dries into a scab. Over the next few days, delicate new blood vessels begin to form in the area under the scab. Meanwhile scavenger cells called macrophages mobilize to clean up clotted blood and fragments of cells that the injury destroyed. Other cells—fibroblasts—begin making collagen, which will be incorporated into the healing wound, including any scar tissue that forms. As the skin regenerates under the scab, the scab gradually detaches and eventually falls off. A scar seals a wound closed, and virtually all wounds result in some degree of scarring once they have healed. Scars have little or no blood supply, and because they are almost entirely composed of collagen they are not as flexible as the skin that surrounds them.

BLEEDING PHASE	INFLAMMATORY PHASE	PROLIFERATION PHASE	REMODELING PHASE
Bleeding	Fibroblasts	Macrophage	Blood vessel
1. The skin is broken.	2. A clot forms.	3. New tissue develops.	4. The scab peels off.

Skin as self

Skin embodies much of the external, physical self we expose to the world, and altering it is a ritual with ancient underpinnings. Since time immemorial humans have gone to great lengths to enhance their skin's appearance, change its color, and otherwise use it as a palette for personal, social and cultural expression. From the South Pacific and Asia to Africa and Alaska, ornate, individualized tattooing has been a traditional means of signifying social status, marking the transition to adulthood, asserting authority or gaining magical protection in battle. Corpses of ancient Europeans, frozen in mountain ice or preserved in peat bogs, still bear tattooed markings received thousands of years ago. Less permanent but equally significant to the wearer, in many societies body paint has been employed to beautify, to identify status or beliefs, or to convey profound social or religious meaning. Changing the size and shape of body parts, especially changing facial features, is a time-honored variation on this enduring desire to shape the impression we make on those around us.

→ **Facial decoration** adorns a girl from Papua New Guinea. In her culture, body paint is used on ceremonial occasions to identify a person's position within a particular social group. Throughout the South Pacific both body painting and tattooing have been practiced for millennia.

↓ **The hands of an Indian bride** display the reddish-orange henna designs that are used to beautify women's skin in India and the Middle East. Obtained from the ground leaves of the shrub *Lawsonia inermis*, henna has been used to dye skin and hair for at least 5000 years.

A PALETTE FOR SELF-EXPRESSION

Painting the skin, or piercing it and injecting pigment, are just two approaches to altering its appearance. A technique common in parts of Africa is scarification—making deep cuts that produce scar tissue in a raised design. Dye may sometimes be used to enhance the effect. Around the globe, each day many of us apply the body paint we know as cosmetics, supporting a multi-billion dollar industry built on the desire to cover defects, minimize flaws and otherwise make our faces look more appealing. Modern make-up reflects a venerable tradition, however. Among others, ancient Egyptians used the black-colored kohl to darken the outlines of their eyes and dyed their skin with henna. Like clothing, modern make-up styles and choices often reflect fashion trends and the acumen of marketers—and high-tech solutions to a fast-paced lifestyle as well. So-called "permanent make-up," including eyeliner and lip color, can be injected into the skin in a process called micro-pigment implantation. The color usually lasts several years.

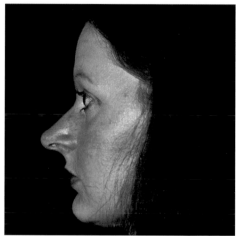

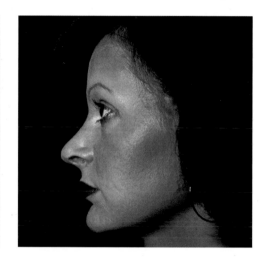

↑ **Tattooing large areas of the body** can take years to complete. Rock star Marilyn Manson wears "full sleeves," tattooing that covers virtually all the skin of his arms. In his stage performances Manson frequently wears striking face paint as well.

→ **The facial tattoos** of this Maori chief, photographed in New Zealand in 1923, would have been created gradually, beginning when he was a youth. Using a technique called *moko*, the tattoo artist would first carve deep cuts into the skin, then apply a natural ink. Undergoing this lengthy, painful process conferred status and prestige, as well as earning the respect of opponents in battle.

↑ **Cosmetic surgery** exploits the malleability of human skin and the tissue beneath it. These before-and-after photographs clearly show how rhinoplasty has produced a shorter, more rounded and less prominent nose.

The three kinds of muscle

Although skin is the body's largest single organ, it is our muscles that form much of the rest of us. An adult man is roughly 40 percent muscle by weight, an average woman just over 30 percent.

The muscles of the body are divided into three distinct categories, each of which performs a particular function. Cardiac muscle is found only in the heart, while smooth muscle lines blood vessels and soft, hollow organs such as the stomach, bladder and intestines. The third, and largest, group of muscles is the aptly named skeletal muscle tissue, which forms the muscles that are attached to bones and move our joints and body parts.

Each muscle type is built of cylindrical cells, and each has another property that no other body tissue can claim: muscle cells can be stimulated to contract—that is, to shorten. When the signal arrives, intricately organized internal parts in a muscle cell can move closer together, pulled by protein "motors." In so doing a muscle cell generates force, which—multiplied millions of times over—moves a body part, such as a leg or an arm, or moves a substance, such as blood pumped by the heart. When a contraction is complete each cell, and ultimately the whole muscle, relaxes. Likewise, the contraction of cardiac muscle cells maintains the normal rhythm of the heart's life-sustaining beat.

The coordination that these operations demand is another feature of muscle action. All three of the body's muscle types contain cells that contract and relax in a rhythm that allows the muscle to perform its assigned task reliably. When a woman gives birth, for example, it is the coordinated contraction of smooth muscle cells in her uterus that pushes her baby out into the world.

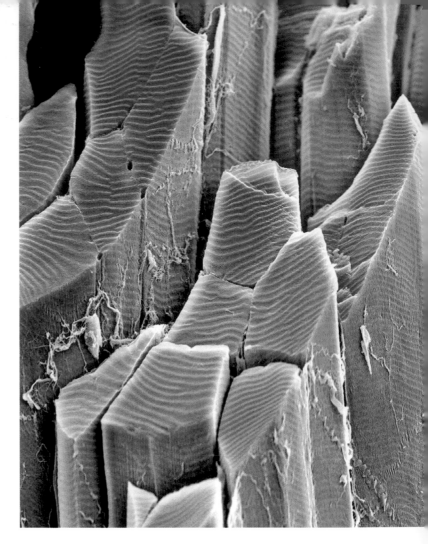

HOW SKELETAL AND SMOOTH MUSCLES DIFFER

Skeletal muscle cells, or "fibers," resemble noodles with tapered ends—they are broad and as long as the muscle they make up. Each fiber is connected separately to the nervous system, so each can contract on its own. Smooth muscle cells are much shorter and slimmer, and they are arranged in sheets in which all the cells contract as one. Whereas skeletal muscle fibers are designed to contract rapidly with great force, smooth muscle cells contract more slowly but for longer periods. A skeletal muscle cell can contract for only a few thousandths of a second before chemical changes cause it to fatigue, but a smooth muscle cell can contract steadily for up to 3 seconds—endurance that assists organs such as a uterus about to deliver a baby, or intestines moving along food to be digested. These and other organs, such as the bladder, can perform their functions because the smooth muscle in their walls is arrayed in layers that are positioned at right angles, with one layer running vertically and the other horizontally. Contractions of the layers alternate, providing the squeezing force that moves material inside the organ to its destination. While we can voluntarily move our skeletal muscles, we have no conscious control over the workings of our smooth muscle.

SKELETAL MUSCLE FIBERS

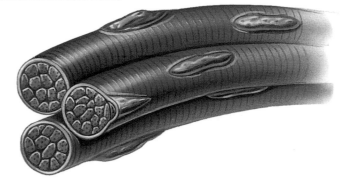

The components of skeletal muscle fibers (top) are arranged in repeating units that appear as alternating dark and light bands. Colored blue-gray by the preparation technique used, the cells are in reality bright red. Skeletal muscle fibers (bottom) are slightly thinner than hairs and up to 12 inches (30 cm) long.

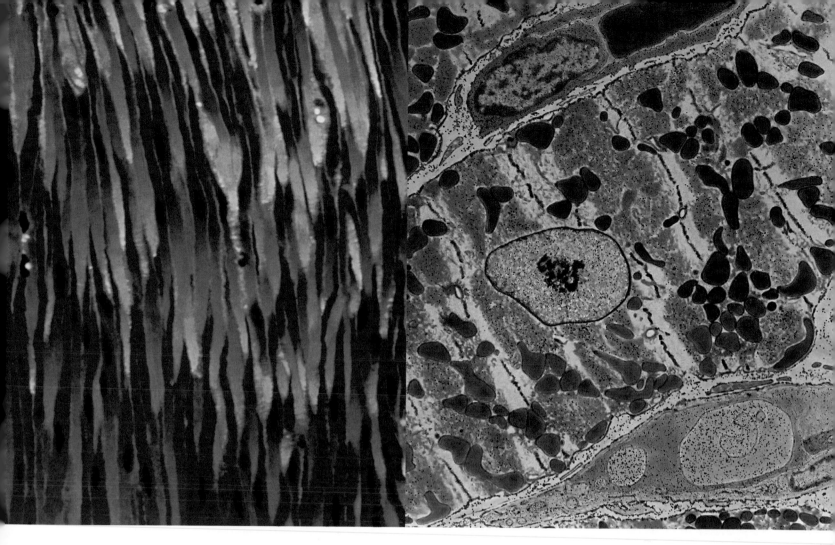

SMOOTH MUSCLE FIBERS

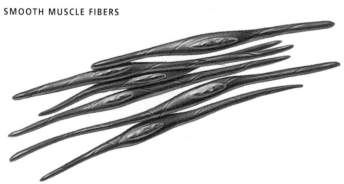

CARDIAC MUSCLE FIBERS

Sheets of smooth muscle cells (top) run vertically in this microscope image. The dark spots are the cell nuclei. Smooth muscle consists of spindle-shaped cells grouped in irregular bundles (bottom). Each cell contains large numbers of mitochondria, organelles that produce a cell's chemical fuel.

A cardiac muscle cell packed with mitochondria (red blobs) runs diagonally across the top image. In the middle is the cell's nucleus (colored blue), which contains a dark, coiled nucleolus. Cardiac muscle fibers (bottom) branch and rejoin in a continuous network that transmits and regulates muscle contraction signals.

Skeletal muscles: 600 strong

Fully one half of the body's tissue is muscle, and most of that is skeletal muscle—the tissue that works with bones of the skeleton to move the body and its parts. This tissue makes up most of what we think of as our flesh, and it is packaged into more than 600 individual muscles, each an organ unto itself.

Unlike the heart's cardiac muscle or the smooth muscles of internal organs, skeletal muscles are under voluntary control. They apply the force to propel us up steps or around a dance floor, to play the piano, lift a heavy bag of groceries or pet the dog. Every eye-blink, nod of the head and wriggle of your toes calls two, three, four or more skeletal muscles into action. In each of those movements, the working muscles never push—they pull in the desired direction as they contract. Many skeletal muscles are arranged in pairs or groups. Some pairs work together, or synergistically, to accomplish a movement. Others work antagonistically, with one opposing the action of the other.

In addition to repositioning body parts, skeletal muscles help stabilize the body's movable joints, from hips and knees to wrists and shoulders. They also generate a significant share of body heat.

↓ **The biceps and triceps muscles** in the upper arm are antagonists; when you raise your forearm, the biceps contracts and the triceps simultaneously relaxes. The two muscles reverse their action when you lower your forearm.

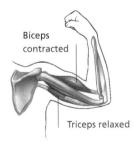

Biceps contracted

Triceps relaxed

Biceps relaxed

Triceps contracted

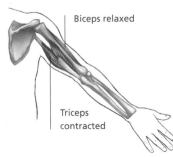

↑ **Ballet dancers** display the human capacity for fine-tuned, voluntary control over skeletal muscles. Ultimately this control relies on the ability of the body's skeletal muscle to respond to exquisitely integrated signals that are sent out from the nervous system.

MAJOR SKELETAL MUSCLES

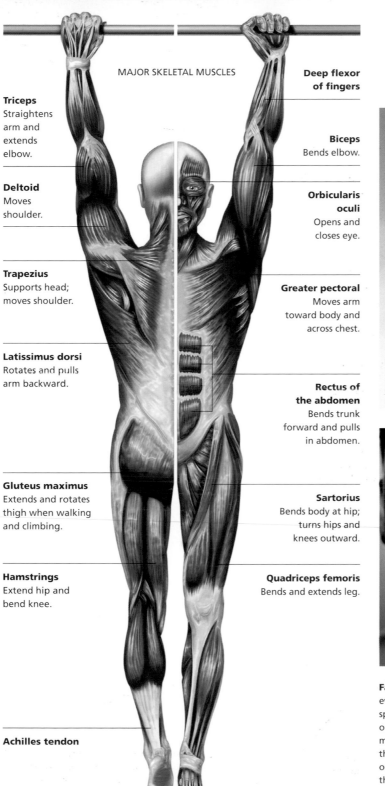

Triceps
Straightens arm and extends elbow.

Deltoid
Moves shoulder.

Trapezius
Supports head; moves shoulder.

Latissimus dorsi
Rotates and pulls arm backward.

Gluteus maximus
Extends and rotates thigh when walking and climbing.

Hamstrings
Extend hip and bend knee.

Achilles tendon

Deep flexor of fingers

Biceps
Bends elbow.

Orbicularis oculi
Opens and closes eye.

Greater pectoral
Moves arm toward body and across chest.

Rectus of the abdomen
Bends trunk forward and pulls in abdomen.

Sartorius
Bends body at hip; turns hips and knees outward.

Quadriceps femoris
Bends and extends leg.

MUSCLE POWER
Some skeletal muscles occur close to the surface while others lie deep in the body. A few attach to skin, but most connect to bones. Regardless of their location, skeletal muscles are arranged for efficient operation, like a system of pulleys that move levers. When a muscle contracts it shortens, and even major movements require the muscles involved to contract only a little. During a movement, one end of a muscle, called its origin, attaches to a bone that remains relatively stationary. The other end, the insertion, is attached to a different bone, which moves more.

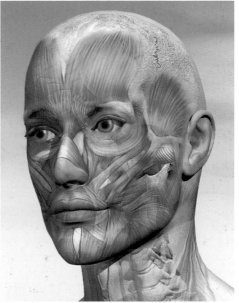

Facial muscles help us express rage, joy and every emotion in between. They operate when we speak, laugh, cry, chew, flare our nostrils or raise our eyebrows. Together with the muscles that move the scalp and neck, facial muscles are among the few skeletal muscles that are attached to skin or other muscles, rather than to bone. As a result the very slightest contraction of a muscle can cause the facial skin to move.

Tendons and ligaments

Tendons and ligaments are nature's solution to the problem of how to hold our muscles and skeleton together, and structurally each is a compromise between strength and flexibility. Tendons join muscles to bones. Built of connective tissue densely laced with collagen fibers, they are strong but relatively stiff, like thick rubber. Some tendons are also quite long. For instance, several of the muscles that operate the fingers are in the forearm and attach to finger bones via long, slender tendons that run down through the hand. If the muscle attachments were in the fingers, the added bulk would drastically reduce our manual dexterity. Ligaments connect bones to one another—for example, uniting the thighbone and the bones of the lower leg at the knee—and they add stability to joints. As well as containing cable-like strands of collagen, ligaments include a greater proportion of elastin fibers than tendons do, making them much more stretchable and vulnerable to tears. While both tendons and ligaments are remarkably strong and usually provide a lifetime's worth of uneventful service, their ability to function can be painfully disrupted by overuse or accidents that strain the tissue beyond what it can bear.

→ **Stretching before exercise** gently flexes tendons and ligaments, reducing the likelihood that they will be injured. This competitive runner is stretching his Achilles tendon, which attaches the calf muscle to the heel of the foot. Sprinters and distance runners alike demand forceful contractions of their calf muscles, which increases the risk that a tendon will rupture.

SPRAINS AND RUPTURES

A sprain is an injury that overstretches or tears a ligament. Partial tears may slowly heal on their own. A completely ruptured ligament is a medical emergency, however, because the body's natural defenses against tissue damage will quickly destroy the severed ends if they are not speedily sutured back together. A severely damaged ligament can be replaced with artificial materials, a piece of tendon from elsewhere in the patient's body, or even one that has been harvested from a cadaver.

← **In running sports,** such as soccer and baseball, knees are the body's most vulnerable joints, as they are not well reinforced against lateral hits and twisting movements.

→ **The knee** has broad tendons that attach the muscles of the thigh and lower leg to the kneecap or to other parts of the joint, and multiple ligaments that connect the various bones.

REDUCING JOINT FRICTION

Muscle fibers, the cells in skeletal muscles, are bundled together by connective tissue that extends beyond them to form the whitish tendons that attach the muscle to bone. In movable joints such as the knees, elbows and shoulders where tendons are subjected to frequent friction, they are enclosed in a sheath equipped with a thin film of lubricating fluid. Likewise, wherever the regular movement of an arm, leg, or other body part can cause a ligament or muscle to rub against bone, it will be separated from the bone by a thin, flattened sac of fluid called a bursa. Tendonitis or bursitis develops when a tendon sheath or a bursa becomes inflamed due to overuse, a blow, prolonged pressure or a bacterial infection.

→ **Typing on a keyboard** for extended periods is a common cause of carpal tunnel syndrome. The tendons under bones of the wrist swell from overuse and press on the median nerve, triggering symptoms from tingling and numbness to debilitating pain.

↘ **To repair injured ligaments and tendons,** the surgeon uses a miniature, lighted device called an arthroscope, which allows a clear view of the damaged area while the repair is made.

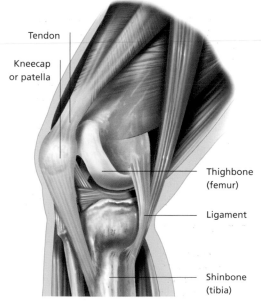

Tendon

Kneecap or patella

Thighbone (femur)

Ligament

Shinbone (tibia)

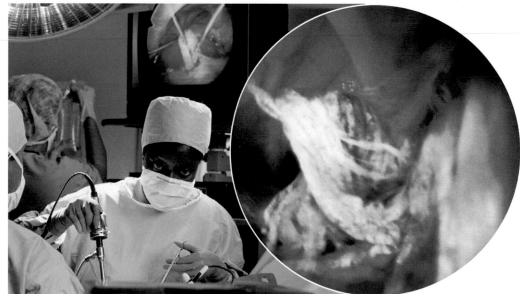

How muscles move bones

Each day most of us perform hundreds of tasks that require the body or some part of it to change position. These position changes are possible because skeletal muscles can contract, or shorten, and pull on bones of the skeleton. Looking deeper, each muscle fiber contains even smaller myofibrils—slender rod-like structures that are lined up in the fiber, like wires inside a cable. Arranged end to end in each myofibril are subunits called sarcomeres—literally, "muscle segments"—which are the microscopic engines of contraction. In a sarcomere, two kinds of protein filaments, called actin and myosin, interact in a way that pulls the ends of the sarcomere closer together, like internal strings drawing in the ends of an accordion. When a contraction stops, the filaments return to their starting positions and the sarcomere lengthens. This sequence of shortening and lengthening goes on in every muscle fiber every time a skeletal muscle contracts and relaxes.

THE ROLE OF CALCIUM

Calcium is key to muscle contraction. Weaving around the myofibrils in a skeletal muscle fiber is a tube-like lacework where calcium is stored. An incoming nerve impulse stimulates contraction when it triggers the release of calcium, setting in motion steps that allow myosin in a myofibril's sarcomeres to attach to actin filaments and pull them toward the center. When signals from the nervous system stop, the calcium is moved back into storage, myosin releases its "grip" on actin filaments and sarcomeres lengthen once again.

→ **The bulging biceps** of a weightlifter hefting a barbell provides a vivid example of muscle shortening: as thousands of its individual muscle fibers shorten, the contracting muscle can lift its load of iron.

STRUCTURE OF MUSCLE FIBERS

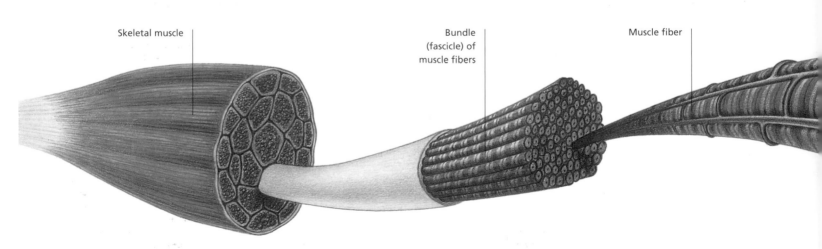

Skeletal muscle

Bundle (fascicle) of muscle fibers

Muscle fiber

Skeletal muscle fibers (colored green here) look striped, or striated, because of the alternating dark and light bands of their myofibrils. The dark bands are the aligned ends of sarcomeres, which shorten when a muscle fiber contracts.

Connective tissue holds muscle fibers together. In this magnified image connective tissue consisting mainly of collagen forms wispy strands that interweave among the long muscle fibers (shown here as orange).

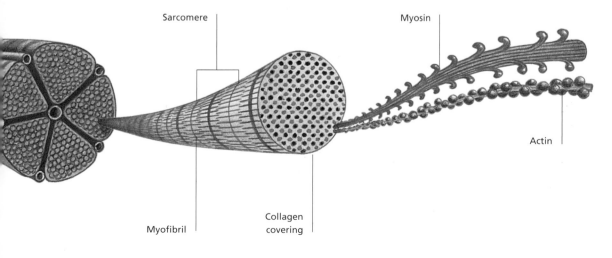

Sarcomere

Myosin

Myofibril

Collagen covering

Actin

MICROSTRUCTURE OF MUSCLE FIBER

Bundles of muscle fibers, called fascicles, make up a muscle. Depending on the muscle, each bundle can include hundreds to thousands of fibers, and each fiber in turn is assembled from myofibrils aligned like strands of uncooked spaghetti. In the myofibrils, filaments of actin and myosin organised in repeating subunits, called sarcomeres, interact when muscle cells—and ultimately whole muscles—contract in response to commands from the nervous system.

Whole muscles at work

Whether a muscle contains hundreds of thousands of fibers, as in the powerful thigh muscles, or only a few hundred, as in the muscles that move our eyes, it is meant to perform work. The mechanical force that allows a muscle to lift a load or do some other form of bodily work is called muscle tension, and it is generated as the steps of a muscle contraction unfold.

Yet because we place a wide variety of demands on our skeletal muscles, they must also be able to function in different ways under different conditions. To help muscles meet such shifting demands, the tension of contracting muscle fibers can be brought to bear in different ways—sometimes lifting a load, at other times holding a steady position, and at others even lengthening a whole muscle. Skeletal muscles (like cardiac muscle) also have varying blends of "fast," or "white," fibers and "slow," or "red," fibers. The pale-looking white fibers have fewer mitochondria, the cell's energy factories, and less of the red pigment myoglobin, which binds the cell's supply of oxygen. They are specialized for short, powerful bursts of contraction. Red fibers, however, have more myoglobin and more mitochondria, and are specialized for endurance.

Although other factors also influence the functioning of our skeletal muscles, one feature is universal: none of our skeletal muscles can do the work of moving body parts unless they can receive and respond to signals delivered by the nerve cells called motor neurons, which reach muscles by way of the spinal cord.

TWO KINDS OF CONTRACTION

Whole muscles can contract in two basic ways, isotonic and isometric. In an isotonic contraction, the force exerted by muscle cells shortens the muscle, while in an isometric contraction the muscle does not shorten overall. Isotonic contractions allow muscles to move loads, such as a book or a dance partner, but the mechanism is versatile. Isotonic contractions in the rings of skeletal muscle called sphincters provide control over the elimination of urine and feces. When you walk down stairs or climb a hill, isotonic contractions in various leg muscles allow the muscles to lengthen even while they generate the force those movements demand. Isometric contractions are common when a load requires more force than a person's muscles can generate. They also help hold joints in stationary positions. Muscles can shift between isotonic and isometric contractions; for example, isometric contractions hold a track athlete in position at the starting line, while at the start signal it is the isotonically contracting leg muscles that launch the runner into the race.

← **Botox® injections** temporarily prevent the muscle contractions that cause wrinkles. This cosmetic form of botulinum toxin paralyzes muscles by halting the release of the chemical signal for contraction. In a person who consumes botulinum-tainted food, the muscles that control breathing stop contracting, often with fatal results.

← **The number of fast or slow fibers** in certain muscles can be increased by targeted training. Sprinters like these athletes competing in a 100-yard (100-m) race follow a training regimen designed to increase the size and contractile strength of the fast muscle fibers in their legs.

→ **This guard** at London's Buckingham Palace can maintain his upright posture and stay immobile for extended periods thanks to isometric muscle contractions. When his postural muscles tire and lose their ability to continue contracting, he may collapse, a risk that is recognized by the ritual "changing of the guard."

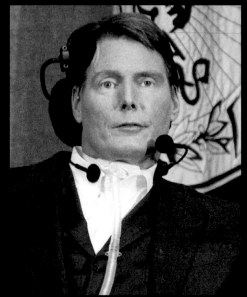

← **Severe spinal injuries,** such as the one that paralyzed actor Christopher Reeve, stop the flow of nervous system signals from the spinal cord so that the injured person loses the ability to control some or all of the skeletal muscles. Researchers today are working to develop drugs that can promote the sustainable regrowth of motor neurons.

← **Different muscles** can function in different ways at the same time. This man is walking thanks to isotonically contracting muscles in his legs and hips, while isometric contractions of muscles in his arms, shoulders and back allow him to carry a heavy stack of books.

Strength, tone and fatigue

Skeletal muscles are extremely versatile, allowing a person to undertake actions as varied as lifting a feather, washing dishes or doing a full workout at the gym. This versatility is due in part to the fact that muscle fibers are grouped into motor units, each of which is a subset of fibers controlled by a single motor nerve cell, or neuron.

When the neuron delivers its signal, all the muscle fibers in the motor unit respond in unison. Large muscles may have motor units that include as many as 2000 fibers, while in very small muscles there may be as few as 10 or 20 fibers in each unit. The strength of a muscle's contraction depends on how many of its motor units are ordered to contract by impulses from the brain to produce a particular movement, what types of cells are recruited and how much force—muscle tension—each fiber develops. Adjusting the number of motor units that are called into action in a muscle is one way the body can fine-tune the muscle power it brings to bear in the ever-changing situations of daily life.

↓ **Muscles bulk up** when certain types of exercise cause "fast" muscle fibers to increase in size. Strength training does not increase the total number of fibers, but the fibers do become larger as they develop more contractile myofibrils. Having unusually bulky muscles represents a compromise: the muscles are stronger than trimmer ones, but do not have much endurance. They also weigh more.

↑ **Intense physical exercise** such as competitive athletics dramatically increases muscle demands for oxygen, which in turn is required to generate the chemical fuel that feeds working muscles. When this fuel is used faster than muscle fibers can produce it, an "oxygen debt" builds up and must be repaid before affected muscles can efficiently contract again.

↗ **Muscles meet the nervous system** at synapses called neuromuscular junctions. Nerve impulses travel down the threadlike motor neurons. Each of the button-shaped end-plates visible in this photograph is a neuron ending that delivers the neuron's signal—a chemical neurotransmitter—to a motor unit of muscle fibers.

HOW MUSCLES BECOME OVERWORKED

An ongoing ebb and flow of neural impulses stimulates our skeletal muscles to contract in a smooth, sustained fashion called tetanus. Usually even relaxed muscles are slightly contracted, a healthy state called muscle tone. The potentially lethal disease caused by a toxin produced by the bacterium *Clostridium tetani*, and also known as tetanus, produces an excruciating form of paralysis by keeping skeletal muscles in a permanent state of unrelenting contraction. When exercise or some other activity works a muscle especially hard, its need for chemical fuel— normally some form of the blood sugar glucose—skyrockets. Gradually, the muscle's energy demands may begin to outstrip the ability of muscle fibers to fuel contractions. At the same time, lactic acid, a byproduct of muscle metabolism, may accumulate and skew the internal chemistry of fibers. The muscle contracts less and less efficiently, and ultimately may be completely unable to respond to the nervous system's signals to contract—a state called muscle fatigue. Even after recuperation fatigued muscles may ache, a side effect of lactic acid build-up.

→ **Messages from the brain and spinal cord** govern the functioning of skeletal muscles. The desire to take a particular physical action is translated into impulses that begin in the brain and are transferred down the spinal cord, then onward to motor units in the muscle or muscles that must move.

FROM IMPULSE TO ACTION

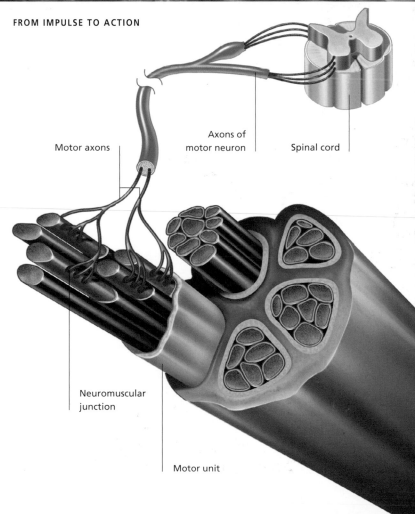

Motor axons

Axons of motor neuron

Spinal cord

Neuromuscular junction

Motor unit

Making the most of muscles

Muscles must be active to be healthy. When a person is inactive, his or her muscles soon begin to degenerate, or atrophy. By weight, 25 to 30 percent of the body may be skeletal muscle, and as much as three-quarters of that muscle mass can waste away if a person is bedridden for an extended period or suffers paralyzing nerve damage. Regular exercise, in contrast, not only increases the size of muscle fibers, it also enhances their capacity to function. For example, even moderate regular exercise increases the number of blood vessels servicing muscle fibers, creating an improved transport network for delivering nutrients and oxygen and carrying away wastes. Active muscle fibers also manufacture more myoglobin, the protein pigment that holds oxygen in the fibers for use in their metabolism. The number of energy-producing mitochondria in each fiber rises as well. The end products are muscles that function more efficiently and are slower to tire.

→ **Aerobic exercise** works skeletal muscles at a level that does not outstrip the rate at which the blood can supply muscle fibers with oxygen for their metabolism. Brisk walking, swimming and cycling also give the body an aerobic workout.

PHYSIOLOGICAL BENEFITS OF EXERCISE

Regular exercise, especially a combination of aerobic and weight-bearing activities, benefits health in the following ways:

1. Helps control body weight by burning more calories.
2. Makes skeletal muscles stronger and more efficient.
3. Increases bone density, retarding or preventing bone loss
4. Makes the heart muscle stronger, so it pumps blood more efficiently.
5. Reduces blood pressure and the level of harmful fats and cholesterol in the blood.
6. Increases energy for daily activities.
7. Leads to more efficient elimination of wastes by the digestive system.
8. Reduces the negative physiological effects of stress.
9. Helps promote restful sleep and prevent insomnia.

← **By age 40,** the natural process of aging means that skeletal muscles do not respond to exercise as dramatically as in a younger individual. Even so, regular physical activity is vital to maintaining muscle tone and a healthy weight, and load-bearing exercise, such as walking and moderate weight training, helps keep both muscles and bones strong.

EXERCISE AND REST

Exercise is a balancing act: to be effective it must tone and strengthen muscles without straining them too much. Even a modest workout results in small tears in muscle fibers, often accompanied by the build-up of lactic acid. Allowing a day of rest between workouts, or alternating days of heavy and light workouts, helps give muscles the time they need to dispose of lactic acid and repair themselves.

→ **A sedentary lifestyle** often adds weight as well as reducing the health and strength of muscles. Disuse weakens abdominal muscles, contributing to chronic lower back pain and other ailments. As fat replaces muscle, body weight has less support, so most of the burden is placed on joints like the knees and ankles.

IMPROVING PERFORMANCE

Extreme resistance training, the forte of bodybuilders, often produces massive muscles. Anabolic steroids, which mimic the natural hormone testosterone, along with chemicals such as human growth hormone (GH) and tetrahydrogestrinone (THG), can artificially stimulate muscle growth. When a person's blood is "doped" with the hormone erythropoietin (EPO), the bloodstream can temporarily deliver more oxygen to muscles, extending their capacity to work before an oxygen debt develops. Although such substances can have serious side effects and are generally illegal in both amateur and professional sports, they and others have been used widely to improve performance.

← **Elite body builders** strive not only to develop well-defined muscles but also to properly balance the development of various muscle groups. Careless training can turn a would-be bodybuilder into a muscle-bound oddity who moves awkwardly, in part because the unbalanced enlargement of some of the muscles has eliminated the person's natural physical flexibility.

The nature of bone

Our bones may seem like solid, lifeless supports for our flesh. In reality, however, each one is a living organ built of bone tissue—a blend of bone cells interspersed with blood vessels, nerves and other structures. Bone is hard because the bone cells are surrounded by a matrix of collagen and other substances that has been infused with crystallized minerals, mostly calcium and phosphorus. From this basic recipe, the body assembles two types of bone tissue. Typically, the outer regions of a bone consist of dense, compact bone tissue, while deeper parts appear deceptively flimsy because their bone tissue has the look of a sponge filled with spaces. The combination of compact and spongy bone reduces the weight of the skeleton, yet makes bones as strong as steel in resisting tension, such as the pull of muscles, and twisting forces. In some bones, the bone shaft and the spaces in spongy bone contain the soft tissue called bone marrow, where blood cells are produced, then released to the bloodstream.

→ **Red blood cells** form from stem cells in red bone marrow, as do white blood cells. Aplastic anemia develops when the bone marrow does not produce enough red blood cells and too little oxygen reaches body tissues.

↘ **In compact bone,** concentric rounds of mineral-rich deposits contain bone cells enclosed in tiny slit-like spaces (colored yellow). Each cell has secreted a surrounding matrix that hardens as mineral crystals are deposited in it. The dark ovals seen here are cavities for blood vessels.

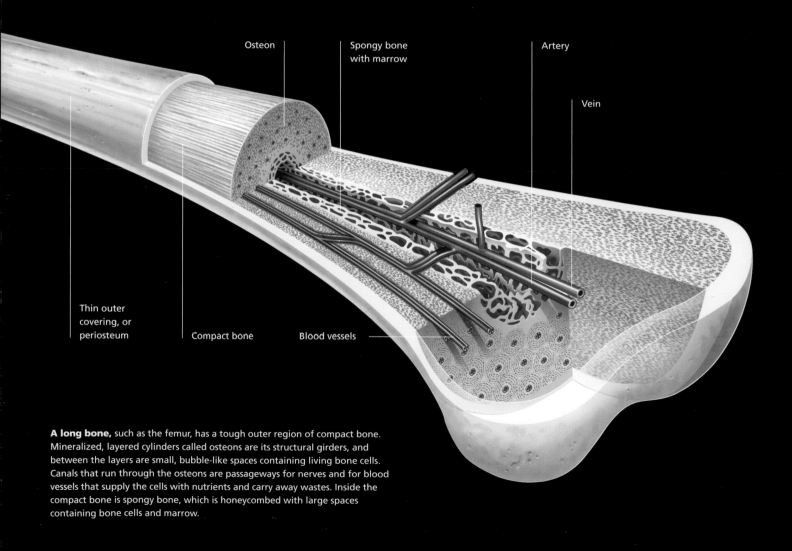

Osteon

Spongy bone with marrow

Artery

Vein

Thin outer covering, or periosteum

Compact bone

Blood vessels

A long bone, such as the femur, has a tough outer region of compact bone. Mineralized, layered cylinders called osteons are its structural girders, and between the layers are small, bubble-like spaces containing living bone cells. Canals that run through the osteons are passageways for nerves and for blood vessels that supply the cells with nutrients and carry away wastes. Inside the compact bone is spongy bone, which is honeycombed with large spaces containing bone cells and marrow.

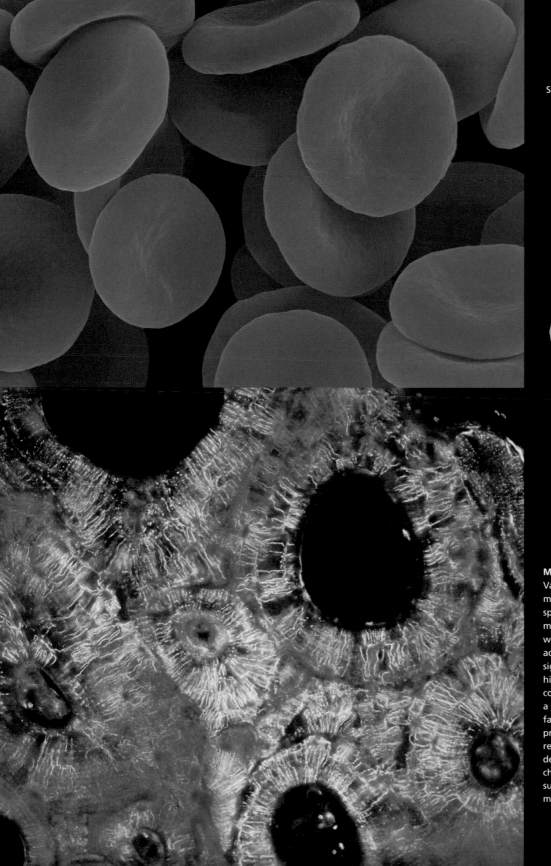

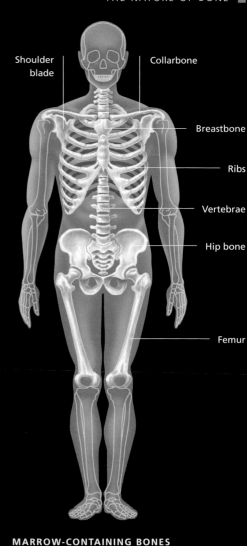

Shoulder blade

Collarbone

Breastbone

Ribs

Vertebrae

Hip bone

Femur

MARROW-CONTAINING BONES

Various bones of the skeleton contain marrow, mostly in the spaces in spongy bone. The marrow spaces of an infant's bones are packed with red marrow, which makes not only red blood cells but white cells as well. By the time a person reaches adulthood, however, only a few bones still contain significant amounts of red marrow. These include the hip bones, ribs, shoulder blades, breastbone, collarbone and spinal vertebrae. Other bones contain a preponderance of yellow marrow, which is largely fat. Even so, the remaining red marrow can usually produce enough blood cells to meet the body's requirements. If a person is anemic or loses a great deal of blood, the kidneys detect the deficit and send chemical signals that convert yellow marrow in bones such as the femur to red marrow, which begins making red blood cells to augment the supply.

The dynamic life of a bone

In the womb, a developing embryo has a delicate skeleton built only of cartilage and flexible membranes. By birth, however, bone-making cells called osteoblasts have replaced this cartilage model with bone, secreting a matrix which is gradually hardened by calcium deposits. First a collar of bone appears around the shaft of the cartilage, which then gradually breaks down and is replaced by bone. During childhood and adolescence, the rounded ends of the long bones in the arms and legs are separated from the shaft by a cartilage growth plate. A young person's limbs can grow until the plate is replaced by bone, usually by the time individuals reach their early twenties. Our bones never truly stop changing, however, for as long as we live the minerals in our bones continually "turn over" in a process called remodeling. This process also allows bones to take on the proper proportions, gain strength and undergo repair when they break. Factors that disturb the balance of this continuing cycle of destruction and renewal can result in debilitating ailments such as osteoporosis.

REMODELING

Processes that add or subtract bone help balance the body's supplies of key minerals. In a sense, bones are reservoirs of calcium and phosphorus that can be tapped if the need arises—for example, if a bone must be repaired or strengthened to adjust to mechanical stress. Bone remodeling operates differently at different times of life, however. In a healthy child, new bone forms faster than existing bone is broken down, reflecting the growth of the skeleton. In early adulthood, the balance between making and removing bone is roughly equal. After age 40, in most people the balance gradually shifts, and bones steadily lose mass and become weaker.

↓ **Mechanical stress** such as load-bearing exercise triggers the deposition of new bone. Weightlifters and dancers often develop unusually dense, strong bones.

↓ **A newborn's skull** is relatively flexible because the bony plates making up the skull have not yet fully fused. In this X-ray, dark lines mark the sutures and fontanels, between plates.

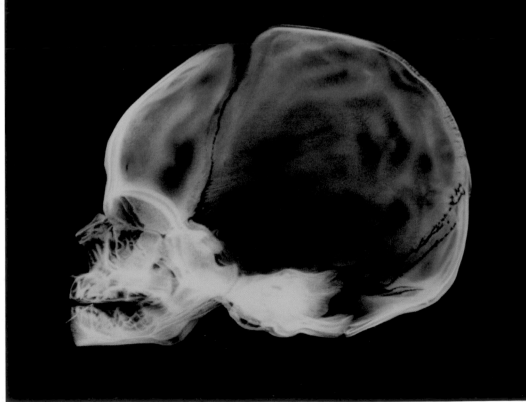

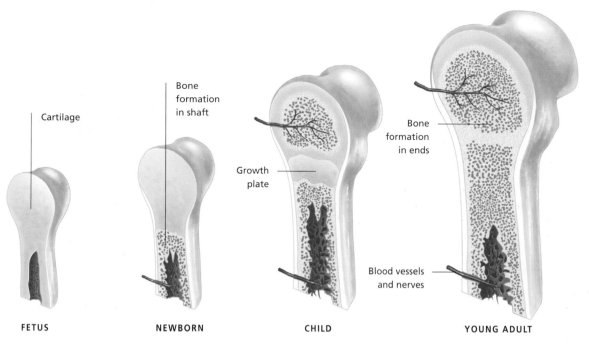

Cartilage

Bone formation in shaft

Growth plate

Bone formation in ends

Blood vessels and nerves

FETUS **NEWBORN** **CHILD** **YOUNG ADULT**

← **The first stage of bone development,** in many bones, is the prenatal formation of the cartilage model for a bone. Next, bone begins to replace the cartilage, so that by birth most of the newborn's long bones are well mineralized. During childhood the bones grow and are remodeled, acquiring a more mature shape. In early adulthood, bones stop growing, but they may add or lose mass.

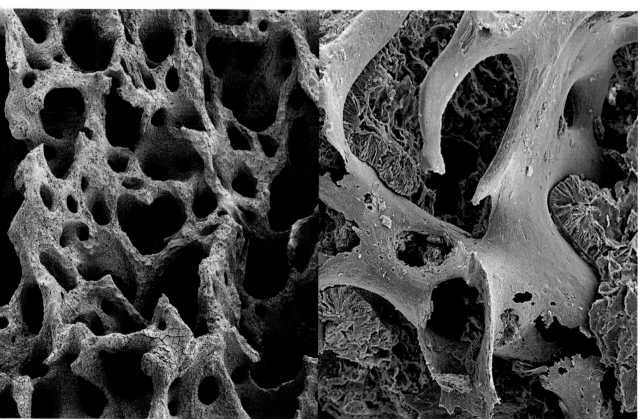

← **The building of healthy spongy bone** (left) is stimulated by sex hormones, adequate Vitamin D and calcium, exercise and other factors. As people age, their sex hormone levels drop, and often they become less active and fail to take in enough calcium. One result can be osteoporosis (right)—literally, holes in the bones—in which the bones become increasingly brittle and fragile. Therapeutic drugs can help stop bone degeneration and even stimulate the creation of new bone.

The skeleton

An adult's skeleton has 206 bones, that of a developing fetus many more. These bones vary in size and shape, from tiny ear bones no larger than a watch battery to long, massive thighbones. Some, such as the vertebrae, are irregular in shape, while others are flat, like the breastbone, or sternum. The ribs and the bones of the skull are curved, those of the wrists and ankles short and stubby. From another perspective, the skeleton's bones also are subdivided into several functional units. The skull, vertebral column and rib cage form the vertically aligned axial skeleton that provides protection for vulnerable soft organs such as the brain, lungs, heart and spinal cord. The so-called appendicular skeleton supports the body's "hanging parts," its arms and legs. Together this collection of bones provides a sturdy framework that both supports our soft flesh and provides hard parts on which muscles can pull to generate movements.

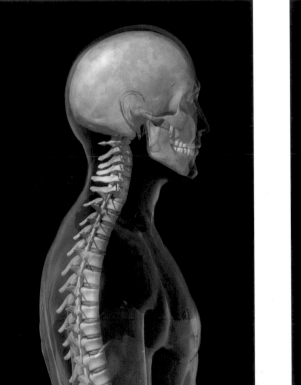

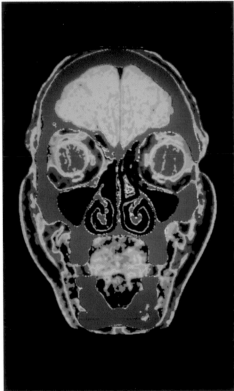

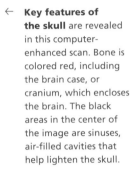

← **Key features of the skull** are revealed in this computer-enhanced scan. Bone is colored red, including the brain case, or cranium, which encloses the brain. The black areas in the center of the image are sinuses, air-filled cavities that help lighten the skull.

← **The spine, or vertebral column,** is both strong and flexible. It is the main bony support for the head, the rib cage and several internal organs. Divided into vertebrae connected by joints, the spine has the flexibility to permit movements. When a person is upright, the spine's natural curves distribute force in a way that helps maintain balance.

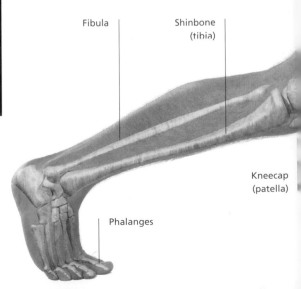

Fibula

Shinbone (tibia)

Kneecap (patella)

Phalanges

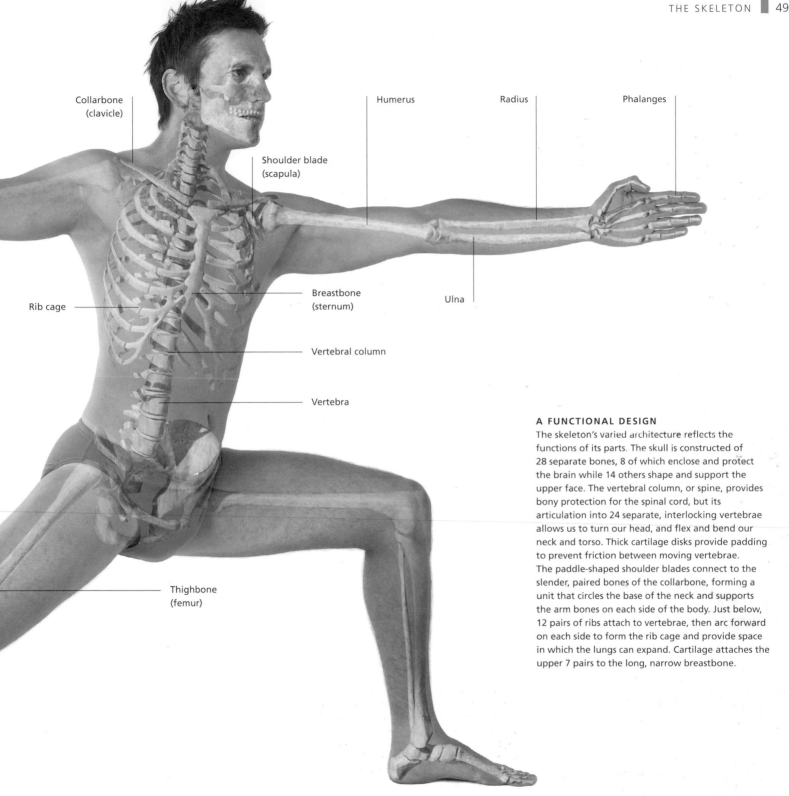

Collarbone
(clavicle)

Shoulder blade
(scapula)

Humerus

Radius

Phalanges

Breastbone
(sternum)

Ulna

Rib cage

Vertebral column

Vertebra

Thighbone
(femur)

A FUNCTIONAL DESIGN

The skeleton's varied architecture reflects the functions of its parts. The skull is constructed of 28 separate bones, 8 of which enclose and protect the brain while 14 others shape and support the upper face. The vertebral column, or spine, provides bony protection for the spinal cord, but its articulation into 24 separate, interlocking vertebrae allows us to turn our head, and flex and bend our neck and torso. Thick cartilage disks provide padding to prevent friction between moving vertebrae. The paddle-shaped shoulder blades connect to the slender, paired bones of the collarbone, forming a unit that circles the base of the neck and supports the arm bones on each side of the body. Just below, 12 pairs of ribs attach to vertebrae, then arc forward on each side to form the rib cage and provide space in which the lungs can expand. Cartilage attaches the upper 7 pairs to the long, narrow breastbone.

Limbs: the "hanging" skeleton

Human beings are mobile creatures, a lifestyle that is reflected in the highly mobile, paired appendages—our limbs—that dangle from our shoulders and hips. Limbs make up the body's appendicular, or "hanging" skeleton, and each is a modular structure with several parts linked by joints. The joints, and the muscles that bind them to limbs, are devised anatomically to facilitate the wide variety of movements we make with our arms, legs, hands and feet. The appendicular skeleton also includes two "girdles" that connect the limbs to the rest of the skeleton. At each shoulder, arm bones attach to the pectoral girdle, consisting of the collarbone, or clavicle, and shoulder blade, or scapula, while a ring-like pelvic girdle, the hip bone, articulates with leg bones. When you put your hands on your hips, your palms and fingers are resting on the sloping upper edges of the pelvic girdle. A person's gender shapes some of these skeletal parts. For example, a man has a pelvis that is long and narrow compared to a woman's wider, shorter and roomier pelvis—the latter an evolutionary adaptation for child-bearing.

STRENGTH AND FLEXIBILITY

Limb bones must absorb a range of physical stresses. Unlike small animals that run on all fours, humans require leg bones—and hip, knee and ankle joints—capable of supporting the body in an upright stance. To help meet this demand the human thighbone, the femur, has become the strongest, largest bone in the body, and both it and the muscles attached to it are engineered so that impact stresses from walking, running and jumping are in line with the bone's long axis. Walking upright allowed our early human ancestors to make free use of their upper limbs and hands. To allow the large human arm bone, the humerus, a wide range of movement, however, muscle attachments at the shoulder joint are rather loose—making it easy to dislocate.

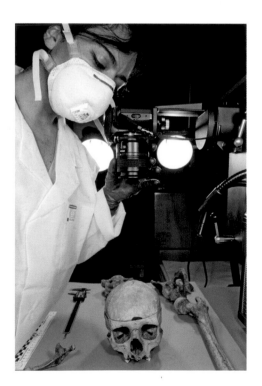

→ **Martial arts moves** exploit the possibilities of jointed limbs. High, leaping kicks and powerful, thrusting arm movements depend not only on muscle power and split-second coordination, but on having shoulder and hip joints that allow limbs to move in a variety of directions

← **Skeletal parts** help forensic scientists determine information about the deceased. In general, males have thicker and heavier bones, with bulkier attachment sites for muscles. The skull of a female will tend to have a longer forehead top to bottom and a smaller jawbone, among other differences.

→ **A ring-tailed lemur** scampers along using all four limbs to bear body weight and absorb impact. In humans, however, the limbs and other parts of the skeleton are knitted together in a way that allows efficient two-footed locomotion. Humans have all but lost a feature that still is important to a lemur, for instead of a tail we have an abbreviated coccyx.

STRUCTURE OF THE ARMS AND LEGS

The internal bony supports for our upper and lower limbs are similar. The thigh and arm are each built around a single, sturdy bone, the leg and forearm each have two long bones, one more massive than the other, while the feet and hands have many small bones arranged to allow flexing and bending.

The shinbone, or tibia, unites at the knee joint with the femur, which is one of the strongest bones in the body. Human feet are strong weight-bearing structures also, despite having small, fragile-looking bones. This strength is provided in part by the arched alignment of bones in the central part of the foot.

Including the finger bones, a human hand has 27 bones joined at 14 joints, an arrangement that gives even the clumsiest of us great manual dexterity. The wrist consists of 8 small, pebble-like bones, called carpals. One of them forms the large "bump" that protrudes under the skin on the little finger side of the lower forearm. Although humans can use their limbs for a striking array of movements, flexibility is especially marked in the thumb, which joins with the wrist in a way that allows it to touch the tips of the other four fingers. Having a versatile "opposable thumb" allows us to manipulate objects such as tools in finely controlled ways.

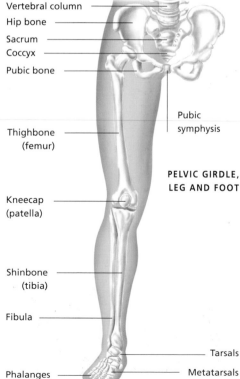

- Vertebral column
- Hip bone
- Sacrum
- Coccyx
- Pubic bone
- Pubic symphysis
- Thighbone (femur)
- Kneecap (patella)
- Shinbone (tibia)
- Fibula
- Tarsals
- Metatarsals
- Phalanges

PELVIC GIRDLE, LEG AND FOOT

PECTORAL GIRDLE, ARM AND HAND

- Phalanges
- Metacarpals
- Collarbone (clavicle)
- Shoulder blade (scapula)
- Humerus
- Carpals
- Radius
- Ulna

A human limb has three major segments. Technically, the "arm" stretches from the shoulder to elbow, followed by the forearm and the hand, which includes the wrist. The thigh extends from the hip to the knee, followed by the leg and foot, which includes the ankle. In all, each of the body's upper and lower limbs has 30 separate bones, including the protective kneecap in the lower limb.

Linking bone to bone

The body's bones come together at joints. Some joints allow expansive movement, others limited movement, and still others, as in the skull, none at all. Movable joints are the essential partners of bones and muscles in all of the highly coordinated movements we ask of our bodies, allowing various parts to extend, bend and rotate in different planes. In fact, many of life's pleasures—eating, giving a hug, strolling in the garden or playing on the beach—would be impossible were it not for our movable joints.

The most common and freely movable joints in the body are synovial joints, which have a capsule that encloses a narrow, lubricated space between the bones. Synovial joints come in various sizes and configurations, from the joint that connects the thumb to the wrist, to the hinge-like knee and elbow joints and the large, ball-and-socket joints at the hips and shoulders.

Instead of being a rigid rod, the spine, or vertebral column, is flexible in part because joints of cartilage unite the central parts of vertebrae and serve as shock absorbers. Other cartilaginous joints that allow limited but important movement are those that link ribs to the sternum, or breastbone. So-called fibrous joints are really not meant to move; instead, they hold bones tightly together by way of bands or a thin layer of connective tissue. Fibrous joints hold our teeth in their sockets and unite the plates of the skull.

↓ **The synovial joint** in the knee, the most complex joint in the body, allows hinge-like movements as well as some rotation. Shown in the front view at left, cartilage plates, each called a meniscus, provide padding between the lower end of the thighbone, or femur, and the top of the tibia. Surrounding the ends of the two adjoining bones is a tight outer capsule, which is lined with a slippery membrane that produces a lubricating fluid. In addition to the joint capsule itself, multiple ligaments and more than a dozen pillow-like bursae provide support and padding, as shown in the side view on the right. Each of them can be a vulnerable point for damage from blows, twisting movements or overuse injuries.

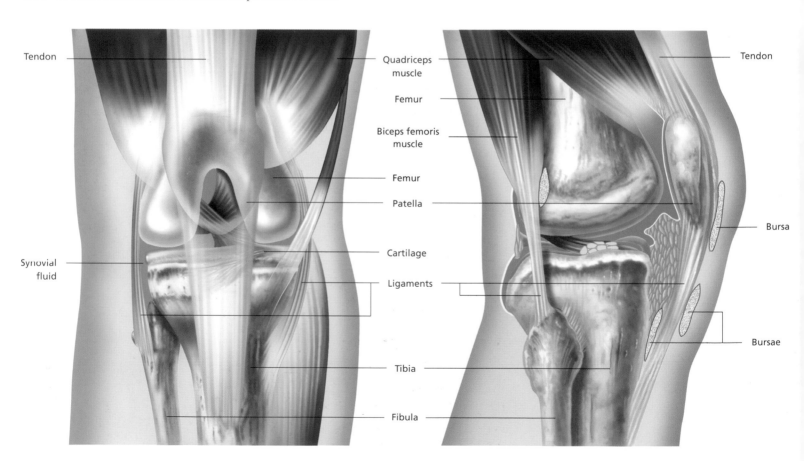

Tendon

Synovial fluid

Tendon

Quadriceps muscle

Femur

Biceps femoris muscle

Femur

Patella

Cartilage

Ligaments

Tibia

Fibula

Tendon

Bursa

Bursae

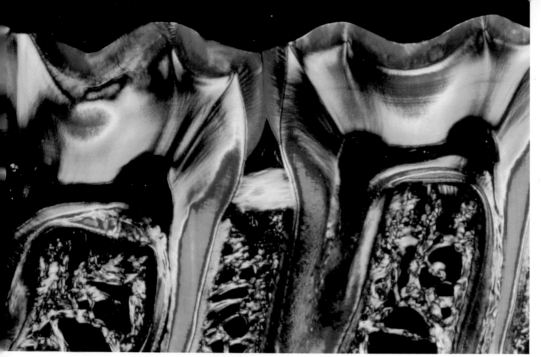

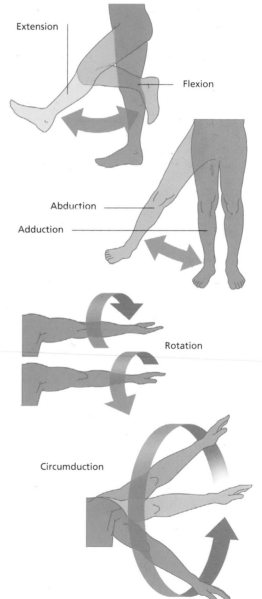

↓ **An array of opposing movements** occur at synovial joints. These include flexion and extension, abduction and adduction, rotation and circumduction.

Extension

Flexion

Abduction

Adduction

Rotation

Circumduction

↑ **Like pegs in sockets,** teeth are held tightly in place in the jaw by a ligament that wraps around the root of each tooth.

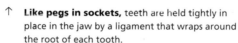

← **In the spine,** pivot-like synovial joints connect the bony, interlocking processes (orange-tipped in this scan), while cartilage disks (blue) provide modestly flexible padding between the round neighboring vertebrae. Each disk has a soft, resilient core. Injury or physical stress can squeeze this material outward, causing a painful slipped, or herniated, disk.

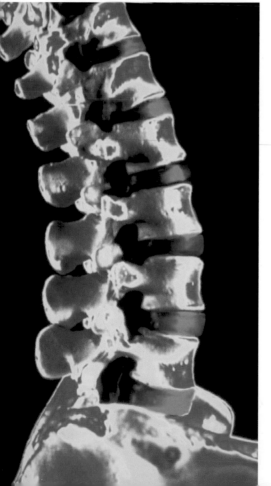

FLEXING AND ROTATION

Many joint movements flex body parts, move them inward or outward in the same plane, or trace all or part of a circle. Raise your heel to the back of your thigh, and you have flexed your knee joint. Kicking a ball extends the knee joint, while looking up at the ceiling or the stars extends joints in the upper spine that move the head. Raising an arm to the side and spreading the fingers apart both are examples of abduction, which moves a body part away from a central point but in the same plane. The motion in reverse is called adduction. Rotation moves a bone around its long axis—for instance, when you move your head from side to side. A dancer twirling her toes is performing circumduction, a movement in which a limb traces an imaginary cone.

The skeleton under siege

The human skeleton is built for durability, withstanding and adapting to a lifetime of physical stresses. When some part of it is injured or impaired by a disorder, body movements are nearly always affected. Millions of people suffer from lower back pain as a result of strain on the muscles and ligaments that help hold the spine more or less erect. Joints are the most vulnerable points, being subject to at least 100 forms of arthritis that inflame the joint or slowly break it down. Either way, the result usually is a swollen joint that is stiff and painful to move. Other common assaults include twisting a movable joint like the knee beyond its natural tolerance, or overstretching the ligaments that stabilize it and help keep bones properly aligned. Powerful blows can dislocate joints, uncoupling the linked bones so that they cannot work in concert; strong as they are, human bones can also break when they are violently struck or twisted, or crumble when they are weakened by diseases such as osteoporosis.

↓ **In rheumatoid arthritis,** the immune system goes awry and attacks the tissues in joints as if they were foreign. Affected joints become painfully inflamed, the cartilage between adjoining bones breaks down and the bones fall out of their normal positions—as this X-ray shows.

OSTEOARTHRITIS

Movable joints fall prey to osteoarthritis. Eventually, nearly everyone develops this degenerative condition as repeated use of movable joints wears away the cartilage padding between the ends of bones. Seen most often in older adults, osteoarthritis can also be a genetic disorder and strike much earlier in life. When the damage is severe, replacing the joint may be the only option for relieving the pain and restoring normal movement. Today, arthritic hips, knees, ankles, shoulders and knuckles all can be replaced with artificial joints made of metal alloys, high-density plastic, ceramics or some combination of those materials.

Arthritic bones also may develop small calcium spurs that reduce the available room for bone ends to move in the joint, making it even stiffer and triggering pain if they press on nearby nerves.

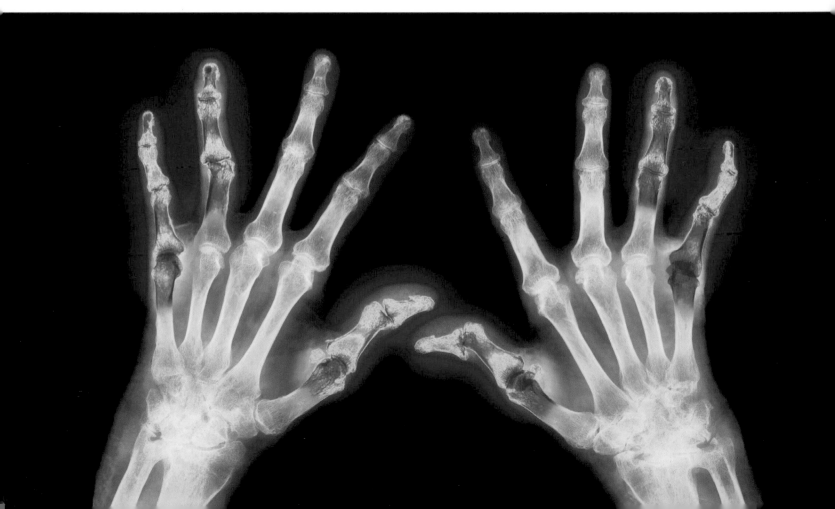

TYPES OF BONE FRACTURE

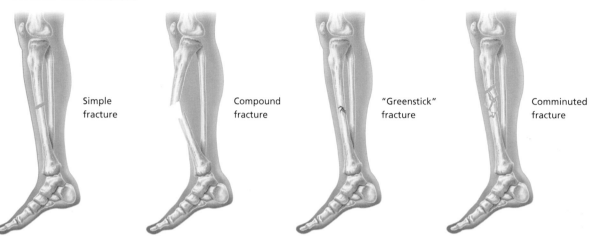

Simple
fracture

Compound
fracture

"Greenstick"
fracture

Comminuted
fracture

Bones break in many ways. In a simple or stress fracture, a crack separates the bone shaft, but the broken ends remain in position. With a more serious compound fracture, the broken bone ruptures the skin and opens the way for infection. In a "greenstick" fracture the bone breaks only part way, while a comminuted fracture breaks the bone into several pieces and can be difficult to repair.

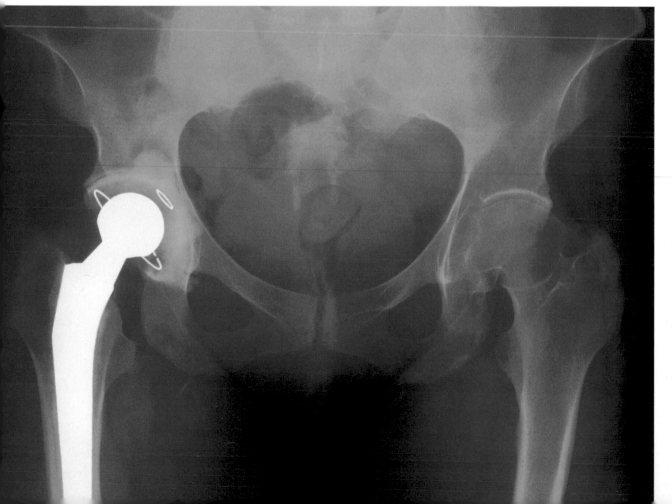

The hip is replaced more than any other joint. In younger patients, the new joint, or prosthesis, usually has all metal parts, which make it more durable. Older patients generally receive a joint that consists of a ball and socket built of metal and high-density plastic. Most prostheses have a porous surface. With remodeling, the patient's own bone cells grow into the cavities and secure the prosthesis in place.

Body cavities and linings

Large body cavities are a major feature of human anatomy. The human body's internal organs—those other than our muscles and bones—are not placed in the body haphazardly, but are suspended in or attached to the walls of these cavities. There are five major cavities that protect and organize the body's vital soft parts. The brain and spinal cord, making up the all-important central nervous system, are enclosed by the cranial and spinal cavities formed by the skull and the bony vertebrae. Inside the chest, or thoracic, cavity are two smaller chambers called the pleural cavities, one for each lung, and the pericardial cavity for the heart. The stomach, liver, pancreas, kidneys, most of the intestines and several other organs rest inside the large abdominal cavity, and the reproductive organs, bladder, lower colon and rectum are housed inside the pelvic cavity below. Membranes line all these cavities, and in the abdominal and pelvic cavities membranes also help hold internal organs in place.

THE ROLE OF MEMBRANES

Membranes line the body cavities and tubes. Thin and moist, they cover or line other body parts, and often lubricate and support them. *Mucous* membranes line the tubes and cavities of the digestive tract, as well as those of the respiratory, urinary and reproductive systems. They secrete mucus or other substances and may also be involved in the absorption of nutrients. *Serous* membranes line body cavities and also cover the outer surfaces of the organs within them. The cranial and spinal cavities have their own distinct lining called the *meninges*, which produce cerebrospinal fluid.

MAJOR BODY CAVITIES

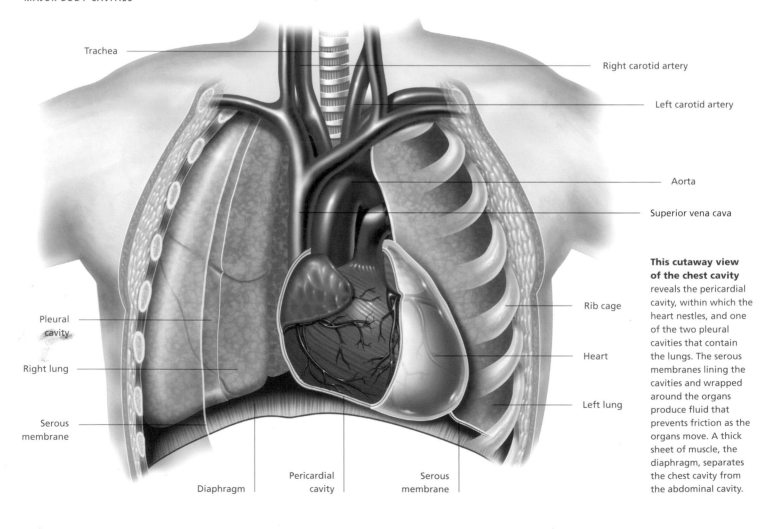

Trachea

Right carotid artery

Left carotid artery

Aorta

Superior vena cava

Pleural cavity

Rib cage

Right lung

Heart

Left lung

Serous membrane

Diaphragm

Pericardial cavity

Serous membrane

This cutaway view of the chest cavity reveals the pericardial cavity, within which the heart nestles, and one of the two pleural cavities that contain the lungs. The serous membranes lining the cavities and wrapped around the organs produce fluid that prevents friction as the organs move. A thick sheet of muscle, the diaphragm, separates the chest cavity from the abdominal cavity.

→ **The pink, moist membrane** lining the mouth is a mucous membrane. Like most mucous membranes, it contains secretory glands—in this case, salivary glands that make saliva.

↘ **The capacity of the cranial cavity,** which encases the brain, was determined by filling skulls with water in this experiment.

↓ **The five major internal compartments** that house and protect our vital organs are all closed to the outside, preventing access by bacteria and other pathogens.

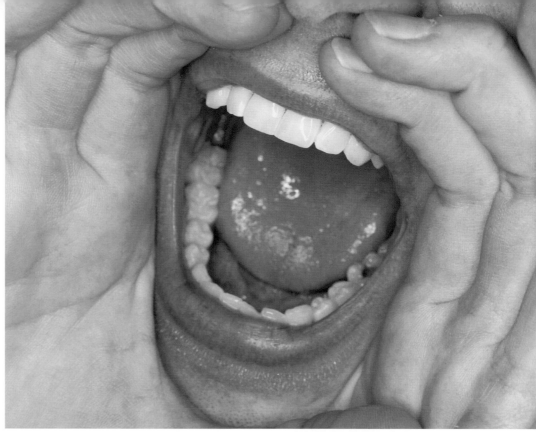

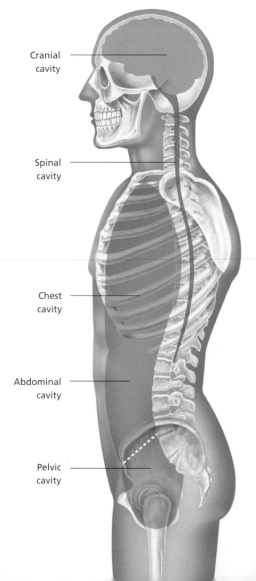

Cranial cavity

Spinal cavity

Chest cavity

Abdominal cavity

Pelvic cavity

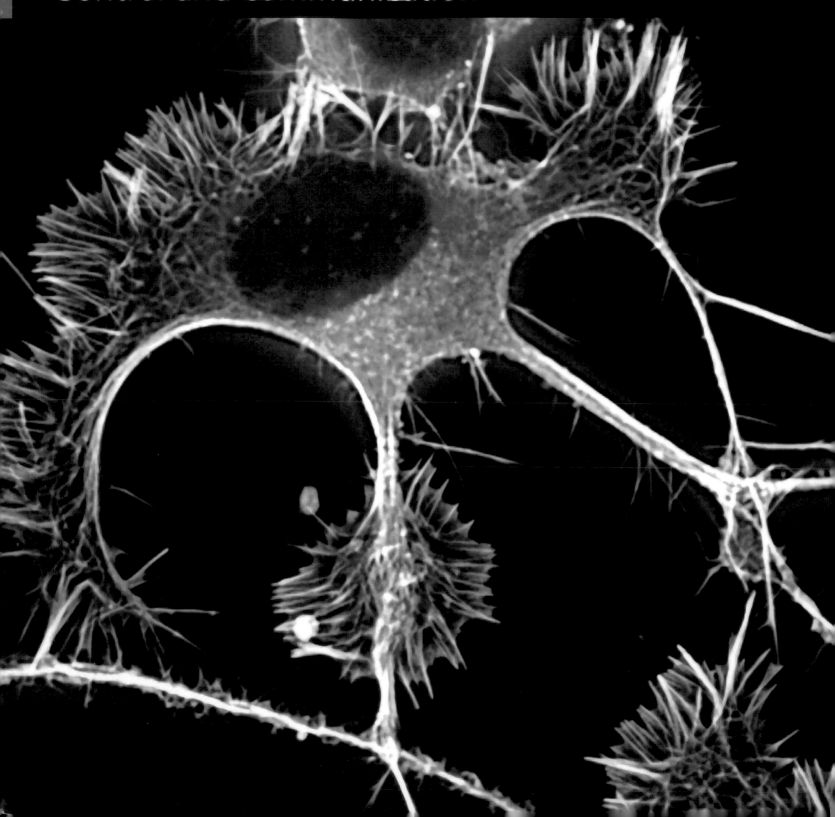

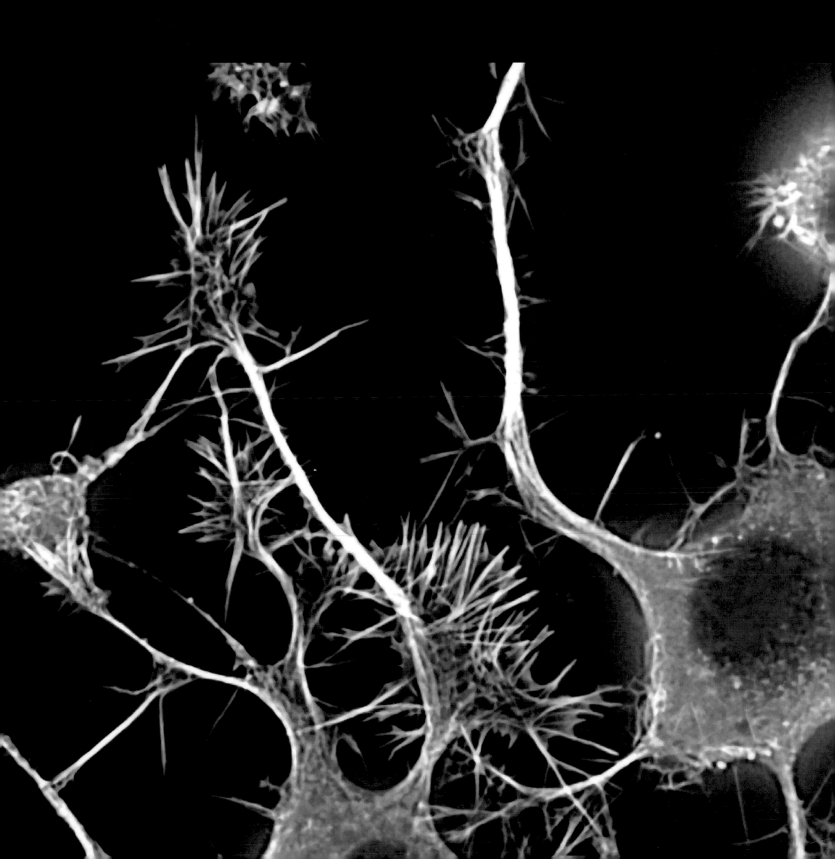

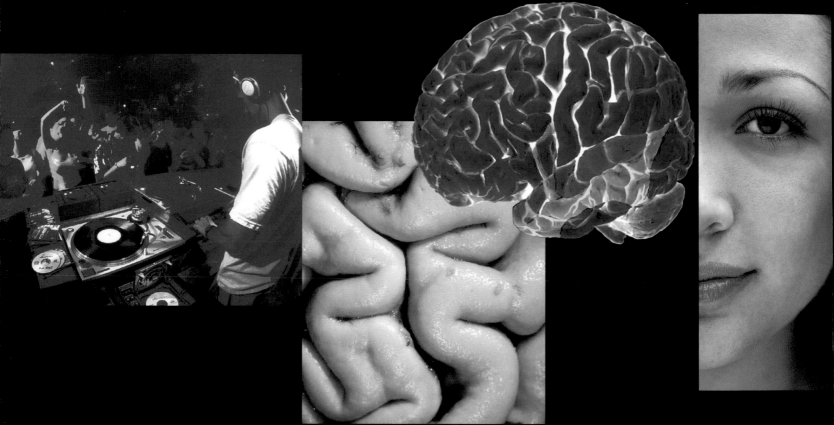

Control and communication

The human body is astonishingly complex, and a prerequisite for even its most basic operations is efficient, effective communication between parts. Two body systems, the nervous system and the endocrine system, provide this "command and control."

The nervous system	62
Master control: the central nervous system	64
Neurons: the great communicators	66
Nerves: long-distance messengers	68
Neural pathways	70
The peripheral nervous system	72
The spinal cord	74
The brain	76
Consciousness, memory and emotions	78
Language and speech	80
Endocrine controls	82
Hormone signals	84
The hypothalamus and the pituitary	86
Insulin and its kin	88
Corticoids	90
Sex hormones	92
The powerful thyroid	94
Hormones from the heart, gut and elsewhere	96
Hormones and the environment	98

The nervous system

Humans have the most complex nervous system in the animal world, consisting of an intricately wired brain, fragile spinal cord and tens of billions of interconnected, constantly "chattering" neurons organized into nerves. Like all high-performance communication systems, the nervous system carries messages between parts that must function in coordinated ways, integrates information arriving from inside and outside the body, and includes controls that allow adjustments in our biological operations as well as our behavior.

Both in structure and function, the two partners of the central nervous system (CNS)—the brain and spinal cord—are the nervous system's organizational center. They receive sensory information from our eyes, ears, skin, taste buds, "smell" receptors and elsewhere, evaluate and integrate that input, and communicate adjustments in body functions. The nerves that serve as the communication lines for this back-and-forth messaging make up the nervous system's other segment—the "outer," or peripheral, nervous system.

↓ **Nerves impulses** travel in two directions in the nervous system. "Incoming" signals about sensory information travel toward the brain and spinal cord, and "outgoing" signals travel away from the brain and spinal cord to muscles and glands, which carry out responses.

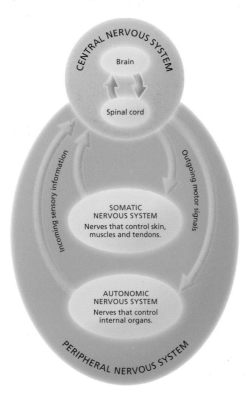

TWO-WAY TRAFFIC
Threading throughout the body, the nerves of the peripheral nervous system carry two kinds of traffic: sensory information traveling to the CNS for evaluation and processing, and commands for motor responses going from the CNS back to muscles and glands. This constant information flow allows adjustment of body functions as conditions vary inside and outside the body. Walking, digesting a meal, answering a telephone and other routine events all demand responses that would be impossible without coordinated interaction of the nervous system's parts.

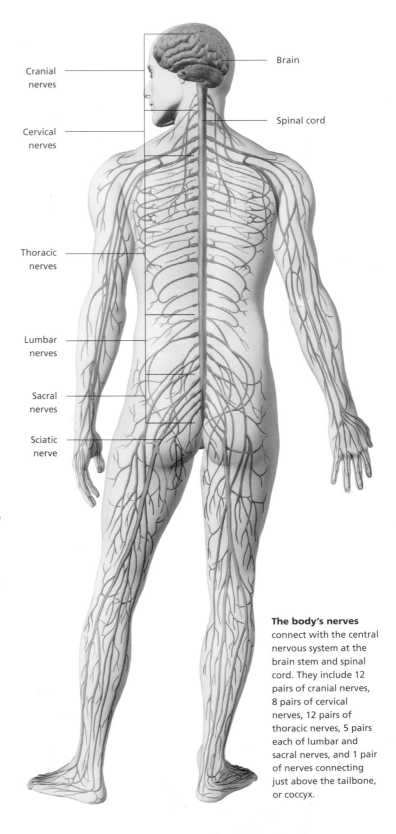

The body's nerves connect with the central nervous system at the brain stem and spinal cord. They include 12 pairs of cranial nerves, 8 pairs of cervical nerves, 12 pairs of thoracic nerves, 5 pairs each of lumbar and sacral nerves, and 1 pair of nerves connecting just above the tailbone, or coccyx.

← **Juggling** requires rapid monitoring of sensory input, coordinated with equally rapid adjustments in physical movements. It triggers sensory and motor signals to and from the eyes and body muscles, as well as demanding keen mental attention.

← **Playing chess** taps central nervous system centers for thought, analysis and planning. The brain must also integrate those mental events with the voluntary muscle actions that are needed to move chessmen around a board.

← **Human nervous systems** are "wired" to respond to the needs of our offspring, a behavior that increases the chances that they will survive and flourish. Human children mature rather slowly compared to the young of other species, in part because it takes a great deal of time for our complex nervous systems to develop fully—perhaps 20 years or longer.

↑ **Animal nervous systems** range from extremely simple to highly complex. A sea star (top) has neither head nor brain. One or more rings of nerve tissue encircle its mouth and short nerves extend into its "arms." A chimpanzee's brain is smaller than ours but still quite complex. As a result, chimps have sophisticated communication skills and problem-solving abilities.

Master control: the central nervous system

The central nervous system consists of the brain and spinal cord. True to its name, the central nervous system (CNS) centralizes the body's master controls in the head, a body part we see in all but the simplest animals. Even worms have a nerve cord, a forerunner of the spinal cord. Spinal nerves and cranial nerves arising from the brain transmit the incoming and outgoing messages that keep the body and its parts functioning. Pairs of these communication lines enter and leave on either side of the cord, each essentially a mirror image of the other. This mirror-image structure reflects the human body's basic bilateral architecture; as with nearly all animals, the body can be divided, more or less, into matching halves dominated by the head, which holds our brain and key sense organs such as the eyes and ears.

OPERATIONAL HEADQUARTERS

A brain that is centrally located at the front of the head correlates with an active body consisting of many parts. Partnered with the spinal cord and receiving input from sense organs, the brain processes incoming information about the outside world as well as about the operations of tissues and organs. It also manages responses, including changes in the way a tissue or organ is operating and changes in behavior—such as eating when the body needs to refuel. We are consciously aware of some of this information traffic, such as sights, sounds, pain, hunger pangs and movements of our skeletal muscles. We are generally not conscious of CNS monitoring that adjusts basic physiological operations such as body temperature, kidney function and blood pressure.

→ **The central nervous system's two parts** are clearly visible in this enhanced scan. Most of the brain is shown as light pink. The brain stem, which monitors many basic bodily functions, and the spinal cord with which it merges are colored blue.

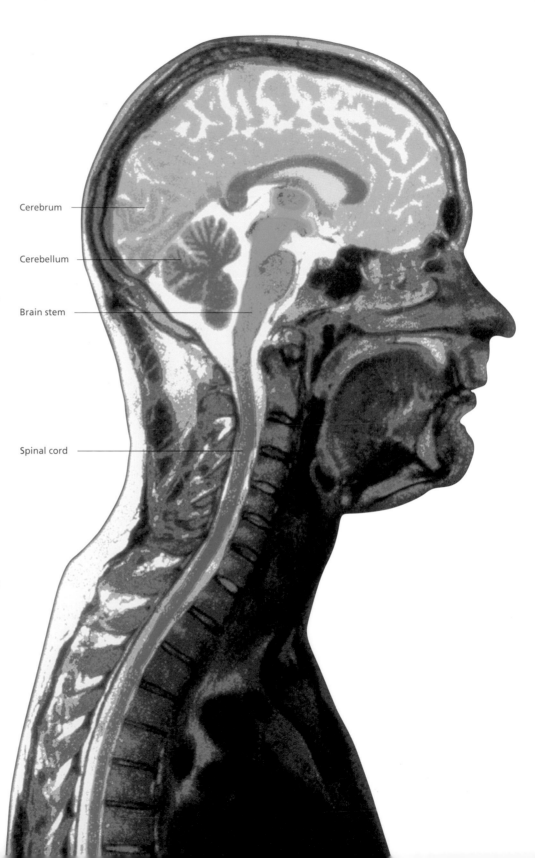

Cerebrum

Cerebellum

Brain stem

Spinal cord

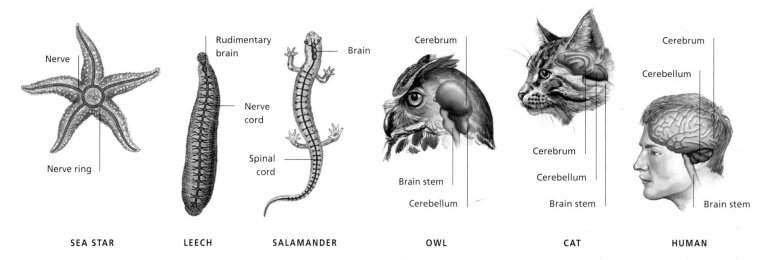

Nerve

Nerve ring

SEA STAR

Rudimentary brain

Nerve cord

LEECH

Brain

Spinal cord

SALAMANDER

Cerebrum

Brain stem

Cerebellum

OWL

Cerebrum

Cerebellum

Brain stem

CAT

Cerebrum

Cerebellum

Brain stem

HUMAN

↑ **Almost all animal groups** have evolved a head and a central nervous system, with nerve cells clustered in a brain-like structure and some form of nerve cord or spinal cord. Radial animals like sea stars are exceptions. The more complex the species and its sense organs, the more complex its CNS, especially the brain.

→ **Like its reptilian ancestors, a lizard** such as the Komodo dragon possesses a relatively small, simple brain capable only of basic responses. Deep within the human brain a primitive region, the limbic system, acts as our "lizard brain." It governs instinctual drives and behaviors, such as eating, rage, sexual responses, fear and fighting.

CROSS-SECTION OF THE SPINAL CORD

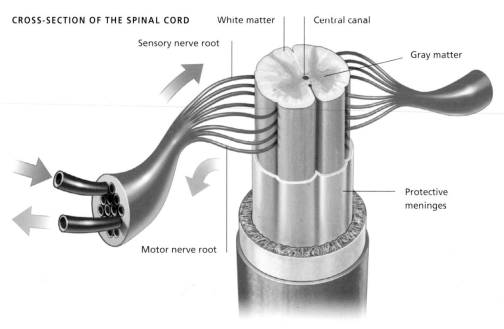

Sensory nerve root

White matter

Central canal

Gray matter

Protective meninges

Motor nerve root

↑ **A slice across the spinal cord** shows the basic pattern of the CNS. Surrounding a central canal containing cerebrospinal fluid are the clustered cell bodies of nerve cells, called gray matter. The next layer, white matter, consists of nerve cell axons—extensions that conduct nerve impulses.

Neurons: the great communicators

Hundreds of billions of nerve cells, called neurons, are the nervous system's workhorses. Motor neurons carry signals to muscles and glands. Likewise, many millions of sensory neurons transmit information about light, sound, smells and other sensory stimuli to the CNS, sending the messages either directly to the brain or via the spinal cord. In the brain and spinal cord an estimated 99 billion nerve cells known as interneurons are the nervous system's crucial go-betweens. They process incoming sensory signals, such as the sight of a child running into the street, and also carry the outgoing response signals—perhaps "orders" for you to slam on the brakes of your car. "Cross-talk" between interneurons in the brain also gives us our memories, emotions, creativity and many other aspects of the human mind.

Every neuron has parts with particular functions. The body of the cell, which includes the nucleus, handles basic housekeeping tasks. Short extensions called dendrites receive incoming information, while the longer axons pass information along to the next neuron in line.

→ **A false-color slice across a neuron** shows its nucleus as green, with a red nucleolus at its center. Axons and dendrites surrounding the nucleus are light blue, while dark blue traces the myelin sheaths of some axons.

↓ **Neural messages** are pulses of electrical signals, as suggested in this artist's rendering. When a neuron "fires," the nerve impulse starts near the cell body and travels down the neuron's long axon.

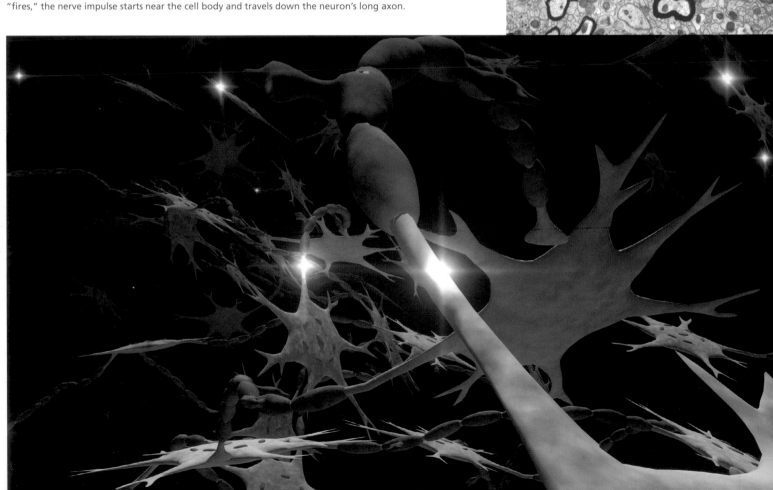

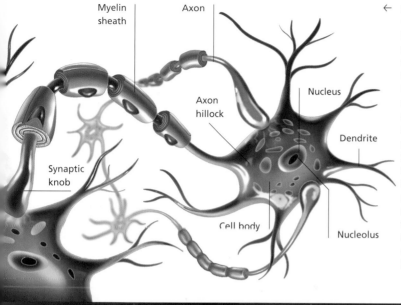

Myelin sheath

Axon

Axon hillock

Nucleus

Dendrite

Synaptic knob

Cell body

Nucleolus

← **Neurons are built** to communicate. Nerve impulses begin at a small mound called the axon hillock and travel outward along the axon. The axons of very long or large-diameter neurons are wrapped with specialized cells, which form an insulating sheath containing a fatty material called myelin. This insulation allows nerve impulses to move rapidly along axons, some at the rate of nearly 400 feet (120 m) per second.

RELAYING THE SIGNAL

Chemicals called neurotransmitters are the means by which neurons throughout the nervous system pass along signals. When a nerve impulse arrives at the ends of an axon—at the synapse, or gap between the sending and receiving neurons—tiny sacs in the axon endings release the chemical neurotransmitter stored inside them. When the receiving neuron takes up the neurotransmitter, it produces another impulse and sends it on to the next synapse, where the process is repeated. Depending on various factors, neurotransmitters can stimulate receiving cells or inhibit their activity. If a neuron is stimulated, it may fire an impulse that travels to yet another neuron, and so on until the original signal reaches its final destination. Antidepressants, cocaine, methamphetamine and drugs used to treat hyperactivity, among other substances, alter the way neurotransmitters are released or taken up by different brain neurons, and so change the way a person's brain functions.

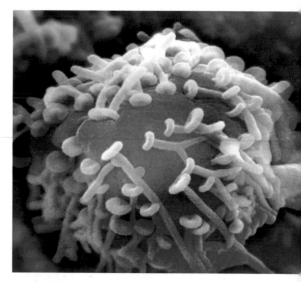

↑ **A large number of axon endings,** colored blue, synapse with a receiving neuron cell body, shown as a pinkish-purple mound.

← **The sound of dance club music** is processed by one set of neurons, while another processes the sight of people dancing. Interpreting a conversation at the same time would involve one or more additional brain regions.

Nerves: long-distance messengers

By itself, the central nervous system can't monitor, assess or regulate many body functions. Nerves extend its reach, serving as the nervous system's long distance communication lines. Taken together, nerves are the cornerstone of the peripheral nervous system, physically linking the brain and spinal cord with virtually every other region of the body. Each nerve is a collection of bundled neuron axons, and most include axons of both motor and sensory neurons. The individual bundles, and the whole nerve, are held together by casings of connective tissue. Twelve pairs of cervical nerves start from the brain, and travel outward through channels in the skull. Most service the muscles and skin of the neck and head, including the face, eyes and tongue. Thirty-one pairs of spinal nerves emerge from the spinal cord and then branch to form smaller nerves that eventually reach to the tips of the fingers and the soles of the feet, and most points in between.

↓ **A cross-section through a nerve** shows the bundled axons of nerve cells, as well as blood vessels and vessels of the lymphatic system. The vessels, which look like dark hollows in this view, line up like pipelines alongside the axon bundles.

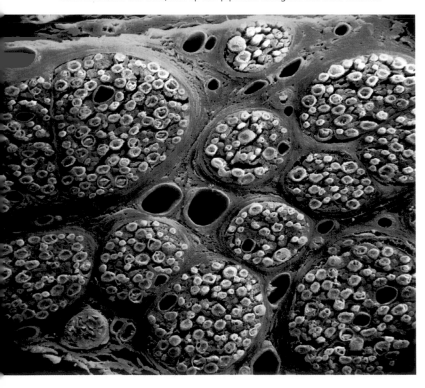

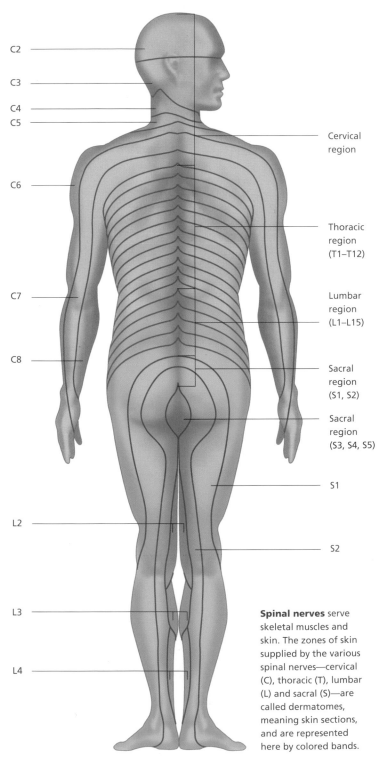

C2
C3
C4
C5
C6
C7
C8
L2
L3
L4

Cervical region

Thoracic region (T1–T12)

Lumbar region (L1–L15)

Sacral region (S1, S2)

Sacral region (S3, S4, S5)

S1

S2

Spinal nerves serve skeletal muscles and skin. The zones of skin supplied by the various spinal nerves—cervical (C), thoracic (T), lumbar (L) and sacral (S)—are called dermatomes, meaning skin sections, and are represented here by colored bands.

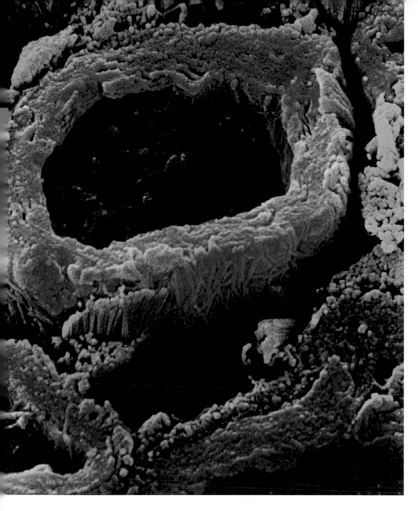

A **close-up view of a single axon** (shown as red) in a nerve reveals its surrounding myelin sheath (lavender) and the connective tissue (green) that holds the axons together in each bundle.

WHEN NERVES ARE DAMAGED

Nerve injuries can run the gamut from inconvenient to painful to lethal. If nerves supplying a certain eye muscle are paralyzed, the patient cannot move the eye sideways but does not lose any vision. By contrast, severe damage to the vagus nerve, which services the heart, lungs, intestines and many other vital organs, can cause death. Inflammation of the sciatic nerve, called sciatica—sometimes triggered by a "slipped disk" in the lower spine—can cause debilitating pain in the buttock, thigh and lower leg. The thickest and longest nerve in the body, the sciatic nerve runs down the back of the thigh and branches into the leg and foot. Sciatica may recur repeatedly during a person's life, and it can be so severe that patients have difficulty walking. Perhaps the most excruciating of all nerve disorders is trigeminal neuralgia ("nerve pain"), which affects the major sensory nerve of the face. The insulating myelin sheath around the nerve disintegrates, leaving the nerve exposed. As a result, even a feather-light touch to the facial skin can cause intense, piercing pain. To relieve the agony, in some patients the affected nerve is severed surgically, a solution that leaves the person with no feeling on the affected side of the face.

↑ **A stroke** is an injury, such as a hemorrhage, that disrupts the blood flow in one side of the brain. If damage affects nerve pathways to the limbs, the person may suffer paralysis. Here a woman, originally right-handed, is paralyzed on her right side, and is learning to write with her left hand.

↘ **A major nerve network,** called the brachial plexus, serves each shoulder and arm. Its nerves include the median nerve, which runs down the inner arm to the palm, the radial nerve running down the back of the arm and the ulnar nerve, which supplies most of the hand muscles.

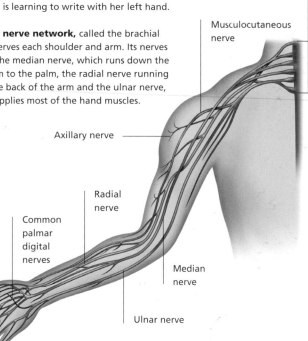

Brachial plexus

Musculocutaneous nerve

Axillary nerve

Radial nerve

Common palmar digital nerves

Median nerve

Ulnar nerve

Neural pathways

Nerves extend throughout the body, providing lines of communication for operating body parts. In the nerves, relays of neurons pass along signals generated in sense organs or in the central nervous system. They function in both awesomely complex arrays and in basic, biologically ancient ones, and life as we know it depends upon both. For example, breathing and the heart's rhythmic beat are reflexes, the simplest of neural operations, while activities such as thinking involve millions of neurons interacting in elaborate circuits.

Despite these variations, information in the nervous system generally follows a simple, overall pattern that we see in many aspects of life: events are noted, forwarded to headquarters for interpretation and processing, and then managers order a response. In the body this sequence unfolds as different types of nerve cells perform their tasks. The neurons in sense organs such as the eyes, ears and skin detect sensory information and relay it along their axons to the nervous system's managers, the spinal cord and brain. This incoming information is delivered by neurotransmitters across synapses. In the central nervous system, interneurons receive the sensory signals, interpret and process them, then issue responses for action—signals that are conveyed across synapses to motor neurons. Bundled in motor nerves, these neurons transport the outgoing signals along their axons to the appropriate muscles and glands.

↓ **Overall, information flows** through the nervous system in a predictable path. Sensory neurons pick up information and carry it into the central nervous system, where they form synapses with interneurons in the spinal cord and brain. After interneurons process the signals, they send responses outward by way of their synapses with motor neurons. Reflexes are the simplest examples of this pathway. Usually, a sensor detects a change, such as stretching muscle fibers or altered blood chemistry. Sensory neurons relay that information to interneurons in the spinal cord which interpret the information and signal a response, producing a rapid, predictable effect. Many basic body functions are controlled by reflexes that occur without us ever being aware of them.

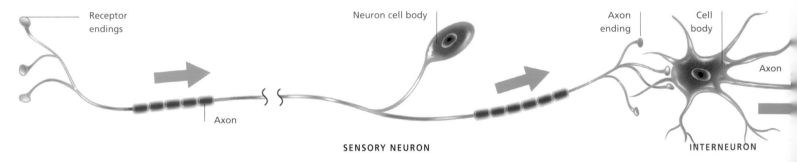

Receptor endings

Axon

Neuron cell body

SENSORY NEURON

Axon ending

Cell body

Axon

INTERNEURON

→ **Inhaling the aroma** of freshly roasted coffee beans, this man is also feeling them in his hands and using various muscles to lift them as he lowers his head. Neurons in his brain integrate these varying inputs into a single pleasurable moment.

→→ **As a dragonfly** feels its way across a young girl's hand, signals from touch and visual sensors are sent to her brain, where they combine to create a sense of wonder and curiosity.

INTERPRETING COMPLEX SIGNALS

One of the brain's amazing features is its ability to sort out and make sense of the vast number of competing signals that arrive each moment. In the brain and spinal cord, incoming signals often are sent to groups of neurons that collectively interpret the information and respond. In areas that must process a large amount of information, some neurons have tens of thousands of dendrites that synapse with the axons of tens of thousands of other neurons feeding in data from many sources. Alternatively, the axon of a single neuron may branch many times over, enabling it to synapse with many neurons. This sort of diverging pathway allows a single neuron to pass signals to numerous others simultaneously.

→ **In the knee-jerk reflex,** a sharp tap on a knee tendon sends signals via sensory neurons to the spinal cord. There, the sensory neurons communicate with motor neurons supplying the quadriceps muscle, and also with interneurons in the cord. The quadriceps contracts, jerking the lower leg upward. At the same time, the cord neurons send signals that inhibit the hamstring muscle on the back of the thigh, so it cannot prevent the quadriceps from contracting.

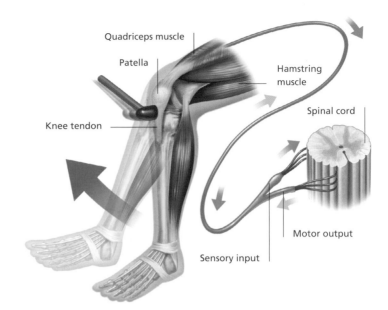

Quadriceps muscle

Patella

Hamstring muscle

Knee tendon

Spinal cord

Motor output

Sensory input

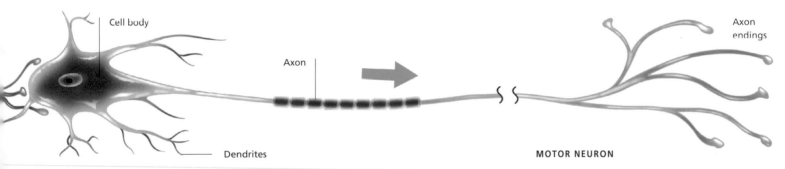

Cell body

Axon

Axon endings

Dendrites

MOTOR NEURON

← **Newborn babies** have an instinctive startle reflex. If surprised by a sudden movement or a loud sound, they extend their arms and spread their fingers wide.

← **A turtle's reflexive response** to surprise—such as the presence of a photographer—is to draw its head and legs up and inward, under the protection of its hard shell.

The peripheral nervous system

The neurons, nerves, and connections outside the brain and spinal cord make up the peripheral nervous system, which meets the needs of our bodies by dividing the labor of communicating inputs and responses. The nerves over which we routinely have conscious control are those that service the skin, skeletal muscles and tendons. Collectively they make up the somatic nervous system, meaning "of the body." Yet many other nerves are not at our beck and call. They make up an autonomic, or "autonomous," system which functions more or less independently of the conscious mind, carrying signals to and from the heart, intestines and other internal organs.

Managing the workings of our internal organs is a hugely demanding task that is parceled out to two subdivisions of the autonomic system, which operate like counterbalances. On the one hand, when a person is excited, upset or stressed in some other way the system's so-called sympathetic nerves muster high-energy responses. On the other, parasympathetic nerves dampen down the activities of internal organs when things are quieter. The result is a finely choreographed "ballet" that maintains biological control over our bodily functions.

→ **Parasympathetic nerves** are most active when basic functions are at the fore, such as the digestion of a meal. Sympathetic nerves dominate when a person is under physical or psychological stress, or giving heightened attention to a task. The heart beats faster, breathing is more rapid and blood sugar levels rise. Depending on the circumstances, a person may develop "jitters," start to sweat or develop a dry mouth. These changes are part of what has come to be called the fight-or-flight response—physiological changes that prepare body systems for maximum exertion.

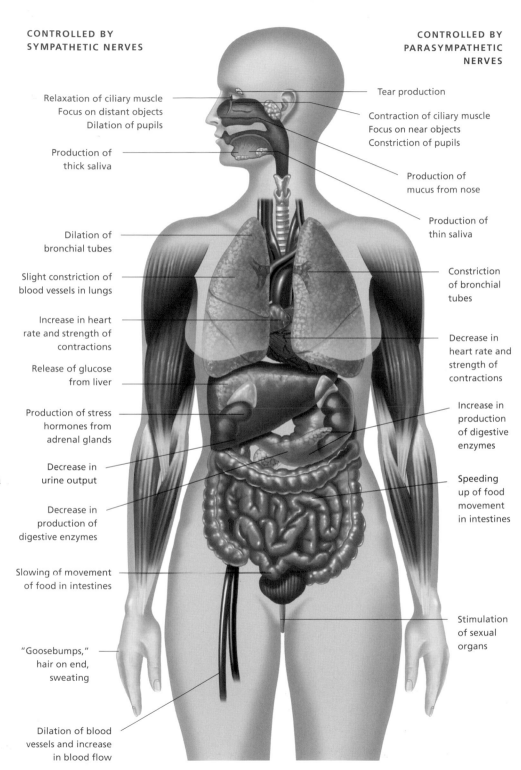

CONTROLLED BY SYMPATHETIC NERVES

Relaxation of ciliary muscle
Focus on distant objects
Dilation of pupils

Production of thick saliva

Dilation of bronchial tubes

Slight constriction of blood vessels in lungs

Increase in heart rate and strength of contractions

Release of glucose from liver

Production of stress hormones from adrenal glands

Decrease in urine output

Decrease in production of digestive enzymes

Slowing of movement of food in intestines

"Goosebumps," hair on end, sweating

Dilation of blood vessels and increase in blood flow

CONTROLLED BY PARASYMPATHETIC NERVES

Tear production

Contraction of ciliary muscle
Focus on near objects
Constriction of pupils

Production of mucus from nose

Production of thin saliva

Constriction of bronchial tubes

Decrease in heart rate and strength of contractions

Increase in production of digestive enzymes

Speeding up of food movement in intestines

Stimulation of sexual organs

HOW STRESS TIPS THE BALANCE

The autonomic system ensures that the operations of body organs are finely calibrated to keep internal conditions within limits that support continued life—a state called homeostasis, meaning "staying the same." Maintaining homeostasis is a key task of the nervous system as a whole. Like crew members on a ship that must maintain an even keel, the opposing signals from sympathetic and parasympathetic nerves cooperate, adjusting the operations of organs as conditions demand.

Stress can unbalance this dynamic system, because it represents extremes—too much or too little in the way of physical or emotional demands. Regardless of whether we perceive a particular stressor as good or bad, the nervous system mobilizes its sympathetic nerves for the fight-or-flight response. The heart works harder, blood pressure rises, digestion slows, the liver boosts chemical reactions that load sugars into the blood, and the adrenal glands increase their production of the hormone epinephrine—which feeds back to prolong and increase these physiological responses even more.

When stress is chronic, some people seem to be naturally able to adapt, but in others physiological and chemical changes can damage the cardiovascular, digestive and immune systems. All things being equal, competitive, impatient "type A" individuals run a much higher risk of heart attack. Some researchers believe that closely timed serious life stressors, such as the death of a loved one and the loss of a job, put people at higher risk of cancer because their immune systems are weakened.

↑ **Impalas flee** at the hint of danger. These nimble, ever-vigilant African animals survive in part because of their well-honed fight-or-flight response.

↗ **After a large meal,** parasympathetic nerves slow down a person's heart and breathing. At the same time, they also increase the secretions of various glands, including the salivary glands—which is why sleeping people sometimes drool.

→ **Sympathetic nerves dilate** a person's pupils in low light, while parasympathetic signals constrict them in bright light. Movements of the eyeball as a whole are under voluntary control.

PERIPHERAL NERVOUS SYSTEM

SOMATIC SUBDIVISION
Controls voluntary functions. Carries signals to and from skeletal muscles, tendons and skin.

AUTONOMIC SUBDIVISION
Controls involuntary functions. Carries signals to and from internal organs.

SYMPATHETIC NERVES
Prepare body for stress.

PARASYMPATHETIC NERVES
Slow activities of internal organs. Predominate during sleep.

← **The peripheral nervous system** consists of two parts—the somatic, over which we have conscious control, and the autonomic, which is responsible for involuntary actions and which is further subdivided into the sympathetic and parasympathetic systems.

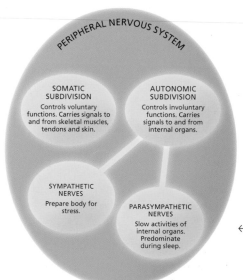

The spinal cord

The spinal cord is the signaling expressway of the nervous system. White and shiny, the cord is both the body's control center for reflexes and a superhighway for signals traveling between the brain and the rest of the nervous system. It begins just below the cerebellum at the base of the brain and travels about two-thirds of the way down the spine. Threading through a channel formed by the stacked, bony vertebrae, the cord stops at the first lumbar vertebra and before the triangular sacrum in the lower back. Inside this channel are layers of protection for the cord, beginning with a layer of fat. Next come three tough membranes, the meninges, which also wrap and protect the brain. The outer one, called the dura mater ("tough mother") has the feel of thin leather. As it extends down the cord, straps (the arachnoid) formed by another of the meninges, the pia mater ("gentle mother") bind the dura mater tightly to the cord. Between the meninges is cerebrospinal fluid. There is only about half a cup (125 ml) of this watery substance, but it provides valuable anti-shock cushioning in both the spinal cord and brain.

The spinal cord itself consists mostly of bundles of the myelin-wrapped axons of neurons, called nerve tracts. Some carry signals up toward the brain, others carry messages away from the brain, and still others move them across the cord. When any of these nerve tracts is injured, compressed or severed, we fully understand how the sensations and movements we take for granted depend on this long, narrow cylinder of nerve tissue.

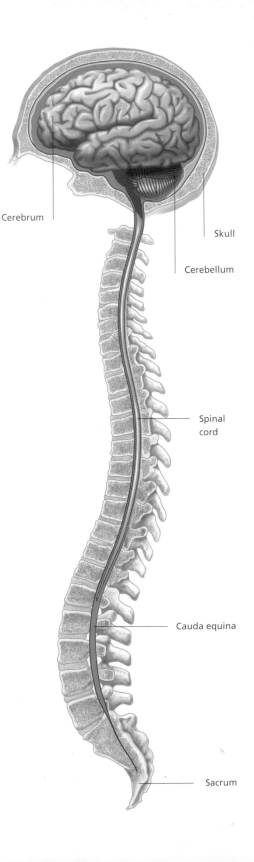

Cerebrum

Skull

Cerebellum

Spinal cord

Cauda equina

Sacrum

← **A ridge formed** by the bony, outer projections of vertebrae is visible on this model's back. Just beneath the projections are the protective meninges, and beneath those layers is her spinal cord. Supportive ligaments on either side of the vertebrae help hold the whole assembly in proper alignment.

→ **End to end,** the spinal cord is about the girth of your thumb and roughly 18 inches long (45 cm) in an adult. Intervertebral disks of cartilage are flex points and help prevent friction between the vertebrae.

↑ **This artificial disk** (orange) was attached with screws in the upper spine to treat arthritic changes that had severely limited the mobility of the neck.

↑ **Grown in a laboratory, these nerve cells** have begun sprouting extensions, shown here as yellow and purple, that are the forerunners of axons—the cable-like carriers of a neuron's electrical signals. Research into nerve cell regeneration may eventually lead to a cure for spinal cord injuries.

REPAIRING DAMAGED NERVES

Each year, accidents that injure or sever the spinal cord leave thousands of people partly or wholly paralyzed. The search for a cure has focused on ways to regenerate nerve tissue by stimulating the unspecialized cells called stem cells to become healthy, functioning neurons. In the brain and spinal cord of newborn animals, chemicals called nerve growth factors can trigger regeneration, but in older animals the process soon breaks down.

Some researchers are taking a different tack by genetically engineering damaged adult nerve cells, inserting a gene that makes proteins that seem to help the cells grow new axons. If this approach succeeds, it will be equally important to get the repaired nerve cells to re-establish connections with other nerve cells. Ultimately, treatments for spinal cord injuries will probably draw on several different strategies that promote new growth from inside and outside the nerve cell.

→ **The spinal cord** is protected by the bony cavity of the vertebrae that enclose it. Spinal nerves exit at side gaps between the vertebrae and travel to either side of the body.

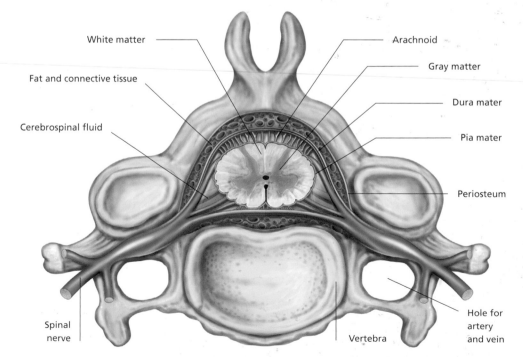

White matter

Fat and connective tissue

Cerebrospinal fluid

Arachnoid

Gray matter

Dura mater

Pia mater

Periosteum

Spinal nerve

Vertebra

Hole for artery and vein

The brain

The Greek philosopher Aristotle believed that the mass inside the human skull served mainly to help "cool" the spirit. Today we know that our soft, wrinkled, grayish brain receives, stores, integrates and retrieves sensory information and coordinates responses by adjusting countless activities throughout the body. Waking and sleeping, hunger and thirst, pleasure and pain are only a few of the fundamentals it governs. Weighing just over 3 pounds (about 1.4 kg) in an adult, the brain also provides the biological mechanics for our capacity to express ourselves through language, to think and reason in complex ways, to learn, to build and savor memories, and to feel and understand complex emotions. We are only just beginning to understand how the brain's many parts manage these feats.

Most of the brain is made up of the cerebrum and its thick, folded outer region, the cerebral cortex. This is the only part of the brain that handles conscious activity. It is divided front to back into two halves or hemispheres, each devoted to certain functions—intuitive and artistic activities on the right, language and mathematical ones on the left. In two slim, neighboring sections of the cortex, the nerve cells are arranged in ways that map the sources of sensory information and the destinations for motor signals. Other parts are reserved for "association" tasks—reasoning, judgment, memory and the like. Below and behind the cerebrum are areas that specialize in the body's housekeeping tasks, from pumping blood and breathing to swallowing food and eliminating wastes.

→   **Deep wavy folds** characterize the cerebral cortex, the outer layer of the cerebrum. Although the cortex is at most one-sixth of an inch thick (about 4 mm), tens of millions of nerve cells communicate within it every moment of our lives.

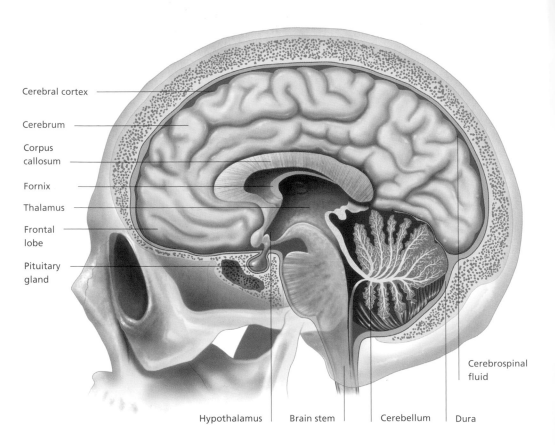

Cerebral cortex

Cerebrum

Corpus callosum

Fornix

Thalamus

Frontal lobe

Pituitary gland

Cerebrospinal fluid

Hypothalamus   Brain stem   Cerebellum   Dura

↑   **Each brain hemisphere** is a blend of structures. The thalamus relays sensory information to the cerebral cortex. The hypothalamus is involved in many functions, including adjustments in body temperature, water balance, growth and metabolism. Cerebrospinal fluid fills brain cavities, cushioning the brain's soft tissues against injury as well as helping to disseminate some chemical messengers.

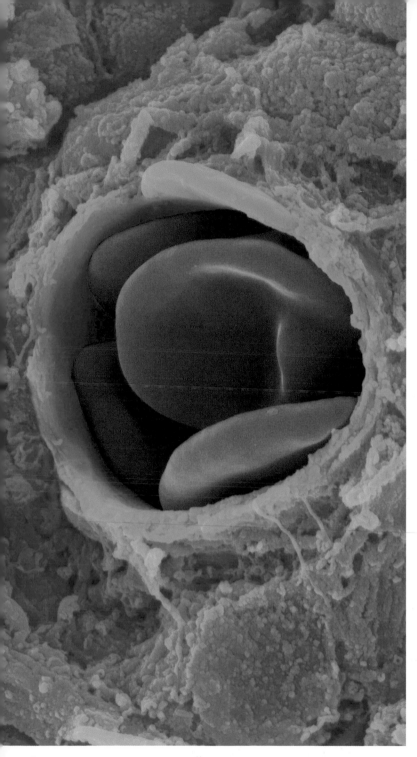

↑ **Red blood cells** are seen inside a blood vessel (colored green) in the cerebellum. Such small blood vessels have unusually thick, tight walls. This "blood–brain barrier" prevents many potentially harmful substances from leaving the blood stream and entering brain tissue, although nicotine and alcohol cross easily.

LEFT AND RIGHT BRAINS

The cerebral hemispheres appear similar but have different functions. The left hemisphere controls most of the actions of the right side of the body, and the right hemisphere most of the left side. Both hemispheres are involved in much of what we do, but each dominates certain functions. In most people, the left hemisphere specializes in language abilities, mathematical acumen and the sequential mental activity we call "logical" thinking. The right hemisphere strongly influences our intuitive, artistic, musical, spontaneous selves. People whose brains work this way are generally right-handed. In roughly 3 to 5 percent of the population, the dominance of cerebral hemispheres is reversed or shared equally. People in this group are usually left-handed or ambidextrous. Despite their differences, the hemispheres interact constantly via the more than 200 million nerves that cross the deep fissure between them, the corpus callosum. Their strengths also complement one another. After hours of analytical work, for instance, a person's more spontaneous side may suddenly trigger the urge to take a break and listen to music. When you stifle an emotional outburst, it is the more logical side of your brain that is exerting control.

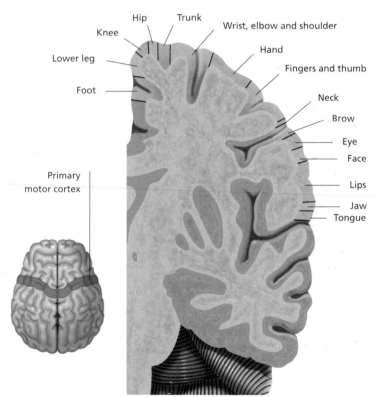

↑ **In the part of the cerbral cortex** devoted to motor functions—the primary motor cortex—proportionately more neurons are dedicated to controlling mouth parts used in speech and movements of the fingers and hands. Control over the trunk and lower limbs is much less specialized.

Consciousness, memory and emotions

Thinking, feeling happy or sad, daydreaming and remembering things past are all aspects of the mysterious brain activity we call consciousness. To a physician, consciousness is a continuum, ranging from full alertness to drowsiness, to various levels of sleep and the unconscious state called coma. Most people experience several states of consciousness in the daily rhythm of life as different areas of the brain come to the fore, then become less active as another state emerges. Mental illness, psychoactive drugs and willed activities such as meditation can all alter "normal" conscious states, radically changing an individual's perceptions of the world around them. Regardless of the state of consciousness, large areas of the cerebral cortex are active, with neurons constantly communicating as sensory information arrives.

Adding immeasurably to the richness of life is the human ability to understand and feel emotions. This ability is centered deep in the cerebrum, where an arc of interacting structures form the limbic system. The system's components—the hypothalamus, thalamus, amygdala and hippocampus—collectively govern emotions, "gut reactions" such as rage, and drives associated with hunger and sex. Several of these structures also connect with circuits that build the repository of experiences we call memory. Studies show that both short-term and long-term memory rely on many parts of the cerebral cortex working in concert.

↓ **The main structures of limbic system,** or the "emotional brain," are seen deep inside the cerebrum in this scan of a healthy human brain.

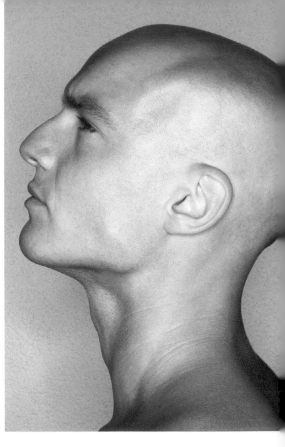

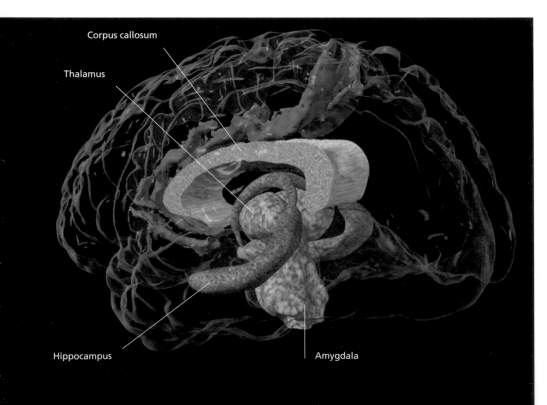

Corpus callosum

Thalamus

Hippocampus

Amygdala

MALE AND FEMALE "WIRING"

Brain scans show that in problem-solving, women and girls use parts of both brain hemispheres, while men and boys use mainly analytical centers. Female infants pay close attention to voices, and as they grow females tend to be more sensitive to verbal nuances and quicker to pick up on emotionally meaningful cues in speech. Male infants focus more on objects, and as they mature, males generally have markedly better spatial skills, which may help explain why males overall are more adept at geometry and mathematics. Research suggests that some brain differences become more pronounced at puberty, when surging sex hormones have an impact on the brain's development. None of the observed male–female differences translates into hard and fast rules about a person's abilities. As with any aspect of human anatomy and functioning, individual differences are common, not the exception.

↑ **Male and female brains** are not only "wired" differently. Men's heads are also larger and their brains weigh about 10 percent more than women's. A larger brain, however, is not necessarily linked to greater intelligence.

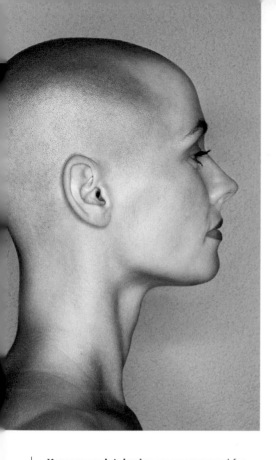

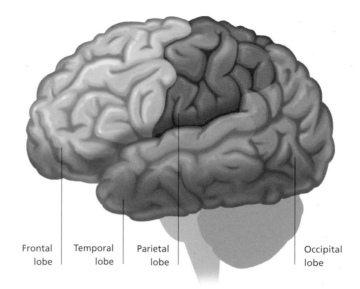

← **The lobes of each hemisphere** have different functions. The frontal lobe manages planning for voluntary movements, some aspects of memory and the inhibition of some inappropriate behavior. The parietal lobe deals with bodily sensations and the occipital lobe with vision. Advanced visual processing and hearing are handled in the temporal lobe.

| Frontal | Temporal | Parietal | Occipital |
| lobe | lobe | lobe | lobe |

↓ **Young people's brains** are programmed for risk. The prefrontal cortex, which controls decision-making and aggression, is not fully developed until humans reach their early 20s.

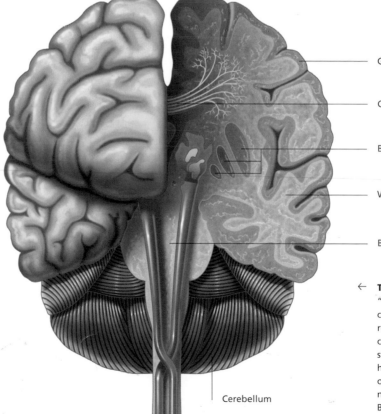

Gray matter

Corpus callosum

Basal ganglia

White matter

Brain stem

Cerebellum

← **The deep folds** of our "thinking brain," the cerebral cortex, are revealed in this front-on cross-section. The brain stem and cerebellum handle automatic operations, such as normal breathing. Basal ganglia neurons coordinate skeletal muscle movements.

Language and speech

Centered in the brain's left hemisphere, language is a vital component of our consciousness and intellect, and especially of our relationships with others. Our brains allow us to think in words, to read, write and speak, to hear and understand the speech of others, and to use physical gestures when spoken communication fails. Brain scans show that these allied skills depend on the flow of information between several specialized areas. The planning required for writing, articulating thoughts and generating words and sentences takes place in the frontal lobe of the frontal cortex. There, too, movements of speech-related tongue, lip and throat muscles are controlled. In the left frontal lobe, a region called Broca's area activates when we speak, while Wernicke's area at the rear of the brain allows us to comprehend the speech of others. Still other areas are concerned with hearing, reading, the act of writing and other language tasks. We are also able to muster all these resources at once—for example, when jotting down a note while listening to the radio and talking to a friend.

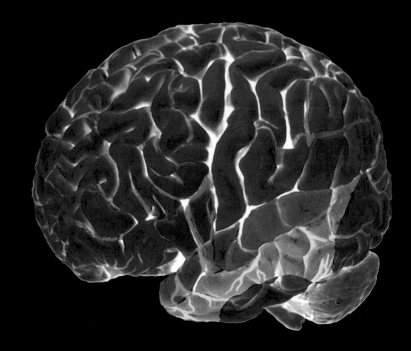

↑ **Language** is a left-brain function. The scan on the left shows several areas in the left brain's temporal lobe that activated while a subject listened to words. The scan on the right shows the same but less active area in the subject's right brain.

← **Using several language skills** at once—here, speaking, listening, writing —demands extremely complex interactions among at least five regions of the left cerebral cortex.

→ **Signing** is language minus sound. Research on people using sign language shows that much the same brain areas are active as if the participants were generating and hearing spoken words.

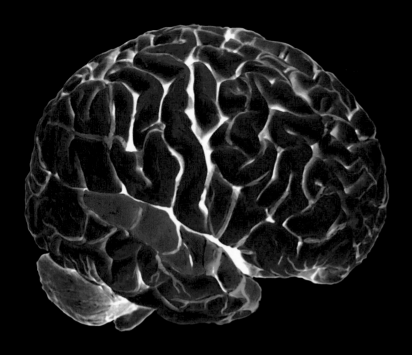

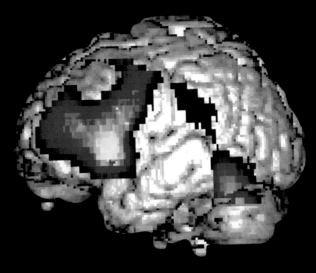

↑ **Speaking** activates several brain areas, shown as red-orange: the frontal lobe (left) for content and muscle movements, and the temporal lobe (lower right) for checking accuracy of output.

CONTROL CENTERS

Much of what we know about different language areas in the brain comes from studies of people who have suffered brain damage. Broca's area, in the left hemisphere's frontal lobe, is vital for normal speech. If it is damaged, the patient may lose the ability to speak at a normal pace, have difficulty enunciating words clearly or be unable to speak at all. However, abilities to grasp both spoken and written language will be intact, and some patients can still sing. Farther back in the left hemisphere is Wernicke's area, which helps govern our ability to comprehend language. When it is damaged, the person cannot understand written or spoken language and may have trouble putting thoughts into coherent speech.

Efforts to treat patients suffering from severe epilepsy shed light on how the two brain hemispheres cooperate in language functions. Hoping to reduce the patients' symptoms, surgeons severed the nerves linking the two hemispheres. While the patients had fewer epileptic seizures, the left hemisphere, which processes language, could no longer respond to information about written words that went to the right hemisphere.

→ **A male robin** sings to attract a mate in the spring. Unlike human language, however, birdsong has an extremely limited "vocabulary" that is used only in certain circumstances and changes relatively little.

Endocrine controls

Together with the nervous system, hormones help control many body functions and even influence behavior. Like the neurotransmitters that pass between nerve cells, hormones are signaling chemicals that serve as the messengers of the endocrine system. The system's components are located throughout the body, from sites in the brain to the intestines and other internal organs. No matter where they are, endocrine cells release hormones that act on specific targets elsewhere in the body. Most travel in the bloodstream to reach their destinations. Some hormones circulate in the bloodstream in predictable cycles while others are secreted only when certain conditions arise and still others circulate more or less constantly.

Some endocrine glands, like the pituitary, make several hormones. A single hormone may also have multiple targets. Insulin is a good example: it acts on muscle, fat and the liver, triggering metabolic activities that store the energy in food in different forms. Target tissues are primed to receive the messages hormones bring. When the hormone arrives, its signal launches a change, subtle or dramatic, in the way the tissue functions. Often, two or more hormones affect the same process. Other signaling chemicals, including the long-distance signalers called pheromones, also act in the body in ways that are much less well understood.

SIGNALS THAT ENDURE

Hormones act more slowly than nerve impulses do, and often they provide controls over more enduring changes, such as growth and reproduction. For example, the pituitary gland makes growth hormone, or GH, throughout life, even after a young adult's bones have stopped growing. At any age, growth hormone helps the body tap its fat stores for energy as it is needed—for example, during prolonged exercise, when the glucose cells normally use as fuel may become depleted. Sex hormones—the androgens and estrogens—are released during prenatal development when the sex organs of a fetus are forming, and then again from puberty until old age.

← **Breastfeeding** requires the interaction of several hormones. In pregnancy, estrogen and progesterone stimulate the development of ducts and glands that will make milk after the baby's birth. Soon after delivery, prolactin triggers the production of milk and oxytocin triggers its release.

→ **Scattered endocrine cells** in the pancreas (islets of Langerhans) secrete the hormones insulin and glucagon, which manage blood sugar levels. Tiny sacs (red spots) contain hormone molecules ready to be delivered to the bloodstream.

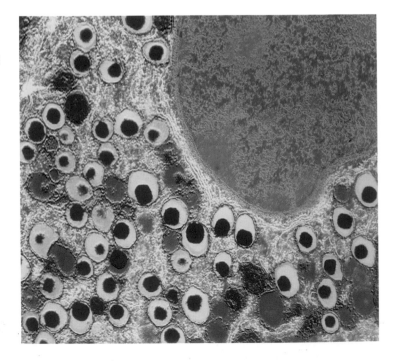

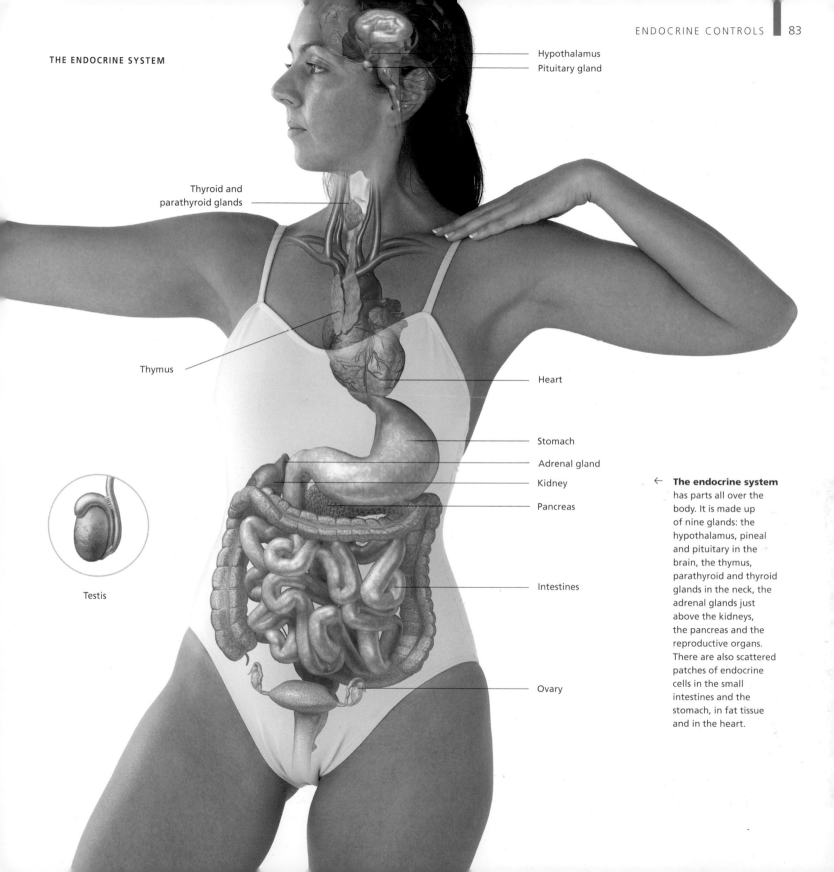

THE ENDOCRINE SYSTEM

Hypothalamus
Pituitary gland

Thyroid and
parathyroid glands

Thymus

Heart

Stomach

Adrenal gland

Kidney

Pancreas

Intestines

Testis

Ovary

← **The endocrine system**
has parts all over the
body. It is made up
of nine glands: the
hypothalamus, pineal
and pituitary in the
brain, the thymus,
parathyroid and thyroid
glands in the neck, the
adrenal glands just
above the kidneys,
the pancreas and the
reproductive organs.
There are also scattered
patches of endocrine
cells in the small
intestines and the
stomach, in fat tissue
and in the heart.

Hormone signals

At any given moment, dozens of hormones are coursing through the bloodstream, each coming close to all of the body's trillions of cells. Yet somehow each hormone manages to interact only with the cells that are its particular targets. Like a key that can fit only one lock, a hormone can interact only with cells that have a matching receptor. Likewise, cells in a tissue or organ can fulfill their biological roles only if they can receive and respond to the signals of particular hormones. Cells have many kinds of receptors, each one akin to a satellite dish that can pick up only limited signals, often only one. Some receptors are built into the cell's outer membrane, others are part of the membrane around the cell's nucleus. When the hormonal key fits, a target cell is open to its signal. The result is a chain of events that produce a change in some aspect of the cell's functioning.

Hormone signals begin operating in the womb, and from then on they are vital to integrating the far-flung changes that take place as the body grows, develops, processes nutrients and deals with the challenges of stress and disease. There are three general types of hormones. The sex hormones and a few others are steroids, synthesized from cholesterol. All other hormones are based on proteins or on an amino acid.

Father and son illustrate how one hormone may operate at different times of life. In males, the hormone testosterone is produced before birth, when it guides the development of reproductive organs in a male fetus. After birth, a boy's body will not produce testosterone again until he enters puberty. A man's testes produce testosterone until the end of his life.

This cell in the pituitary gland manufactures growth hormone (GH), which has widespread effects in the body. Genetic instructions from DNA in the cell's nucleus (pink) have guided the production of GH, which is stored inside the many small vesicles (brown) in the cytoplasm (yellow) outside the nucleus.

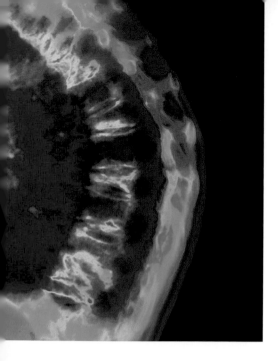

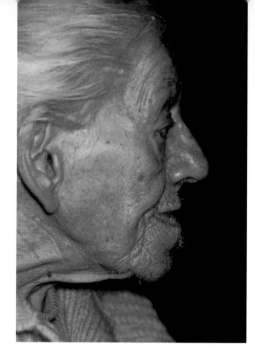

← **Acromegaly**—abnormal growth of cartilage and other tissues in the jaw, lips, eyelids, hands and elsewhere—is caused by a pituitary defect. This elderly woman shows signs of the disorder, which develops in adulthood when the pituitary gland overproduces growth hormone.

↞ **Hormone replacement therapy,** which replenishes a woman's estrogen supply, is one strategy for limiting the bone degeneration of osteoporosis, shown here in the bowed spine of an elderly woman. Although men sometimes develop osteoporosis, the disorder is most common in women past menopause, whose bodies have all but stopped making estrogen.

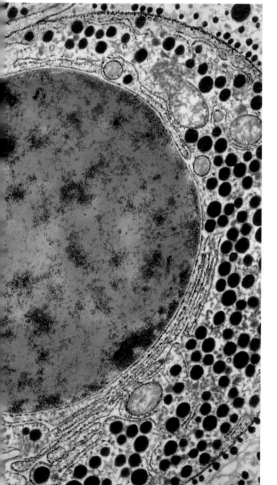

DUAL ROLES

Some hormones do double duty in the body's communication networks. Prime examples are epinephrine and its chemical cousin norepinephrine, hormones produced by the adrenal glands that promote the fight-or-flight response. In the brain, certain nerve cells use these chemicals as neurotransmitters to convey messages to other neurons. Another example is cholecystokinin, a hormone that helps regulate food digestion. It too has been discovered in brain tissue, where it may signal to the brain that a person has eaten enough.

Two recently discovered signaling chemicals are gases, nitric oxide (NO) and carbon monoxide (CO). Both act in the peripheral nervous system, among other functions. When NO is released by neurons that supply smooth muscle cells in the walls of blood vessels, the vessels dilate, so more blood flows into the tissue. CO slows the activity of muscles in the small intestine. CO also operates in the brain, helping to control the operations of the hypothalamus.

The discovery of dual roles for so many chemicals shows that, contrary to previous ideas, the nervous and endocrine systems share many of the responsibilities for controlling and regulating body functions.

→ **Hormones act** inside body cells. Once a steroid hormone has connected with its receptor inside the cell (lower right), its signal is transmitted to the cell's DNA in the nucleus. Protein-derived hormones bind with receptors on the cell's outer membrane (top left). In both cases, the hormone prompts a change in the way the target cell operates.

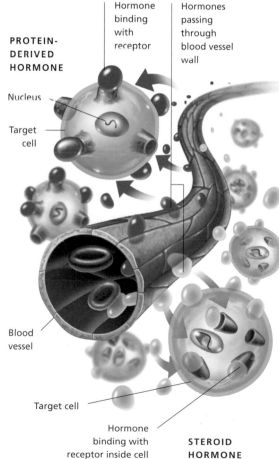

PROTEIN-DERIVED HORMONE

Hormone binding with receptor

Hormones passing through blood vessel wall

Nucleus

Target cell

Blood vessel

Target cell

Hormone binding with receptor inside cell

STEROID HORMONE

The hypothalamus and the pituitary

Deep in the brain, the hypothalamus and the pea-sized pituitary cooperate to make and release eight hormones that influence processes as different as general body growth and contractions of the uterus during childbirth. They jointly make up what physiologists call the "neuroendocrine control center," and share the tasks of manufacturing hormones and delivering them into the blood. The hypothalamus is the overseer, the pituitary a powerful middle manager.

Hormones made in the hypothalamus follow one of two pathways. Some are sent directly to the pituitary via nerve axons, which release the chemicals into small blood vessels there. The hormones are then transported by the bloodstream to tissues elsewhere in the body, such as the kidneys. Other hormones travel in the bloodstream from the hypothalamus to the pituitary, where they relay signals that spur or halt the release of the pituitary's own hormones, such as growth hormone. Once released, the pituitary hormones also enter the bloodstream, destined for other endocrine glands or elsewhere. In endocrine glands, the pituitary's signals may promote the release of yet another hormone. In this way, the pituitary directly or indirectly influences most other endocrine glands and many body structures and functions.

↑ **Gigantism and pituitary dwarfism** result from pituitary malfunctions, which can abnormally boost or reduce the amount of growth hormone in a growing child's body. Here a giant and dwarf flank a man of normal size.

← **A child's body grows** in response to growth hormone, which begins to be produced during fetal life and continues to influence many tissues, especially bone and muscle, throughout childhood.

A KEY COORDINATING CENTER

Not much larger than a kidney bean, the hypothalamus has neurons that coordinate processes involved in regulating body temperature, blood pressure, heart activity and other functions of internal organs. Some of its neurons are concerned with responses to biological drives, such as hunger and thirst, and others with emotional states. At the same time, the hypothalamus makes hormones that spark or shut down the secretion of hormones by the pituitary gland. Incoming signals from the nervous system and endocrine glands feed back to the hypothalamus, giving notice that the flow of hormones requires adjusting. For example, the adrenal glands make the hormone cortisol, which triggers several physiological responses to stress. Cortisol is pumped into the bloodstream when the anterior pituitary, stimulated by a releaser hormone from the hypothalamus, sends a chemical signal to the adrenals. Later, as circulating blood brings cortisol to the brain, receptors in the hypothalamus detect it and reduce the releasing hormone. Feedback loops such as this make possible this small organ's substantial contributions to "whole body management."

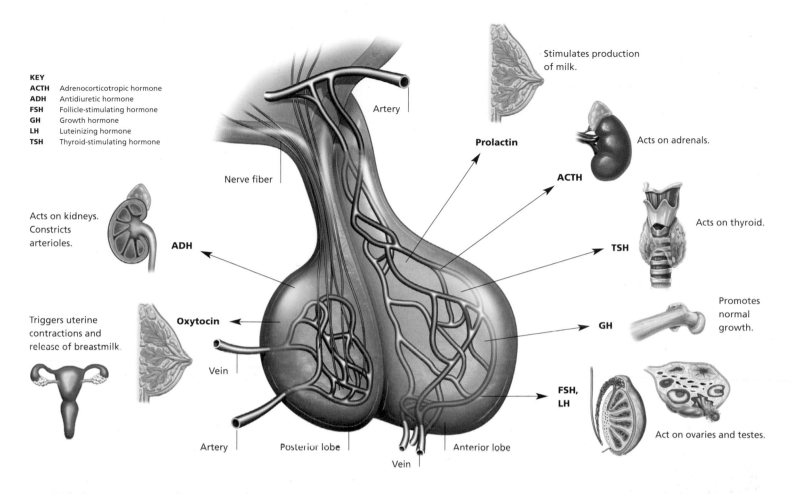

KEY
ACTH Adrenocorticotropic hormone
ADH Antidiuretic hormone
FSH Follicle-stimulating hormone
GH Growth hormone
LH Luteinizing hormone
TSH Thyroid-stimulating hormone

Artery

Nerve fiber

Stimulates production of milk.

Prolactin

Acts on adrenals.

ACTH

Acts on kidneys. Constricts arterioles.

ADH

Acts on thyroid.

TSH

Promotes normal growth.

GH

Triggers uterine contractions and release of breastmilk.

Oxytocin

Vein

Artery | Posterior lobe | Anterior lobe

Vein

FSH, LH

Act on ovaries and testes.

↑ **In the pituitary,** the smaller posterior lobe is a pass-through point for two hormones from the hypothalamus. The larger anterior lobe releases six hormones. Several of these control the activity of other endocrine glands and two have key roles in the reproductive organs.

← **Immediately after a meal,** insulin prompts cells to take up the sugar glucose from the bloodstream. Later, when less blood glucose is available, glucagon, cortisol and growth hormone help the body convert its stored carbohydrates, fats and amino acids to forms that cells can use as fuel until the next meal.

→ **Hormonal signals** from the hypothalamus trigger reduced metabolic activity during the Belding's ground squirrel's long hibernation in its Californian mountain home, when its core body temperature is just above freezing.

Insulin and its kin

Located behind the stomach, the pancreas has a dual role. One part of it is devoted to manufacturing various enzymes used in digestion, but it also contains several million scattered clusters of endocrine cells, called islets of Langerhans. In each cluster are cells that produce insulin and glucagon, two hormones that have a major influence on how the body uses the sugars we ingest in food. The interplay between insulin and glucagon regulates how much sugar, or glucose, is in the blood at any given moment, regardless of how often and how much we eat. Our survival depends on this fine-tuning, in part because glucose is the only substance that the brain can use for fuel.

Closely regulated blood sugar also is a key factor in weight control and maintaining overall health. Excess sugar may be converted into fatty acids and stored as fat, which can put a serious strain on the body's other organ systems. When the system for managing blood sugar breaks down, the result can be diabetes mellitus, the most common of all endocrine disorders. People with this form of diabetes may have sugary urine. Mellitus comes from the Greek word for honey, and ancient physicians who suspected diabetes reportedly would sometimes taste a patient's urine to confirm their diagnosis. Today, the form of diabetes called type 2 has become a global health epidemic. It seems often to be related to lifestyle choices, which makes it one of the most preventable serious diseases in the world.

→ **Endocrine "islands"** in the pancreas, called islets of Langerhans, make insulin and glucagon. In this magnified view, dark pink cells on the left are glucagon-secreting alpha cells. To the right, cells that appear green and yellow-brown are beta cells, which make insulin. Seen at the bottom are other cells (orange) that manufacture somatostatin, a hormone that normally shuts down the production of insulin and glucagon at appropriate times.

↓ **Obesity can be a sign** of insulin resistance and is a major risk factor for diabetes. Genes can also be a factor. Studies show that, other things being equal, people of American Indian, Asian, African and Hispanic descent are more likely to have insulin-resistant cells and thus to develop diabetes. For all ethnic groups, exercise and a healthy diet can help prevent the onset of diabetes and limit its progression.

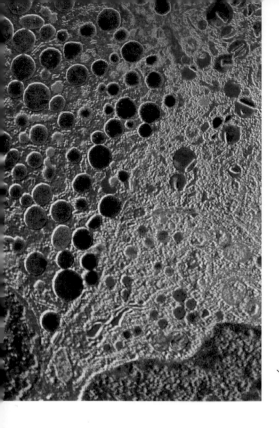

THE DIABETES THREAT

Diabetes has two main forms. Type 1 more commonly begins in childhood, developing when the insulin-making cells in the pancreas are destroyed, possibly by an autoimmune response to a viral infection. In type 2 diabetes, which usually occurs later in life, insulin receptors on body cells gradually stop working and cells do not respond to insulin's signal to take up glucose from the blood. As abnormally high levels of the blood sugar circulate, the smallest blood vessels, or capillaries, are damaged and ultimately tissues and organs suffer from an impaired blood supply. People with advanced type 2 diabetes may become blind, suffer a heart attack, lose a foot or leg to gangrene, or have other complications. Treatment may include insulin injections and drugs that make cells more receptive to insulin or stimulate the pancreas to make more of the hormone.

↓ **Insulin lowers** blood sugar and glucagon increases it. Together they interact to keep blood glucose levels at a set point.

↑ **Using a pen-like device** that meters out the correct dose from a cartridge, this girl is giving herself one of the regular insulin injections that are essential to her survival.

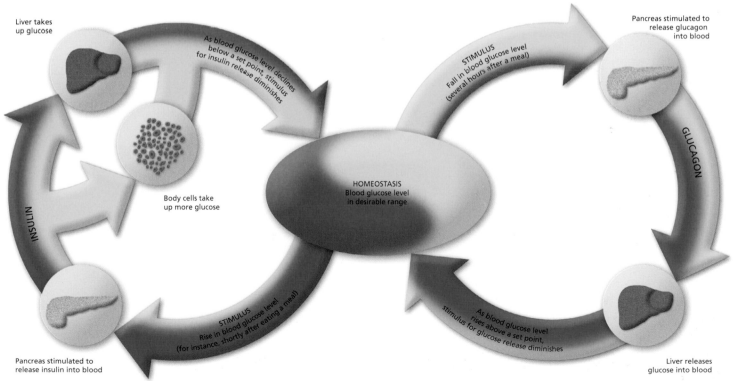

Liver takes up glucose

As blood glucose level declines below a set point, stimulus for insulin release diminishes

STIMULUS
Fall in blood glucose level (several hours after a meal)

Pancreas stimulated to release glucagon into blood

Body cells take up more glucose

INSULIN

GLUCAGON

HOMEOSTASIS
Blood glucose level in desirable range

STIMULUS
Rise in blood glucose level (for instance, shortly after eating a meal)

As blood glucose level rises above a set point, stimulus for glucose release diminishes

Pancreas stimulated to release insulin into blood

Liver releases glucose into blood

Corticoids

Just above each kidney is a cone-shaped adrenal gland. Its outer layer, called the adrenal cortex, secretes potent steroid hormones, including two families of chemicals called corticoids. The family of glucocorticoids includes cortisol, which can swiftly raise blood sugar levels—an important response when the body is stressed and tissues such as muscles may need extra energy. Cortisol also is a potent anti-inflammatory agent. Its chemical kin, the drugs cortisone and hydrocortisone, are used in anti-inflammatory remedies. The adrenal cortex also makes DHEA, a male sex hormone. In males, DHEA is overshadowed by testosterone, but in females its roles include helping promote the growth spurt at puberty. The other corticoids, termed mineralocorticoids, adjust the levels of minerals such as sodium and potassium. Aldosterone, the most abundant of these, helps the kidneys conserve water. Inside the medulla region of each adrenal gland, neurons release the hormones epinephrine and norepinephrine, which stimulate the flight-or-fight response.

HORMONES AND STRESS	
Physical or emotional stress can trigger one or more of the following responses from different parts of the endocrine system:	
Hormone	**Source and effects**
Cortisol	Adrenal gland hormone that recruits energy sources by helping boost blood levels of glucose, fatty acids (from fat) and amino acids
Epinephrine	Adrenal gland hormone that helps sustain the fight-or-flight response mobilized by the sympathetic subdivision of the nervous system
Vasopressin	Hormone of the hypothalamus that raises blood pressure and is also involved in water conservation
Glucagon	Pancreatic hormone that increases blood glucose
Aldosterone	Adrenal gland hormone that is part of a sequence of reactions that conserve body water and so elevate blood pressure

← **Mineralocorticoids** are mobilized in a firefighter's body, generating the classic fight-or-flight response to heavy physical demands, anxiety and fear.

↑ **Asthma inhalers** may contain steroids based on cortisol, which help tame the potentially lethal swelling of airways.

↑ **Crystals of the hormone cortisol** reflect a rainbow of colors when viewed with polarized light under a microscope.

ADRENAL GLANDS

Like the hypothalamus, the adrenal glands embody a blend of hormonal and nervous system controls. The cells of the cortex are specialized to make and release their steroid hormones, while the gland's interior, the medulla, is populated by neuron-like cells. These release chemical signals, mainly epinephrine, that function in both realms—as neurotransmitters in the brain, and as hormones elsewhere in the body. The adrenal medulla is actually an outpost of the sympathetic nervous system, the subdivision that operates during times of heightened awareness, excitement or danger. Malfunctioning adrenal glands underlie a variety of disorders. Because cortisol influences the body's use of glucose, a cortisol deficiency can lead to severe hypoglycemia, in which blood sugar levels are too low. A deficiency of the hormone aldosterone may result in abnormally low blood pressure and life-threatening changes in the heart's normal rhythm.

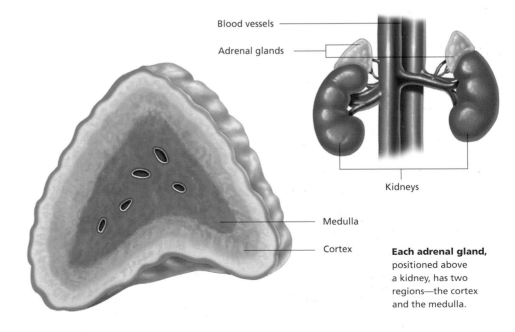

Blood vessels

Adrenal glands

Kidneys

Medulla

Cortex

Each adrenal gland, positioned above a kidney, has two regions—the cortex and the medulla.

Sex hormones

Sex hormones—estrogens and androgens—influence the development of the bodily features and behaviors we associate with femininity and masculinity. They also orchestrate the development of sperm and eggs, the reproductive cells that ensure the continuity of our species. In females, two key sex hormones are estrogen and progesterone. Not only do they influence external aspects of the body, they also regulate a woman's menstrual cycles and help sustain pregnancy. In males, testosterone both shapes physical features and stimulates sexual behavior. Males tend to be more aggressive than females, another difference that may be related, at least in part, to sex hormones. Although such effects may seem neatly divided between the sexes, androgens and estrogens are produced by both males and females. When a young adult's long bones stop growing due to the hardening of the cartilage growth plates, estrogen is responsible in both men and women. A woman's ovaries make small amounts of testosterone, which may influence her sex drive, as it does in men. Connections like these may hark back to the early weeks in the womb. Although genetic sex is determined at conception, the reproductive parts that make a baby recognizably male or female develop only after the respective genes begin operating, and sex hormones take on their many body-shaping tasks.

SECONDARY SEXUAL TRAITS

The most important functions for sex hormones are stimulating the making of eggs and sperm, but hormones from the ovaries and testes also play central roles in shaping a person's secondary sexual traits—the varied features that we associate with maleness and femaleness. In females, estrogen and progesterone promote the accumulation of fat in the breasts, hips and buttocks, helping give women their rounded body contours. In males, testosterone promotes the growth of facial hair, growth of the larynx that deepens the voice, and the development of more massive bones and skeletal muscles. In both sexes, hormonal changes increase the activity of oil glands in the skin or the face and back. They also promote the growth of underarm and pubic hair—sure signs of impending sexual maturity.

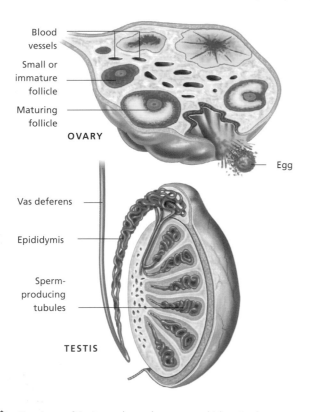

OVARY

Blood vessels

Small or immature follicle

Maturing follicle

Egg

Vas deferens

Epididymis

Sperm-producing tubules

TESTIS

↑ **Ovaries and testes** make sex hormones, which not only stimulate the release of eggs, or ova, and sperm but also influence the development of the physical traits that mark us as male or female .

↑ **Joan of Arc** reportedly had several "male" characteristics, such as a lack of menstruation and aggressive behavior. Although a variety of factors can underlie such traits, a few researchers have speculated that the legendary heroine was a genetic male affected by testicular feminizing syndrome— the development of external female features due to skewed hormonal signals during prenatal life.

← **The faces of this couple** clearly show sex-based differences. In addition to his beard, the man has relatively thick facial skin. Because the skin of a woman's face is thinner, it may also begin to show signs of aging sooner.

↗ **At puberty,** both estrogen and testosterone stimulate the enlargement of reproductive organs and external genitals, as well as the development of secondary sexual traits. They also promote the growth of long bones, so youngsters undergo a growth spurt, and influence developmental alterations in the brain as well.

→ **Successful pregnancy** depends on hormones. The placenta that nourishes a fetus produces hormones that prevent the uterus from shedding its lining after an embryo has implanted. The hormones also stimulate the enlargement of the woman's breasts in preparation for milk production.

The powerful thyroid

The thyroid sits at the base of the neck in front of the windpipe, or trachea, looking like a lumpy butterfly with outspread wings. How we develop and grow, and how the body's metabolism operates are directly related to the way this gland functions. With few exceptions, two of the hormones it makes—T_3 and T_4—affect every cell in the body. Cells use the blood sugar glucose as their basic fuel. T_3 and T_4 stimulate cells to remove glucose from the circulating blood and directly regulate the processes by which they transform it into usable chemical fuel. T_3 and T_4 have major impacts on other body functions as well. Unless they pave the way, growth hormone and several other hormones cannot act on their target cells. They are also essential to the processes that assemble proteins and that trigger the growth and maturation of bones, the development of the fetal nervous system, and many other fundamental events.

Another thyroid hormone, calcitonin, interacts with parathyroid hormone (PTH), made in four tiny parathyroid glands tucked behind the thyroid. PTH promotes the release of calcium into the blood from bones, while calcitonin slows it. Jointly these substances provide enough calcium for neurons and muscle fibers to function efficiently, without drawing so much from bones that the skeleton weakens.

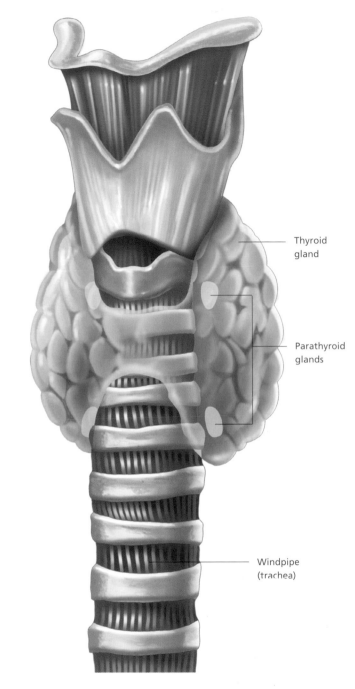

Thyroid gland

Parathyroid glands

Windpipe (trachea)

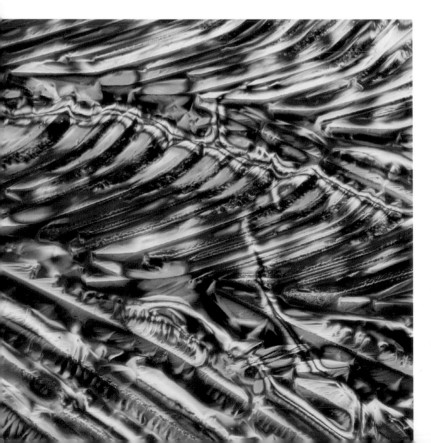

↑   **The thyroid gland** is the body's largest endocrine gland. Its two lobes consist of many round, hollow sacs where hormones are produced. Each of the four parathyroid glands is the size of a lentil.

←   **Calcitonin,** seen here as crystals under the microscope, is released by some of the cells of the thyroid when calcium-rich blood flows through the gland. The hormone slows the withdrawal of calcium from bone.

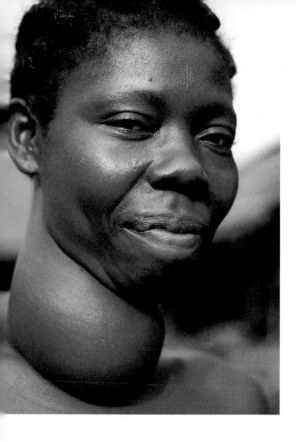

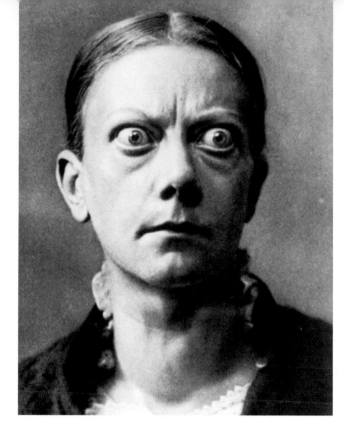

THYROID DISORDERS

Various factors can cause the thyroid gland to malfunction. Having too little of the thyroid hormones T_3 and T_4 is called hypothyroidism. People who suffer from this deficit tend to feel cold and gain unwanted weight because their cells are not receiving the necessary signals for proper metabolism. They may also become depressed and develop a goiter. The hypothalamus monitors the thyroid's output, and orders an increase in pituitary TSH if the blood levels of thyroid hormones fall below a set point. If the thyroid is unable to secrete enough T_3 and T_4, the signals keep coming and ultimately the overstimulated gland grows larger. Too little T_3 and T_4 in children stunts growth and can lead to mental retardation. If the deficit is caught in time, thyroid hormone replacement therapy can replenish the missing hormones and prevent these complications from occurring.

The opposite problem, having surplus thyroid hormones, is called hyperthyroidism. Causes vary, and some people are genetically predisposed to the disorder, which can lead to Graves' disease or a "toxic" goiter. An affected person may feel jittery, lose weight even though they eat well, and sweat profusely as overworking cells give off excess heat. The heart works overtime as well.

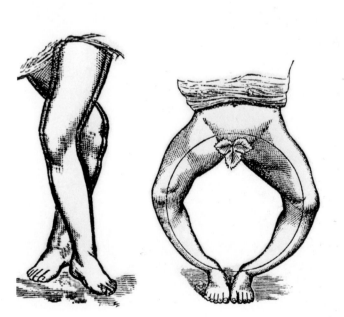

↑ **Rickets** is a childhood bone disorder caused by vitamin D deficiency, producing misshapen legs (left and center). Activated in part by parathyroid hormone, vitamin D acts as a hormone, promoting the digestive tract's ability to absorb calcium from food.

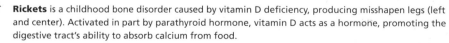

Hormones from the heart, gut and elsewhere

Although we do not usually think of our skin, heart or digestive organs as endocrine glands, in fact there are hormone-producing cells and tissues throughout the body. Human skin manufactures vitamin D, which then is chemically altered into a form that partners with PTH, the parathyroid hormone that ensures an adequate supply of calcium for building bones and teeth. In the heart, clumps of cells in the atria, its two smaller chambers, make a hormone called atrial natriuretic peptide (ANP). When blood pressure rises above a set point, ANP signals to cells in the kidneys to excrete salt. Water follows the salt, moving out of the blood and into urine to be expelled. As the blood volume falls, so does blood pressure. The thymus, situated behind the breastbone, or sternum, makes hormones that guide the development of invasion-fighting cells of the immune system. This gland is most active in children, helping to build a mature immune system. Patches of cells located in the stomach and small intestine release hormones that convey many messages related to feeding—including appetite, hunger and the "full" feeling that tells us to stop eating. Nearly all our body cells have the precursors of prostaglandins, signaling molecules that are among the most biologically active substances science has discovered.

→ **At least half a dozen hormones** govern the chemical processing of food, as well as producing cues related to appetite, hunger and satiety. Some of these hormonal signals are sent by the brain, others are generated in the digestive tract. Jointly they have a profound affect on a person's eating patterns.

↓ **Soaking in a hot spring pool** can have the same effect as a rise in blood pressure. The bloodstream carries excess body heat to the skin where it can dissipate. When the heart pumps this heated blood, the heart's left atrium may stretch slightly, interpret the change as a rise in blood pressure and release ANP. The kidneys then increase the flow of urine to the bladder, and the bather has to heed a "call of nature."

↓ **Headache pain** is often relieved by aspirin and other nonsteroidal anti-inflammatory drugs (NSAIDs). These drugs work in part by inhibiting the synthesis of prostaglandins that constrict blood vessels. The same drugs are useful for relieving heavy menstrual cramps and bleeding—both of which are caused by prostaglandins that act on smooth muscle in the uterus—and for reducing inflammation in overworked muscles.

PROSTAGLANDINS RULE

Every cell in the body has the chemical ingredients needed to make prostaglandins, and prostaglandins are virtually everywhere in the body. To date at least 20 have been discovered, and they produce a wide range of effects. Prostaglandins are "local" signaling chemicals that act in the area where they are released. There, they may boost the sensitivity of small blood vessels to other chemical changes that are unfolding in the defensive counterattack to injury, called inflammation. They are part of the signaling pathway that raises body temperature and produces a fever. They cause menstrual cramps, but also help the uterus contract to deliver a baby. Elsewhere, different prostaglandins can help open up or close down airways to the lungs, shut down the release of stomach acid or stimulate the movement of food through the intestines, worsen pain and even stimulate the muscle contractions that help propel sperm.

→ **Skin irritated by poison oak** erupts in painful, itchy blisters. As with poison ivy and some other "poisonous" plants, the culprits are oils in the leaves that trigger an allergic reaction in human skin. Prostaglandins fan the inflammation; corticosteroid drugs help keep it in check.

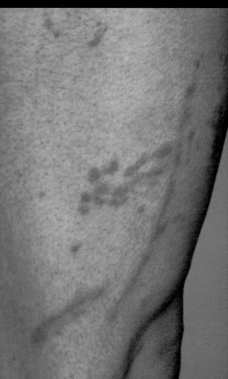

Hormones and the environment

Chemical signals tell us a great deal about the world we inhabit. Many creatures are extremely attuned to light, and in particular to the shifts that mark day and night. Like other complex animals, humans have light-sensitive eyes that send nerve impulses to the brain. The pineal gland in the brain responds to these impulses and so is sometimes called the "third eye." It secretes melatonin, a hormone that operates as part of the body's internal biological clock. In a general way melatonin "keeps time," increasing in the blood as night falls and decreasing with the morning light. Signaling molecules also convey other environmental cues. Some of the most mysterious communication signals in the animal world are pheromones—airborne chemicals that bear messages about territorial boundaries, readiness for mating and other information. Human activities are adding other chemicals to the environment, such as pesticides and human hormones, that can disrupt the normal functioning of many species.

→ **"Good chemistry" between people** may be pheromones in action. Possible pheromone sensors have been discovered about half an inch (just over 1 cm) inside each nostril, which may account for responses such as "love at first sight."

↓ **The bushy antennae** of the male silkworm moth detect a pheromone released by females ready to mate. The male can sense the pheromone as far away as 1 mile (1.6 km) and follow its chemical trail back to the source.

↑ **Sleep/wake cycles** fall prey to changing time zones. The pineal gland makes melatonin in the dark, so levels rise in the evening, preparing us for sleep. Until the pineal gland adjusts to day and night in a new time zone, jet-lagged air travelers typically find their sleep/wake cycle disturbed.

↑ **Light treatments may offset** some physiological responses to extended darkness. In parts of the world where winters are long and dark, humans may become depressed, crave sleep and binge on carbohydrates—all symptoms of seasonal affective disorder, or SAD. Deprived of adequate sunlight, some people also develop a deficiency of the vitamin D needed for healthy bones and teeth. In this school in Stavropol, Russia, girl students are being exposed to strong, full-spectrum fluorescent light to help stave off these potentially harmful effects.

THE BIOLOGICAL CLOCK

Many animals have an internal timekeeping mechanism that is influenced by the hormone melatonin. In humans, this timekeeper is a cluster of nerve cells in the hypothalamus called the suprachiasmatic nucleus (SCN). The SCN is studded with melatonin receptors, and its operations include regulating daily rhythms such as appetite and cycles of sleep and waking. In animals melatonin helps set the biological schedule for seasonal breeding. Reproduction in humans may also be subtly shaped by changing levels of melatonin. Youngsters whose pineal gland does not produce melatonin may start to mature sexually during childhood, suggesting that the hormone influences puberty.

Many other, "non-clock" roles have also been proposed for melatonin. It is a potent antioxidant, removing harmful free radicals that can damage a cell's DNA. This and other discoveries have led some researchers to speculate that melatonin might be harnessed to retard aging and enhance the functioning of the immune system.

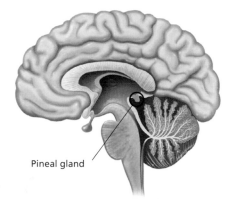

Pineal gland

↑ **The tiny pineal gland,** which releases melatonin, nestles deep in the brain. Although its workings are still not fully understood, the pineal gland probably monitors changing light levels by way of sensory signals from the eyes.

↑ **Deformed frogs,** like this five-legged African clawed frog, have become a worldwide phenomenon. Although several factors may be at work, researchers have evidence that some deformities are related to pesticides that slow or halt the production of thyroid hormones. In frog tadpoles, as in human embryos, these hormones play key roles in the proper development of body parts.

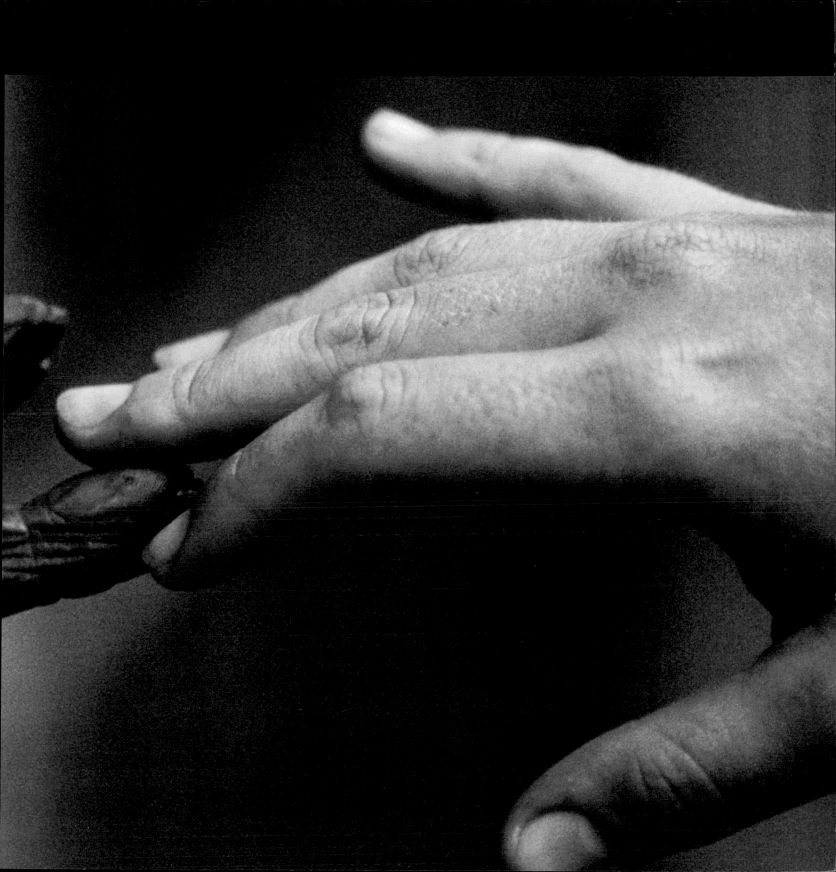

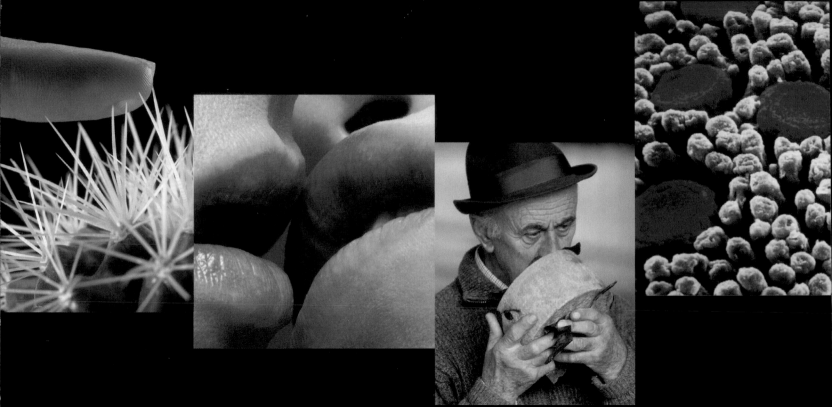

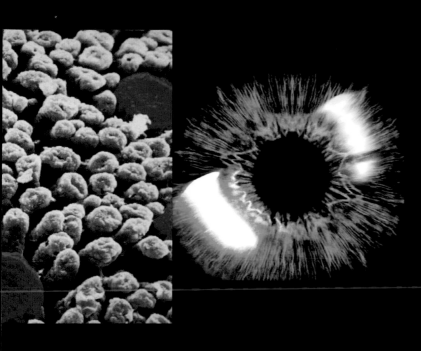

Sensing the world

Our senses are portals for gathering essential information about the ever-changing world around us. Several, such as vision, hearing and balance, are attuned to external events, while others, such as the sense of pain, feed back information about the internal workings of tissues and organs.

Humans as sensory creatures	104
Responding to a stimulus	106
Sensory pathways	108
Touch and pressure	110
Pain and its perception	112
The sense of smell	114
The sense of taste	116
Sound waves and the ear	118
Hair cells and hearing	120

Humans as sensory creatures

Nearly everything we know about our world we discover and experience through our senses. Exquisitely attuned sense organs located in various parts of the body allow us to perceive colors and shapes, tastes and smells, sounds and temperatures, body positions, pressure and pain. Yet vital and wondrous as these sensory capabilities are, they have limits. No human can see in the ultraviolet spectrum, as bees do, nor can we discriminate the hundreds of millions of odors that a bloodhound can smell. We do not sense the Earth's magnetism, which helps guide newborn leatherback sea turtles on their first ocean journey, nor do we sense the high-frequency sound waves that lead a bat through the dark to its supper.

Each species has senses that are suited to the opportunities and challenges of its environment. In combination with the brain's sophisticated centers for processing and interpreting sensory signals, the array of human senses dramatically expands our ability to respond to our surroundings. Survival—the ability to hear an approaching automobile, see and smell smoke, taste spoiled food, or feel the heat of a stove burner—depends in large measure on these inputs, and so does our psychological health. Studies show that people deprived of sensory stimuli for long periods begin to hallucinate and break down mentally. Those vital stimuli include interactions with other people, from a warm hello to a handshake or hug.

→ **Young babies are** exquisitely attuned to touch—including caresses and cuddling that provide comfort and security in the strange new environment they have entered. Even before birth, a fetus hears sounds in the womb ranging from its mother's heartbeat to voices and music, and feels the amniotic fluid around it.

→ **Smell, or olfaction,** is one of the most ancient senses. Recent studies have revealed that humans smell "in stereo," with signals from each nostril traveling to a different processing region in the brain. This arrangement may help a person determine if a smell is coming from a particular direction.

→ **Hearing is a sense** that detects sound waves of different frequencies and intensities. Headphones filter out "noise"— unwanted sound signals that complicate the task of discriminating more important or desirable ones.

→ **A gravity-defying snowboarder** depends on his sense of balance. This sense, which conveys information about the body's location in space, relies in part on signals from sensors in the inner ear. It also requires the brain to rapidly integrate simultaneous inputs from the eyes, skeletal muscles and skin, and to coordinate the appropriate responses.

MIXED PERCEPTIONS

In a small number of people, the brain processes one type of sensory signal—such as sound—in a way that produces multiple sensory experiences. This unusual condition is called synesthesia, meaning "perceived together." For example, the person may both hear a given sound and "see" it as a particular color, or taste flavors and also "hear" them as sounds. Pain may be perceived along with colors or sounds as well. In all, researchers have confirmed at least 20 different combinations of such mixed perceptions, and some believe there are more. Synesthesia is not a form of mental illness, but it is as puzzling as it is remarkable. Neurologists do not understand why the brain of a synesthete handles the affected sensations in different processing areas simultaneously. As a result of this two-track processing, however, both the actual sensory experience and its "partner" perception are equally real and vivid.

↑ **As he climbs the mountain,** this hiker takes in the expansive view of snowcapped peaks thanks in part to his senses of vision and balance. Smelling the fresh air and feeling the breeze enrich the experience.

→ **Invisible to the human eye,** dark lines and blotches on an evening primrose can be seen by insects with eyes sensitive to ultraviolet light, directing them to the parts of the flower with nectar and pollen.

Responding to a stimulus

Perception is what our brain makes of a sensory stimulus—how we consciously interpret its meaning. By contrast, a sensation is the conscious awareness of a physical event that takes place when a receptor detects a stimulus. A receptor can be a specialized cell that communicates with the ending of a sensory neuron, or the branched terminal extensions of nerve cells may receive incoming signals. Most receptors are specialized to detect a particular type of stimulus and send that information back to the brain. The stimulus provides the energy this communication process needs to get under way. It can take the form of a mechanical force, a chemical, or electromagnetic and thermal energy—light and heat. Some receptors are scattered in a number of locations and are known as "somatic" receptors, meaning "of the body." Other receptors are restricted to specific locations, such as the eyes, ears, nose and taste buds, and are the basic elements of the body's special senses, including vision, hearing, smell and taste.

There are six major families of sensory receptors. Four of them encompass receptors that respond to light, sound and chemicals in the air or in the mouth. Our sense of touch and that of the position of the body and its parts are provided by pressure-sensitive "mechanoreceptors" in the skin and inner ear respectively. A sixth group, osmoreceptors, lie deep in the brain and respond to changes in the volume of water in the surrounding tissue fluid. They enable the brain to monitor and adjust the body's balance of water and salts.

↓ **Pain is communicated** by nociceptors, bare nerve endings that respond to intense mechanical pressure, such as from a fall, as well as to temperature extremes and noxious chemicals. Any sensory neuron that is overstimulated may send signals that the brain also interprets as pain.

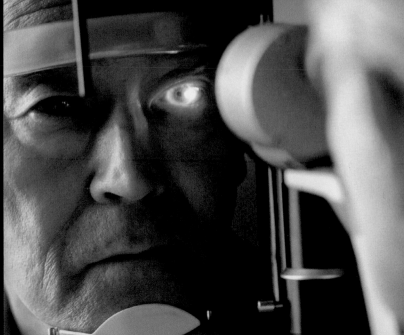

← **Eating a lemon** (left) triggers special chemoreceptors in the taste buds that respond to chemicals associated with a sour taste. A *salty snack* (center) stimulates osmo-receptors in the brain that monitor the water content of blood. To prevent too great a water loss as the kidneys filter excess salt, the hypothalamus stimulates thirst and water-conserving mechanisms in the kidneys. A *massage* (right) activates a subset of mechanoreceptors that respond to touch.

← **Photoreceptors sense light.** Located in the retina at the back of the eyeball, human photoreceptors are packed in dense arrays that feed signals to nerve cells. The axons of some of these neurons are bundled to form the optic nerve, which carries visual signals to the brain for processing.

→ **Thermoreceptors in the skin** register heat and cold. Bare nerve endings monitor shifts away from core body temperature. In the brain, thermoreceptors directly monitor core temperature by tracking the temperature of blood flowing through brain tissue.

FROM STIMULUS TO SENSATION

Some of our sense organs pick up stimuli from the outside world, others sense internal stimuli related to blood chemistry, pain, muscle stretching, body temperature and so forth. Regardless of where a stimulus arises, it does not travel directly to the brain to be processed. Instead, the stimulus first must be converted into nerve impulses. Channel-like proteins spanning the outer membranes of a sensory cell permit select substances to cross into it. If the chemical change is substantial enough, it may set up the conditions that generate a nerve impulse. All nerve impulses are equal in strength, but the stronger the incoming stimulus, the more often the sensory neuron will fire. A very weak stimulus may be ignored because its effect on the receptor is minor. Ultimately, the body responds only to sensory inputs that are forwarded to the central nervous system. Countless signals arrive in the CNS at any given moment, but only certain ones will become conscious sensations. Some, including many signals from our muscles, joints and internal organs, are processed in the spinal cord or elsewhere below the level of consciousness. Sensory receptors may be the "front doors" of our sensory systems, but we see, taste, smell, hear and feel touch entirely in the brain

Sensory pathways

Messages from receptors are transmitted via sensory nerves to the brain, traveling in the form of nerve impulses. This raw data, however, acquires meaning only after it is processed in a series of steps. The brain learns the general nature of the stimulus from the type of sensory nerve that communicates its signal. Optic nerves from the eyes bear only information about light, auditory nerves from the inner ear carry only information related to sound, and so on. Other information indicates the extent and intensity of the stimulus. A sensory nerve contains axons of many sensory neurons. When a drop of water falls on your knee, only a few neurons may fire, while an ice pack applied to the knee may trigger nerve impulses in dozens or hundreds of them. The strength of the stimulus is measured by how often neurons fire off impulses. Once the sensory signal reaches the brain, the information can be integrated with memories and sensory inputs arriving from elsewhere in the body. Only when all this information is assembled does the brain organize a response.

SENSORY ADAPTATION

When some sensory neurons are stimulated continually, they may suddenly slow or even stop firing off nerve impulses. In effect, the receptors become accustomed to the unvarying stimulus and will trigger a new volley of nerve impulses only if something changes. If you put on socks, this sensory adaptation is at work when you stop being highly aware of the pressure of the socks on your skin. Some of the mechanoreceptors in the skin of your feet have adapted and shut down "pressure" signals to your brain. Remove the socks and the receptors will again be activated, so for a few moments you become consciously aware of having bare feet.

Nerve impulses

→ **Gentle pressure** from the tip of a cactus spine activates only a few pressure receptors in the skin. Press harder, however, and receptors for pain may be stimulated—triggering "orders" from the brain to pull your finger away.

⇢ **This kiss recipient's brain** processes information about the sustained, gentle pressure of another's lips, while his eyes send nerve impulses generated by a different stimulus, the bright light of a camera's flash bulb. His brain can clearly differentiate between the two.

← **Awareness of a stimulus** such as the surface of a basketball begins at sensory receptors in the hands. These generate nerve impulses that travel via sensory neuron axons in the arm to the spinal cord and then via inter-neurons to the brain.

Clasped hands stimulate sustained, moderate pressure over a fairly large area of skin, and trigger sensory impulses from hand muscles and tendons as well. The result is many separate nerve impulses that the brain integrates and interprets as a firm handshake.

↑ **Even a glancing touch** can activate many sensory receptors if it involves a relatively large area.

↓ **Kicking a ball** is a complex sensory and motor event. Motor signals to leg muscles trigger the contractions needed for running and kicking. The foot-on-ball impact elicits a brief, strong sensory message from his foot to his brain. Sensors in his ears, eyes and elsewhere monitor the positions of his head and other body parts, enabling his brain to order compensatory movements—like outspread arms—that help prevent him from falling.

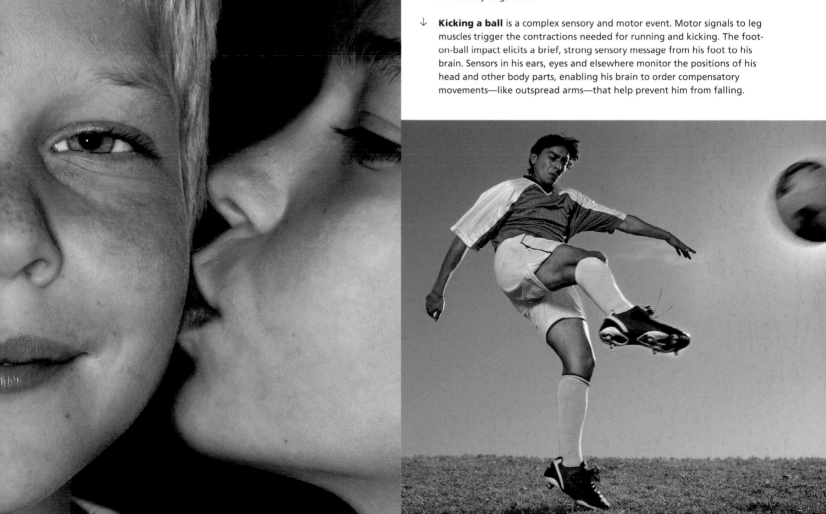

Touch and pressure

The skin covering each of your fingertips has roughly 300 sensory receptors, and countless others are packed into other areas of your skin, in skeletal muscles and tendons and in the muscular walls of organs like the stomach and bladder. Sensory receptors are the entry points for somatic sensations, which are consciously recognized when signals from the receptors reach the brain's "body sense" region, a layer of gray matter in the cerebrum called the somatosensory cortex.

Bare nerve endings—thin, terminal branches of sensory neurons—can be found both close to the surface and deep within the body. They can be sensitive to heat, cold, pressure or pain. Other nerve endings twine around hair follicles. Any movement of the hair—whether from a breeze wafting over the skin or an insect traversing it—will relay a message. Also close to the body surface are receptors in capsules, together with other special non-sensory cells that augment the sensation or play a supporting role. They include structures that are attuned to fine textures, vibrations or steady touching, heat or cold. Large numbers of receptors called muscle spindles are found in skeletal muscles, and there are also sensory nerve endings in the tendons that strap muscles to bones. These receptors register the stretching of muscle fibers during limb movements. Along with other mechanoreceptors in muscles and movable joints, muscle spindles help convey information on the position of limbs and the body as a whole.

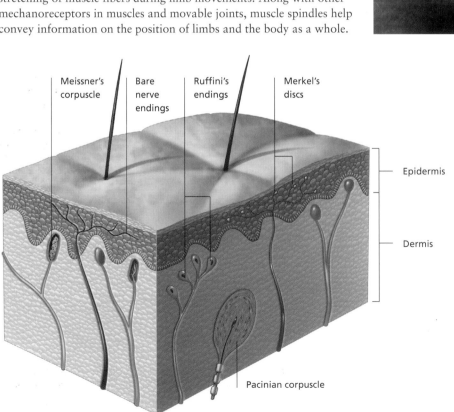

Meissner's corpuscle | Bare nerve endings | Ruffini's endings | Merkel's discs

Epidermis

Dermis

Pacinian corpuscle

↑ **Lips and the tip of the tongue** have more sensory receptors per unit of area than most other parts of the body. Embedded just beneath the skin's surface, the receptors help convey the sensations of a kiss by detecting light pressure, the duration of a touch, the area it covers and other tactile information.

← **Sensory receptors** pack the skin. They are found in and under the dermis as well as at the boundary of the epidermis and dermis, while some penetrate the epidermis. Bare nerve endings serve as pain receptors and mechanoreceptors that respond to the movements of hairs. They also provide much of the sensory feedback we interpret as an itch or a tickle. The main tactile receptors are Merkel's discs, which steadily monitor the "touch" of things that are in contact with the skin. Meissner's corpuscles specialize in detecting changes in touch stimuli, such as repeated taps on a keyboard. Ruffini's endings respond to low-frequency vibrations and skin stretching. Deeper in the dermis, Pacinian corpuscles keep track of stronger vibrations.

← **The delicate skin of the eyelids** contains many sensory receptors, particularly ones specialized for touch and low-frequency vibrations. Most of the body's light-touch sensors are in facial skin, fingertips, nipples and other hair-free areas.

← **A dancer's acrobatics** require constant monitoring of body position. Key sources for this information are muscle spindles and other sensors that provide feedback on the relative locations and movements of a person's limbs.

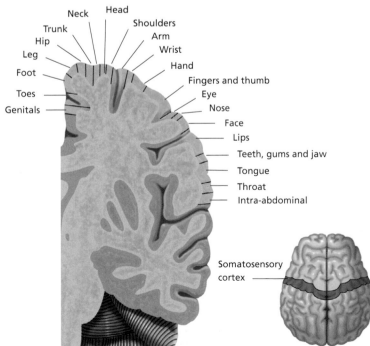

Neck
Head
Shoulders
Trunk
Arm
Hip
Wrist
Leg
Hand
Foot
Fingers and thumb
Toes
Eye
Genitals
Nose
Face
Lips
Teeth, gums and jaw
Tongue
Throat
Intra-abdominal

Somatosensory cortex

KNOWING WHERE YOUR PARTS ARE

Almost any kind of athletic movement requires the central nervous system to make non-stop adjustments in the movements and relative positions of body parts, including the limbs. A variety of receptors help supply the necessary information, including muscle spindles and other specialized receptors in muscles and joints. Muscle spindles and receptors in tendons report almost constantly to the brain about the length of muscles and the tension, or force, they are exerting. In joints, other receptors detect changes in the angle and direction of movement.

Muscle spindles are particularly important in feeding back information to the CNS. Embedded deep in the muscle, each one is a unit of several specialized muscle fibers plus a nerve fiber from a sensory neuron near the spinal cord. The neuron communicates directly with a motor neuron in the spinal cord that supplies the same muscle. When the spindle is stretched, its sensory signals rapidly trigger a corresponding contraction in the muscle. A golfer's backswing or a basketball player's pre-jump crouch greatly stretches the spindles in the muscles concerned, and the muscles respond to the signals with extra-strong contractions.

← **Brain neurons** that process touch sensations physically "map" different areas of the body surface. The neurons are arrayed in a strip of gray matter called the somatosensory cortex. Proportionately larger areas of the cortex are devoted to very sensitive parts of the body such as the fingers and lips.

Pain and its perception

Stripped to its essentials, pain is a biological warning. It notifies the brain that tissues elsewhere in the body have been damaged or are under duress. This unpleasant but vital message begins with signals registered by the bare nerve endings called nociceptors—a term that comes from an apt Latin word meaning "to do harm." Various nociceptors respond to extremes of pressure or temperature or to chemicals that can damage the skin or internal tissues. We have millions of them located virtually everywhere in the body, except in the brain. The central nervous system itself deals with pain in intriguing ways, and has a built-in mechanism for dampening pain. When pain signals reach the spinal cord, interneurons there release substance P, a neurotransmitter that launches the release in the brain of endorphins and enkephalins. These natural pain suppressors adjust the activity of brain neurons that process pain signals, so the person perceives pain as less intense. They seem to be released at times when reducing pain has a survival benefit, such as during childbirth or when an injury must be temporarily ignored while a person escapes from danger. Endorphins also are credited with triggering euphoria during prolonged exercise. Other chemicals that bind endorphin receptors and mimic its effects include euphoria-producing narcotic drugs such as opium, heroin and morphine.

↓ **Headache** is one of the most common types of pain. Triggers may range from hormonal changes, stress and changes in diet to allergies and even the weather. Pain-sensing neurons respond to the swelling of blood vessels in the brain's meninges or in muscles of the scalp, face or neck. Today so-called "tension headaches" and migraines are considered part of a continuum that can be treated with drugs ranging from aspirin to corticosteroids. Men are particularly susceptible to agonizing cluster headaches, which are experienced as a piercing pressure in one eye and can recur several times a day for weeks or months.

→ **Referred pain** from various internal organs is felt in certain skin areas. A physician can use this information to help diagnose illness.

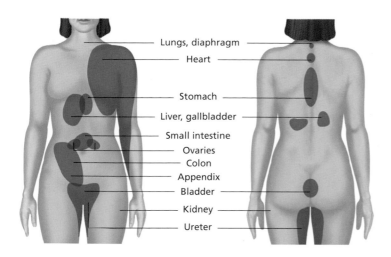

Lungs, diaphragm
Heart
Stomach
Liver, gallbladder
Small intestine
Ovaries
Colon
Appendix
Bladder
Kidney
Ureter

PHANTOM AND REFERRED PAIN

For you to be aware that some part of your body hurts, your brain must have identified the source of the damage and sent a "pain" signal back to it. When an amputee experiences "phantom pain" from a lost body part, the severed sensory nerves continue to send pain signals long after the amputation.

The phenomenon of "referred pain" results from a different quirk in the wiring in the nervous system. Pain sensations from internal organs sometimes are assigned not to the organ, but to some part of the skin. A classic example is the pain from a heart attack, which victims often feel in the skin over the heart, in the left shoulder and arm, or in the upper back, between the shoulder blades. Likewise, pain emanating from an inflamed appendix—a fingerlike projection at the start of the large intestine—may be experienced over a large area on the right side of the abdomen.

The confusion may develop because nerves carrying pain signals from the skin enter the spinal cord at the same level as nerves from internal organs. It is also possible that the skin is the brain's "default setting" for pain signals, because in daily life most painful stimuli come from areas of the body surface.

← **Sunburn** signals major injury to the skin. The inflammatory response includes the release of histamine and prostaglandins that stimulate pain receptors. This is why many sunburn remedies, like pain treatments in general, incorporate anti-inflammatory agents.

→ **Acupuncture** has been a staple of pain relief in China for more than 2000 years. A trained practitioner inserts slender needles that activate the neurons of specific sensory nerves. The resulting nerve impulses travel to the central nervous system, where the signal is passed to centers that release endorphins, the body's own pain reducers.

The sense of smell

Olfaction, the sense of smell, is one of the most ancient senses. For our ancestors a capacity for sensing airborne chemicals was a crucial tool in locating food, detecting threats and even finding mates. The importance of a versatile smell sense is reflected in the fact that between 400 and 1000 of our genes are dedicated to it.

We can distinguish thousands of different odors by way of the estimated five million olfactory receptors clustered in patches of the membrane that lines the upper portion of our nasal passages. Each receptor is a sensory neuron that responds to one or a few odorants, or smellable chemicals. Unlike most other sensory triggers, these substances have a direct line to the brain. This is because the neuron axons run in olfactory nerves directly to the olfactory bulbs, the paired brain centers that process odor cues.

The sense of smell is closely allied with taste, the other major chemical sense. In fact, odors make up much of what we think of as taste. This is why a head cold that clogs nasal passages temporarily deprives the sufferer of the ability to fully enjoy the flavors of a meal.

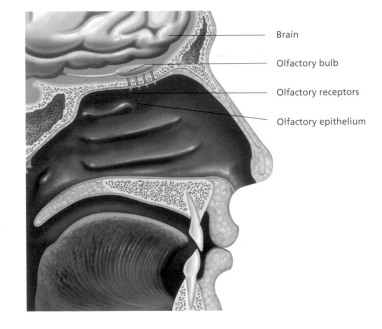

Brain

Olfactory bulb

Olfactory receptors

Olfactory epithelium

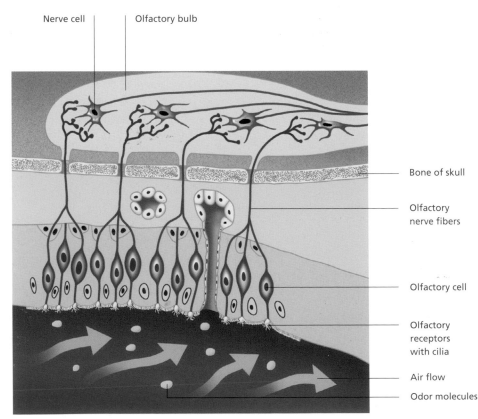

Nerve cell

Olfactory bulb

Bone of skull

Olfactory nerve fibers

Olfactory cell

Olfactory receptors with cilia

Air flow

Odor molecules

↑ **Smell is a chemical sense.** Deep in the nasal passages a thin layer of mucus covers the exposed surface. Odorant molecules from a source must diffuse through the watery mucus before binding with olfactory receptors.

← **Olfactory receptors** are neurons that connect the nose with the brain. In the nose, their tips protrude slightly from the surface of a membrane known as the olfactory epithelium and bind odorant chemicals. The resulting nerve impulses travel along the neuron axons, which pass through channels in the skull to reach the olfactory bulbs.

HOW THE NOSE KNOWS

With at most 1000 working smell sensors, how can humans distinguish millions of different odors? We know that odor molecules are chemically complex, consisting of several different components, just as a sentence consists of a number of different words. Each smell receptor detects one of these elements and sends this piece of information on to the brain. There, in the olfactory bulbs, all the incoming smell "words" are reassembled into the full messages, such as "garlic" or "freshly mown grass." The possible combinations of receptor inputs number in the millions, so allowing us to detect many scents. From the olfactory bulbs this message is sent on to the cerebral cortex, where the scent becomes part of our conscious world.

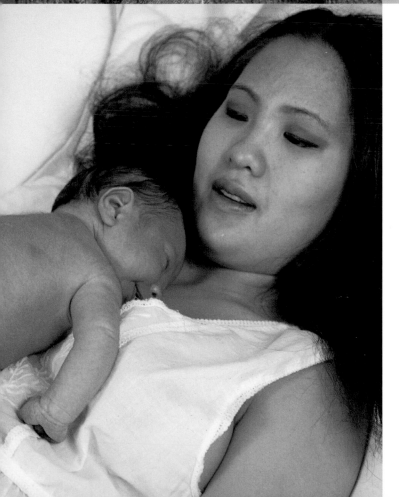

↑ **Sharks** have an extraordinary sense of smell. Captive reef sharks have been able to detect one part prey fish extract in a billion parts of water.

↖ **Sniffing** enhances the sense of smell. It draws odor molecules into the upper reaches of the nose more rapidly than normal quiet breathing.

← **Newborns** are thought to recognize their parents partly by smell.

→ **A bloodhound** detects odors through more than four billion olfactory receptors—hundreds of times the "smelling power" of a human.

The sense of taste

Some 10,000 tiny organs scattered in the surface tissue of the tongue, throat and roof of the mouth, the taste buds are responsible for gustation—our sense of taste. Like smell, taste is a chemical sense. It relies on chemoreceptors in taste buds that can respond to chemicals from food, or other substances, that are instantly dissolved in saliva. Each of the thousands of tastes we perceive is some combination of five primary tastes: sweet, sour, salty, bitter and umami—the savory taste associated with meats.

Although the ability to taste food is one of life's enduring pleasures, it also has survival value. When we first taste (and smell) food, the digestive system boosts its secretion of digestive enzymes that will be needed to break down ingested food into nutrients. Many poisons and other noxious substances taste bitter, and spoiled, potentially harmful food often has a characteristic "off" flavor. Inborn taste preferences shape our predisposition for certain foods. Even newborn infants show a particular liking for sweet tasting food, an observation that helps explain why modern humans often develop a waistline-expanding affinity for sugary treats.

WHAT TASTE BUDS TASTE

The five basic taste categories each relate to receptors that are attuned to particular chemicals, or tastants. Sweet tastants often contain sugars, but some are alcohols or amino acids—the building blocks of proteins. Anything that tastes sour contains some type of acid. Receptors that signal a bitter taste are responding to any of the substances in a family of more than 100 chemicals, including plant alkaloids such as quinine, caffeine, nicotine and the poison strychnine. A salty taste results when receptors for sodium chloride, the stuff of common table salt, are stimulated. Umami is a response to certain amino acids, especially glutamate—the main component of monosodium glutamate, or MSG, which is often used in Asian dishes. Two, three or more different basic taste receptors may be clustered in each taste bud. In combination with messages from olfactory receptors, the signals they send are integrated into the symphony of flavors humans can perceive.

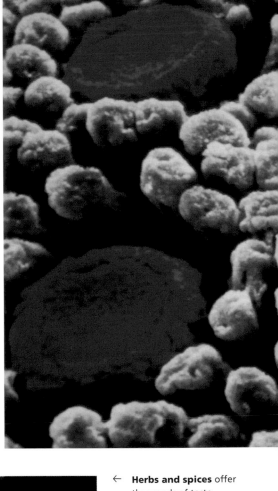

← **Herbs and spices** offer thousands of taste experiences. Before refrigeration and other food preservation methods were widely available, meats, fish, and other fresh foods were highly spiced to mask the tastes and odors of impending spoilage, or were salted or pickled for long-term storage. Today the spicy variations of different cuisines please the palates of diners everywhere.

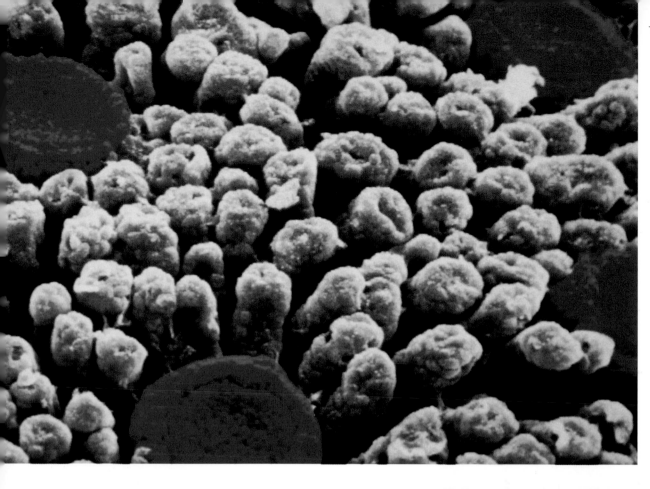

← **The surface of the tongue** is covered in nodules, called papillae (red and pink in this close-up). Inside these are minute taste buds, each containing from 50 to 100 chemoreceptors. A pore lets in fluid, such as saliva, that contains tastant molecules from food. When a receptor binds a tastant, it stimulates a sensory neuron, which transmits the message to the brain, where the signal is interpreted.

↓ **A housefly tastes** with its feet. When the fly walks on a substance, chemoreceptors in hairs on its feet sense whether the substance is suitable for eating.

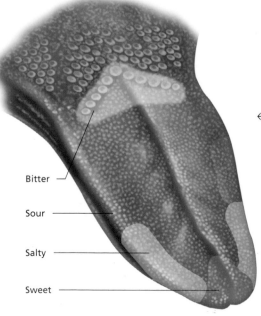

Bitter

Sour

Salty

Sweet

← **Receptors** for sweet tastants are most abundant at the tip of tongue, while those for salty tastants are more prevalent around the forward edges. We detect sour tastants farther back along the sides and bitter flavors in the center back of the tongue. The umami taste is registered by some of the taste buds in the upper throat.

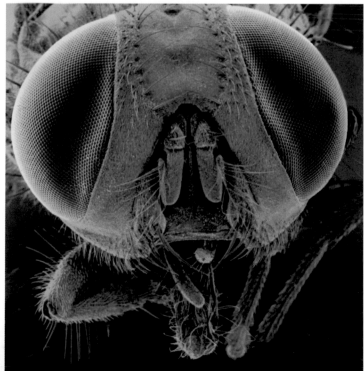

Sound waves and the ear

Our sense of hearing depends on the ear's ability to capture sounds traveling through air and to convert them to nerve impulses that can be sent to the brain. This process brings together the physics of sound waves and the biological functions of the ear's many parts.

A sound wave is a physical force not unlike a ripple moving across the surface of a pond. Instead of water, however, the sound wave consists of compressed air. Although invisible, air molecules—oxygen, nitrogen and carbon dioxide gas—are compressed when some force pushes them together, even if briefly. This is what happens, for example, when a baseball player's bat strikes a ball or a person's vocal cords vibrate and generate speech. If we visualize the waves with vertical peaks and troughs, the distance between the highest and lowest points is a sound's amplitude. The higher the waves, the louder the resulting sound.

The frequency of sound waves—the number of waves created in a given slice of time—determines a sound's pitch. Loud, deep tones of a cello and loud soprano notes from a violin may have the same amplitude, but they vibrate the air at different frequencies. Hearing is actually one of the less acute human senses. Bats, whales, dogs, the tiny mouse-like shrews and many other animals can hear sounds at much higher and lower frequencies than our human ears can detect.

↑ **A bat pinpoints potential food** by producing ultrasonic sound waves inaudible to humans and interpreting the returning echoes.

↗ **The Pacific Science Center's fountain** in Seattle, Washington, provides a multifaceted sensory experience with jets of water, music and changing colors.

→ **Musical instruments** create sounds by vibrating objects, such as guitar strings. Amplifiers can increase their loudness to levels that cause hearing damage.

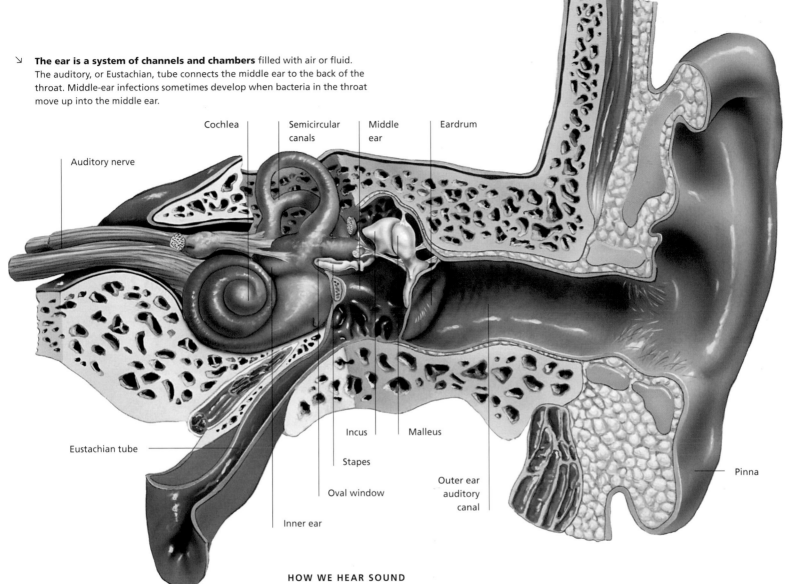

↘ **The ear is a system of channels and chambers** filled with air or fluid. The auditory, or Eustachian, tube connects the middle ear to the back of the throat. Middle-ear infections sometimes develop when bacteria in the throat move up into the middle ear.

Auditory nerve

Cochlea

Semicircular canals

Middle ear

Eardrum

Eustachian tube

Incus

Malleus

Stapes

Oval window

Outer ear auditory canal

Pinna

Inner ear

HOW WE HEAR SOUND

The ear is built to receive and amplify sound, and each of its parts—outer, middle and inner—has a specific role to play. The flaring flap of the outer ear, called the pinna, collects sound waves and funnels them inward through the auditory canal to the eardrum, or tympanic membrane. Now begins a sequence of events that will transfer the energy of sound-wave vibrations to nerve cells that can fire off impulses to the brain. The sound vibrations cause the wafer-thin eardrum to vibrate. These vibrations are transferred to the three minuscule, interlocking bones of the middle ear, the malleus ("hammer"),

incus ("anvil") and stapes ("stirrup"). These are the smallest bones in the body, and they serve as a movable bridge between the outer and inner parts of the ear. As they vibrate, their back-and-forth motion is focused on an equally tiny, flexible membrane called the oval window. This in turn amplifies the force of the eardrum's vibrations, so that when they are transferred to the inner ear, they have the physical force necessary to launch vibrations in the fluid there. That moving fluid generates signals that travel along the auditory nerve to the brain, which interprets the information as sound.

Hair cells and hearing

Our hearing sense depends on a rapid sequence of cause and effect. When the vibrations generated by sound waves reach the inner ear, they set up pressure waves in the fluid-filled chambers of a structure called the cochlea. Coiling back on itself like the shell of a garden snail, the cochlea is the size of a small pea. In a compartment inside it is a strip of tissue called the organ of Corti. This tiny organ contains the ear's mechanoreceptors—flexible structures called stereocilia that protrude from hair cells. Like many of the body's mechanoreceptors, stereocilia initiate sensory signals when a membrane pushes on them and causes them to bend. When the membrane pressure lets up, the hair straightens again. The pattern of bending and straightening precisely tracks the distortions caused by waves of fluid in the cochlea, which in turn track the physical properties of the original stimulus. As a result, we perceive sounds as they were first generated in air.

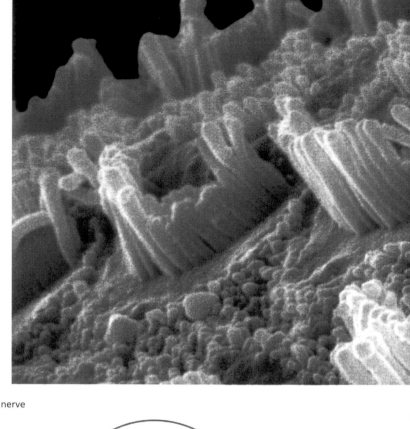

→ **Stereocilia, the ear's sensory hairs,** project up from hair cells. The upper surfaces of the cells are visible here as round, tan-colored areas and the hairs look like crescents of pink bristles. Stereocilia bend easily when a pressure wave arrives. Each human cochlea has about 16,000 hair cells, and about 1.6 million stereocilia.

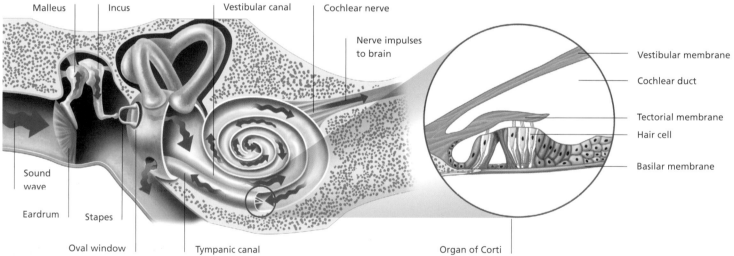

Malleus   Incus   Vestibular canal   Cochlear nerve

Nerve impulses to brain

Sound wave

Eardrum   Stapes

Oval window   Tympanic canal

Vestibular membrane

Cochlear duct

Tectorial membrane

Hair cell

Basilar membrane

Organ of Corti

FLUID WAVES AND THE SOUND PATHWAY

Sound waves arriving from the outer ear strike a membrane at the start of the cochlea. With each wave the oval window bows inward, then out again. This motion produces pressure waves in the fluid in the vestibular canal. Pressure is then transmitted into the cochlear duct, where waves are translated into nerve impulses in the organ of Corti. There, hair cells are sandwiched between two membranes, called the basilar membrane and the tectorial membrane. When fluid waves cause the basilar membrane to vibrate, their force bends the stereocilia of

hair cells against the tectorial membrane and triggers a nerve impulse. Different parts of the basilar membrane vibrate at different frequencies. Higher-pitched sounds vibrate the membrane at its thinnest end, while lower-pitched sounds vibrate the thicker parts. A sound signal only produces a nerve impulse when fluid waves arrive at the part of the basilar membrane that vibrates at a corresponding frequency. The vibrations then pass down the tympanic canal, at the end of which another membrane, the round window, flexes and relieves the pressure.

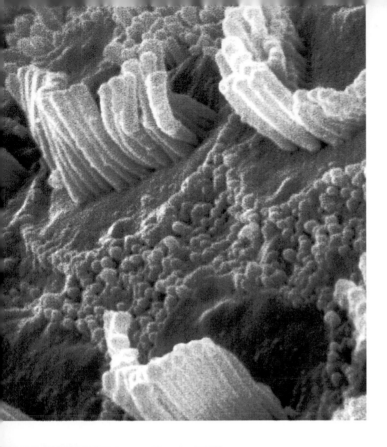

← **Sound becomes painful** at 130 decibels. People who are exposed to such "ear-splitting" auditory threats can suffer a permanent hearing loss after only a few hours. Government health agencies estimate that worldwide tens of millions of people are at risk of hearing loss related to high noise levels on the job, at home or in recreational settings. Normal aging also takes its toll as the ears' hair cells break down. By age 70, most people have lost close to half of the stereocilia they were born with. The sensory hairs in parts of the cochlea that detect high-pitched sounds usually are the first to degenerate.

↑ **Rustling leaves** generate low-amplitude sound waves, about 20 decibels (dB) on the loudness scale where 0 dB is the lower limit for normal hearing.

↑ **The roar of jet engines** on the ground, reaching up to 150 dB, is an intense sound that can do irreparable harm to the ears' sensory hairs over time.

↓ **Starting from 0 dB,** every increase of 1 dB represents a ten-fold increase in loudness. For example, a sound that measures 100 dB, such as music on a personal stereo, is about a billion times louder than the ticking of a clock at only 10 dB.

LOUDNESS OF COMMON SOUNDS	
Sound	**dB**
Ticking clock	10
Rustling leaves	20
Normal conversation	50–60
Urban street noise, busy restaurant	75–80
Kitchen blender (high speed)	90
Boom box stereo (high volume)	100+
Jet engine (on runway)	110–150
Rock concert	120–130

A sense of where you are

Balance is a sense of the body's natural position. The structures that support this fundamental sensory ability are located in the inner ear, but they have a different function from the organs of hearing. Collectively called the vestibular apparatus, they consist of a closed system of fluid-filled chambers and channels that assess changes in the position of the head. In one part of the apparatus, the fluid-filled semicircular canals, rotation of the head is monitored. Straight-line movements of acceleration and deceleration are tracked in another part of the vestibular apparatus, called the vestibule. The result is a steady stream of nerve impulses sent to reflex centers in the brain stem, where the signals are integrated with information from our eyes and muscles. The brain then orders movements that help you maintain your balance when you walk, run, dance, ski or otherwise move your body.

MONITORING EQUILIBRIUM

A position that "feels natural"—not upside down or tilted sideways—is called the body's equilibrium position. Dynamic equilibrium is determined by sensory hair cells in the semicircular canals, which monitor tilting or rotation of the head. The hairs of these sensory hair cells project up into a soft mass that deforms under pressure like bread dough poked by a finger. When head movements propel fluid in one of the semicircular canals against this soft substance, the bending of hairs triggers nerve impulses. Static equilibrium is our unconscious "sense" of gravity and monitoring of linear head movements. It draws on sensory hairs inside sacs in the vestibule, which trigger nerve impulses when head movements shift fluid and hard bits of calcium carbonate called otoliths.

→ **As a dancer twirls,** balance organs in her ears' semicircular canals assess dynamic equilibrium. Her brain coordinates this input with signals from tendons and joints, and adjusts contractions of her eye muscles to keep her eyes focused as she moves.

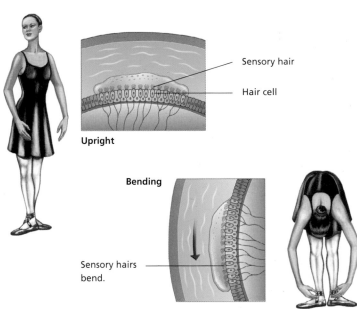

Sensory hair

Hair cell

Upright

Bending

Sensory hairs bend.

↑ **Sensory hairs,** shown here in blue bundles of short threads surrounding a single longer one, are the key to detecting directional movements. Together with otoliths (purple), the hairs function in static equilibrium.

↑ **Hair cells that monitor linear movements**—walking, then stopping, or bending and straightening—are found in patches of tissue called maculae in the sacs of the vestibule. By signaling the brain about the head's position relative to the ground, they also help us maintain posture.

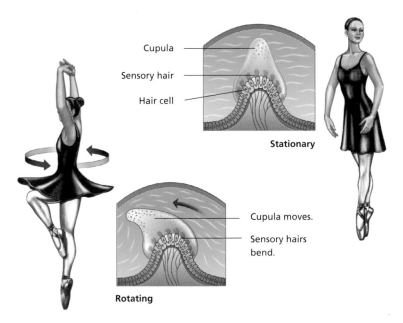

Cupula

Sensory hair

Hair cell

Stationary

Cupula moves.

Sensory hairs bend.

Rotating

↑ **The cristae in the semicircular canals** sense rotational movement. As the head turns in a certain plane, fluid called endolymph moves in the corresponding canal and bends a jelly-like mass called a cupula. As the cupula bends, so do the sensory hairs inside it.

↑ **The brain's ability to assess** changes in static equilibrium allows this gymnast to maintain his posture. As hair cells in the vestibular apparatus send signals about the position of his head, he is able to balance briefly on one hand by using muscle power to adjust the position of other body parts.

Light and vision

Light is one of the most pervasive elements in our world. No matter where you are on the Earth's surface, the sun rises and sets in a predictable 24-hour cycle that brings with it periods of light and dark. Light is a form of energy packaged in units called photons, and almost all living things are equipped to detect and respond to it in some fashion. In one form or another, creatures as varied as animals, plants and tiny microorganisms have photoreceptors—simple or complex structures that are stimulated by light energy.

In green plants, photoreceptors capture sunlight and immediately harness it to make the plant's food. An earthworm has photoreceptors all over its body, so its simple nervous system can respond to light even though the worm has no eyes. More complex animals, however, do have eyes of one type or another. Eyes provide what vision requires—an organized array of photoreceptors that are linked to brain centers capable of receiving and interpreting the patterns of nerve impulses those photoreceptors generate.

↓ **Eyes are complex organs** that receive and organize light rays. Eyebrows, eyelids and eyelashes have important functions as well. They help shield the eyes from intense light, perspiration and foreign objects. The lining of the eyelid secretes lubricating fluid over the eyeball.

↑ **A rainbow displays light** that has passed through millions of tiny prisms—water droplets suspended in air. When light passes through a prism, its wavelengths bend at different angles and produce a band of colors like this one arcing over a field of macadamia nut trees in Hawaii.

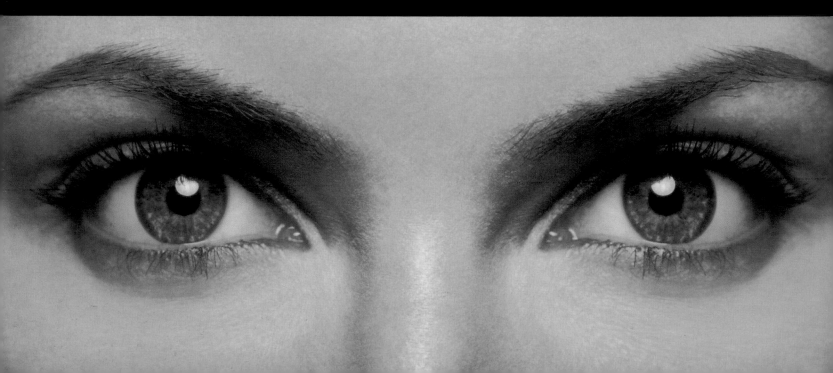

↑ **A moth's compound eye** consists of many separate units called ommatidia. All insects have this type of eye, which produces a mosaic-like image because each optic unit is aimed at a slightly different part of the visual field.

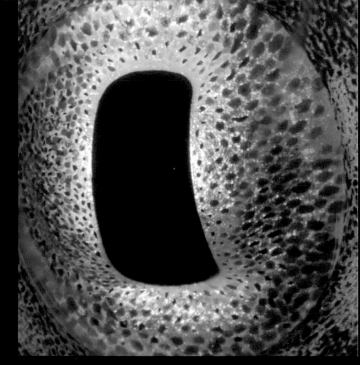

↑ **The eye of an octopus** is built much like a human one. Although they lack color vision, octopuses and their relatives can discriminate different shapes, such as squares and circles.

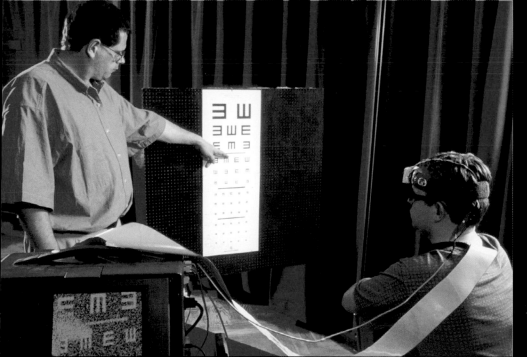

WAVELENGTHS AND COLOR

Light normally travels in a straight line at the astonishing speed of 186,000 miles (300,000 km) per second. When it encounters an opaque surface, however, it "bounces off"—is reflected—at an angle. Most of the light that our photoreceptors detect comes from this reflection of light by the objects around us. Visible light—the part of the electromagnetic spectrum to which human photoreceptors respond—is a blend of light of different wavelengths. It begins with deep purple at one end and moves on through blue, green, yellow, orange and bright red at the far end of the spectrum. Objects appear colored because they reflect some wavelengths and absorb others. A leaf looks green because it reflects mostly light in the green part of the spectrum, a blue sock reflects mostly blue wavelengths, a banana yellow wavelengths, and so on. Black and white are not colors at all. Things appear white if they reflect all light wavelengths, and look black if they absorb all wavelengths.

← **A blind boy tests** new technology that converts images from a video camera on his head to tactile sensations that allow him to "see" the image using his sense of touch.

Eyes: built to manage light

Our eyes are versatile organs for detecting and managing light. In some ways they operate like an old-fashioned film camera. Light enters through the cornea, a thin, transparent sheet of tissue at the front of the eyeball that slightly bends the incoming light as it passes through on its way to a small opening, the pupil. The colored iris—a muscle around the pupil— can widen or narrow, adjusting the diameter of the pupil to increase or decrease the amount of light that can enter. From the pupil light travels through a lens, then bends again at a sharper angle. It ultimately strikes a layer of photoreceptors in the retina at the back of the eyeball. Called rods and cones, these receptors are specialized neurons, whose signals travel along axons that join at the back of the eye to form the optic nerve.

To ensure that light casts a clear, sharp image on the retina, the lens becomes rounder to focus on near objects or flatter to view those that are farther away. As a result of the way the cornea and lens bend light rays passing through them, the rays arrive at the retina reversed top to bottom and left to right. Photoreceptors send this message to the brain, which restores the original orientation. Only after light signals are processed in the brain do we "see."

INSIDE THE EYE

The human eye is only about 1 inch (2.5 cm) in diameter. Padded by fat and set deep into bony orbits, each eye is fairly well protected from external harm. The transparent cornea is lubricated by tears and covered with a tissue that protects against airborne dust, hairs and other objects that may come in close contact with the eye. More protection for light-processing structures is provided by the white of the eye, a tough, fibrous layer called the sclera, which extends around the eyeball. Inside the eyeball, a watery aqueous humor (body fluid) lubricates both sides of the lens, and a jellylike vitreous humor fills the region behind it. Light passes through these substances, bends, and finally converges at the back of the eye on the retina.

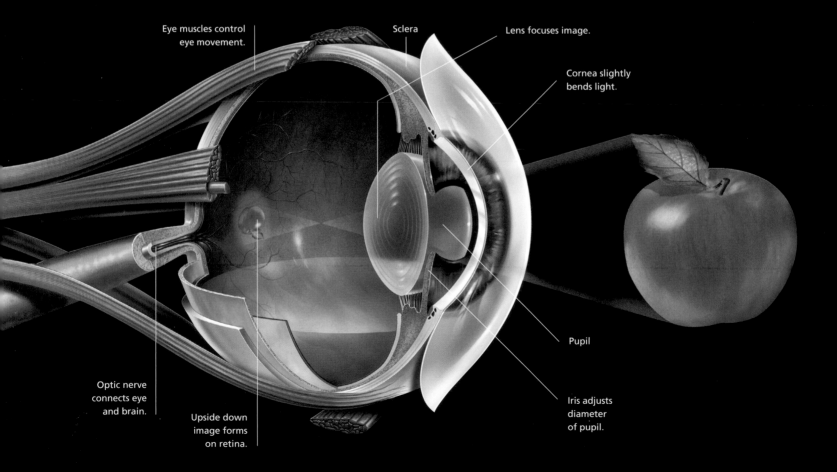

Eye muscles control eye movement.

Sclera

Lens focuses image.

Cornea slightly bends light.

Optic nerve connects eye and brain.

Upside down image forms on retina.

Pupil

Iris adjusts diameter of pupil.

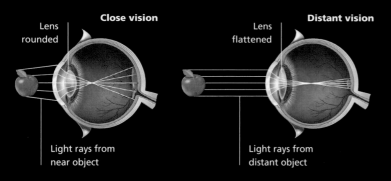

Close vision

Lens rounded

Light rays from near object

Distant vision

Lens flattened

Light rays from distant object

↑ **In accommodation,** ciliary muscles contract to round up the lens when a viewed object is close (left), and relax to flatten out the lens when the object is distant. Without this adjustment, light from near and distant objects would be focused either in front of or behind the retina, respectively.

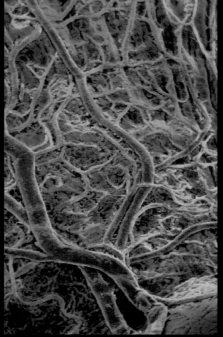

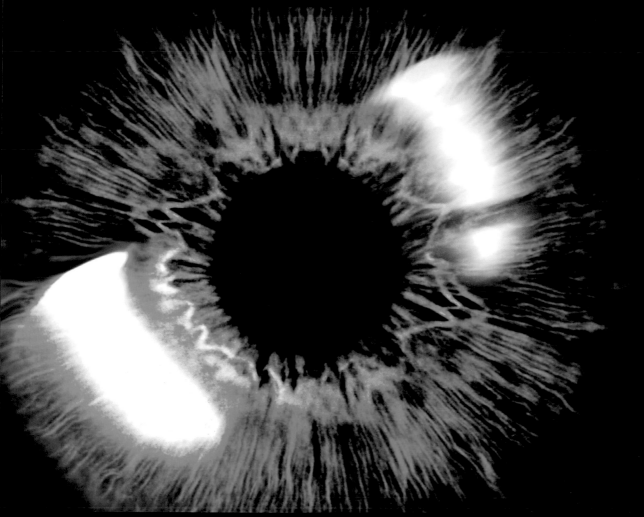

↑ **A straw in water** illustrates refraction, the bending of light when it passes from air to water. Light entering the fluid chambers of the eye is refracted, bending toward a precise focal point at the back of the retina.

↖ **The choroid,** a layer just under the white sclera, is laced with blood vessels that service the wall of the eyeball. This layer also contains pigments that help prevent light from scattering before it reaches photoreceptors in the retina.

← **The eye's iris** is equal to a fingerprint. Its smooth muscle fibers radiate out from the center in a fixed pattern that differs in each one of us. Iris scanning security systems can be used to confirm a person's identity.

The photoreceptors

Each of your eyes has an estimated 130 million photoreceptors—rods and cones—that provide the raw material for vision. Named for their shapes, these modified neurons make up roughly two-thirds of all the sensory receptors in the body. Both contain light-absorbing visual pigments. Yet rods and cones have very different functions, in part because their "wiring" to other neurons is organized differently.

Rods do not detect color but are highly sensitive to light. They are responsible for our ability to see in a dimly lit room or at night, if only fuzzy outlines of objects in shades of gray. Rods also sense shifts in light that provide peripheral vision—movements you detect "out of the corner" of your eye.

Cones detect only bright light, but they endow us with acute color vision. Cones are concentrated around the fovea at the back of the retina, which is why visual images are sharpest in a well-lit situation if they are viewed straight on.

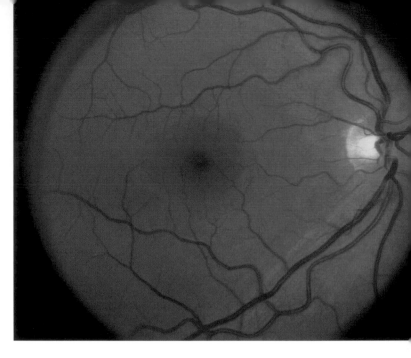

↑ **Different lighting** can change our perception of things. Cones allow us to see noticeably sharper images by day than after dark, when rods give us less precise vision, as represented by these day and night views of the Eiffel Tower.

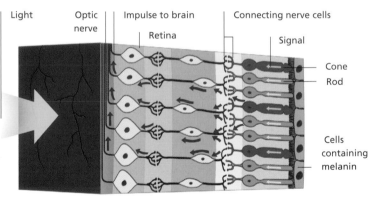

COLOR-CODED VISION

Our color vision depends on three pigments in cones, each one suited to absorb a particular wavelength of light. Hence "red" cones are most sensitive to wavelengths in the red part of the spectrum, "green" cones to green wavelengths, and "blue" cones to blue. Humans can distinguish thousands of colors, however. We perceive yellow, orange, pink, brown, tan, turquoise and countless other hues when several types of cones are stimulated, some more strongly and others more weakly. Rods, the photoreceptors specialized for dim light, contain a single, purple-black pigment. If you switch off the lights in a room at night, it takes a while for your eyes to adapt to the dark because their cones suddenly have little or no light to stimulate them and the rod pigment has not yet been mobilized. Move from a dark room to a brightly lit one, and the situation is reversed.

↑ **Signals from stimulated rods and cones** flow to layers of interneurons in the retina. Axons of these nerve cells are bundled into the optic nerves, which convey the signals to the brain's visual cortex, and then an image is perceived.

← **The retina** is served by a network of blood vessels. It covers almost the entire inner surface of the eye, with photoreceptors clustered at the rear. The light-focusing fovea is in the darker area near the center of this image from an opthalmoscope.

→ **Like all dogs,** a Labrador retriever sees only in black and white, but many predators have good to extremely acute color vision. Great white sharks see color, and birds of prey such as hawks have so many millions of cones in their eyes that they can spot a mouse in a field from high in the air.

↓ **Rods and cones** are packed tightly in the retina. The tips of the slender rods (white in this image) and the tapered tips of the plumper cones (pale green) contain the visual pigments that respond to light.

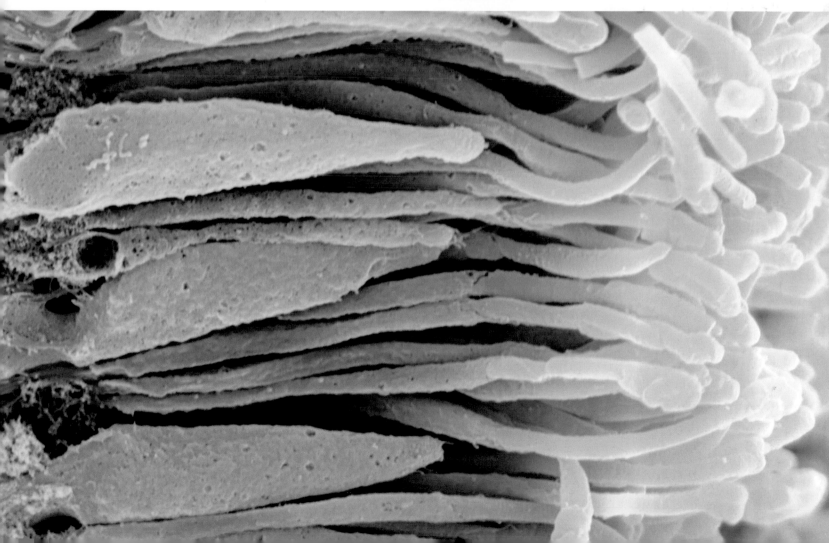

Interpreting visual signals

Our sense of vision includes an awareness of the shape and brightness of visual stimuli, as well as their orientation, texture, color, location and movement. The brain can interpret so many parameters because the retina is organized into a variety of receptive fields, areas in which photoreceptors feed signals to different sets of sensory neurons. The neurons in each set respond best only to certain stimuli, such as motion, hard-edged bars or lines, or a small spot of light ringed with dark. This specialization helps the brain sort out the barrage of visual signals it receives. Because we have two eyes set a short distance apart, visual signals arrive at the brain from slightly different vantage points. Visual processing normally blends the different images in a way that leads to accurate depth perception.

Optical illusions result from the brain's efforts to sort visual stimuli into standard categories, such as "a tilted line," "a sharp edge" and "a shadow." When incoming visual stimuli are confusing or don't fall neatly into one of these set categories, the brain tries to manipulate the input in ways that will make it fit with visual patterns the brain "expects" to encounter. In this sense, optical illusions are visible proof of how much our visual processing system does to help make sense of the world around us.

↑ **Contrasting tiles** on a café wall create an optical illusion. Due to the way retinal neurons and the brain process light signals of dramatically different intensity—here, black and white—the parallel gray grout lines between the tiles appear to be sloping and jagged.

VISUAL FIELDS AND PATHWAYS

The "visual field" is everything that our eyes can see—either sharply focused in the center or "fuzzier" to the sides. Light entering through the lens of each eye is reversed, so signals from the left side of the visual field are detected by the right side of the retina and signals from the right side of the field are detected by the left side of the retina. The right and left visual fields overlap a great deal, omitting only a small area blocked by the nose.

Signals from each side of the retina are carried deep into the brain by an optic nerve. Each of these nerves then "splits," just like a dividing highway: some of the nerves' fibers travel into the same side of the brain, while others cross over into the other hemisphere. As a result, the left hemisphere receives a complete "picture" from the right visual field, and a full set of signals from the left visual field arrives in the right hemisphere.

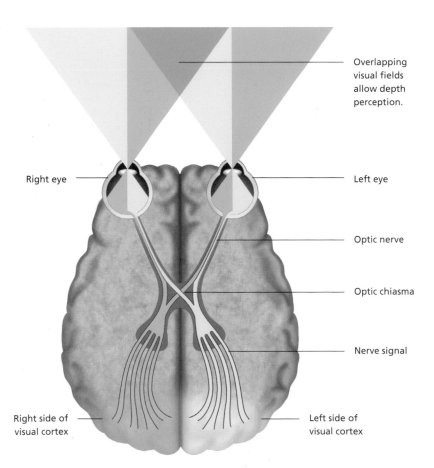

Overlapping visual fields allow depth perception.

Right eye

Left eye

Optic nerve

Optic chiasma

Nerve signal

Right side of visual cortex

Left side of visual cortex

← **This view of the brain from below** shows how optic nerve fibers cross over to the opposite brain hemisphere at the optic chiasma ("crossing"). Farther back in the brain, inputs from the right and left visual fields are combined so that the visual cortex processes a unified signal, producing a single image.

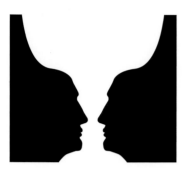

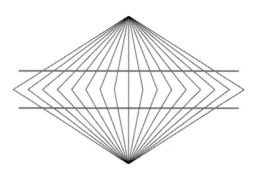

↑ **Is this image a goblet** or two faces in profile? Despite the brain's virtuoso ability to process visual stimuli, when there are two (or more) possible interpretations of an image the brain may well switch from one to the other—but it cannot present both at once.

↑ **Lines or bars** in different orientations stimulate neurons in different receptive fields in the retina. The lines radiating from two points create an illusion of depth, so some people may perceive the green bars above as curving inward, although they are straight and parallel.

↑ **Which red circle** is larger? The brain equalizes perceived size differences between adjacent objects, in this example "seeing" each circle as close in size to its square. However, while the square on the left is larger than the one on the right, the circles are identical in size.

← **Light photons** stimulate photo-receptors, which then launch sensory signals that the brain translates into visual images. There are limits, however, to how well a receptor can track a light stimulus. If a photoreceptor is stimulated by a slowly flickering light, we can perceive the separate flickers because there is a substantial delay between nerve impulses. If the rate of flickering increases too much, the sensory neurons send signals constantly, and the brain interprets this as a steady light. Standard light bulbs and some projected film images flicker, but too rapidly for photoreceptors to keep up, so we see images as smooth and uninterrupted.

Sight under siege

Our eyes are our most important source of information about the outside world. If we didn't know this simply from experience, we have indisputable evidence from experiments: more than half of the brain's cerebral cortex is devoted to processing visual signals. Another major indication of the importance of vision is how much injury, disease and genetic conditions that disrupt normal vision may hamper our ability to navigate the world, read a recipe or road sign, or perform some other routine function. Minor vision problems may only require us to use eye drops or wear corrective lenses, and many new technologies are now available to treat some vision ills. More serious maladies can lead to blindness. In developed countries, increasing attention is focusing on common age-related disorders such as cataracts and macular degeneration. In some parts of the world, highly contagious eye infections annually rob thousands of their sight.

AGING'S TOLL ON VISION

Over time, nearly all people experience a reduction in the acuity of their vision. With the passing years, the lens becomes thicker and less flexible, so it does not focus light rays as precisely. This is why so many people, even those with once-perfect vision, need glasses in their 40s. With cataracts, changes in proteins of the lens make the once-transparent tissue cloudy. In severe cases the lens becomes opaque like milk glass, and must be replaced with an artificial lens. Macular degeneration leads to partial blindness, most often in the elderly. Once-healthy parts of the macula, in the central part of the retina, degenerate and are replaced with scar tissue.

↗ **To see someone clearly close up** requires small muscles to contract sufficiently to "round up" the lens so that light rays converge on the retina.

→ **Viewing people across the room** in sharp focus requires muscles to relax so that the lens flattens, so allowing light rays to form an image on the retina.

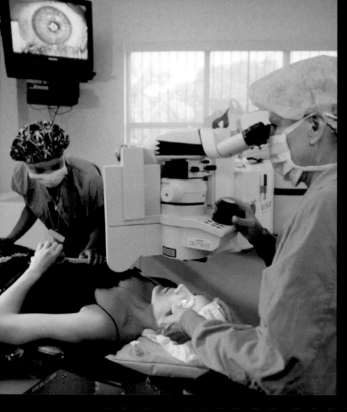

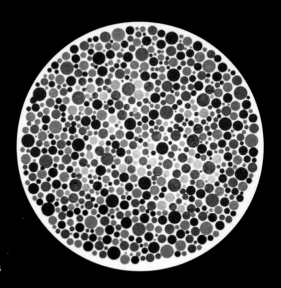

← **Painless LASIK surgery** uses a laser to reshape the cornea, a strategy for correcting nearsightedness.

→ **The number 74** in this circle of colored dots may be invisible to people with red–green color blindness.

↘ **The bacterium** *Chlamydia trachomatis* can invade eye tissues and trigger secondary infections that scar the cornea. The microbe is a major cause of blindness in North Africa and the Middle East.

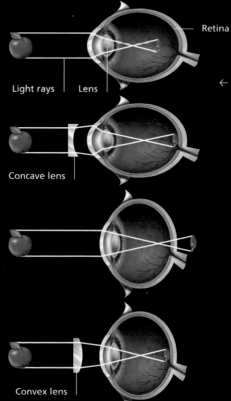

Nearsightedness

Light rays | Lens

Nearsightedness corrected

Concave lens

Farsightedness

Farsightedness corrected

Convex lens

Retina

← **Nearsightedness,** or myopia, results when the eyeball is too long or when the muscles that move the lens over-contract. Both problems cause distant images to be focused in front of the retina (top). Corrective lenses or surgery shift the focal point farther back by flattening the cornea (second). In farsighted people the problem is reversed. Their eyeballs are too short, or the lens muscles are weak, so images of near objects are focused behind the retina (third). Corrective lenses are the usual solution (bottom).

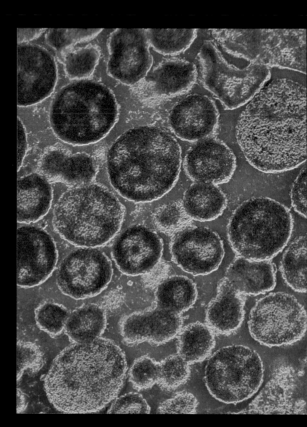

Breath, blood and defense

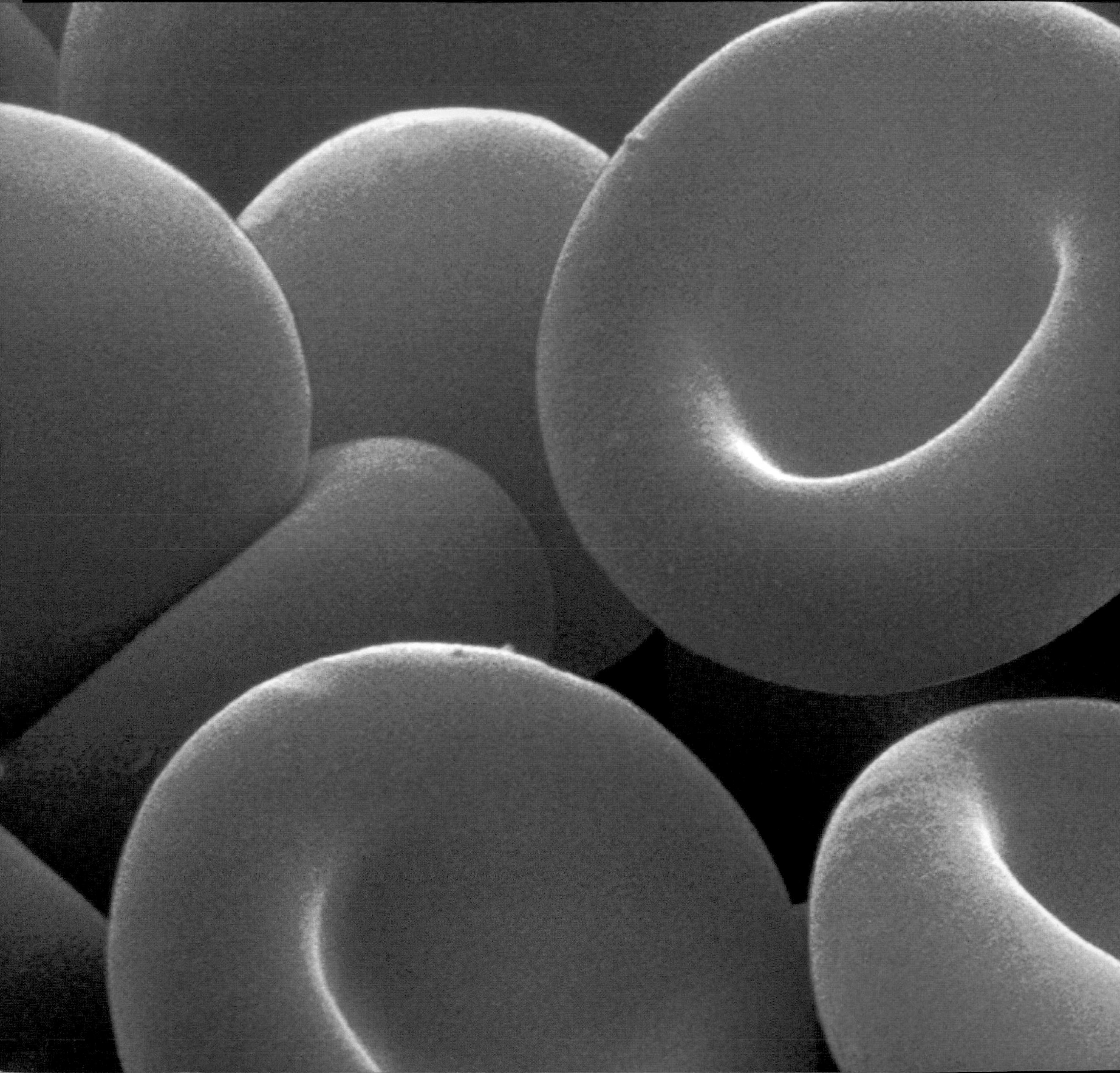

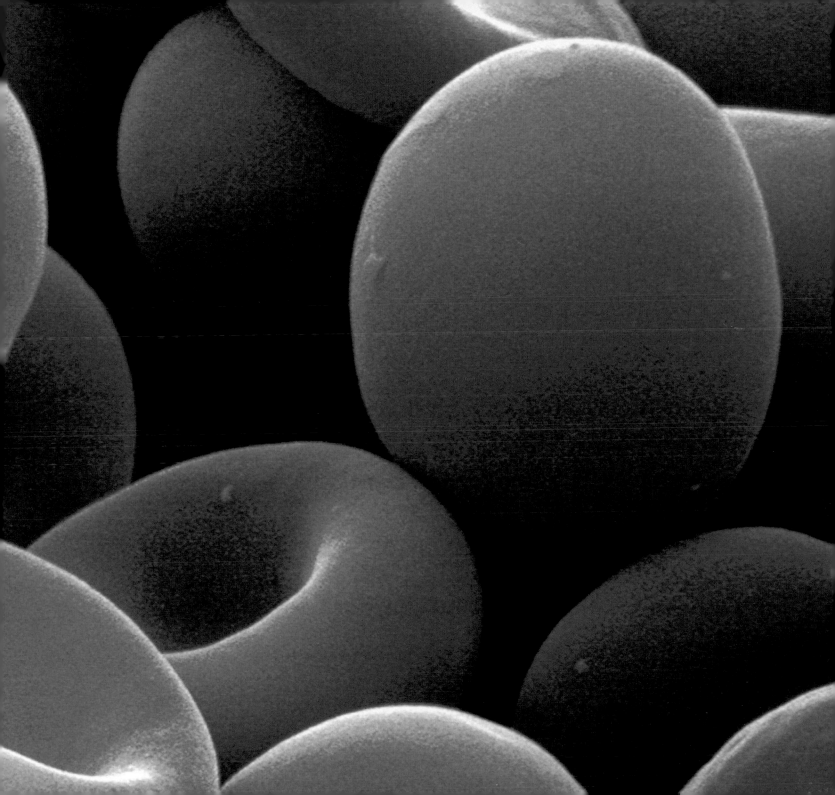

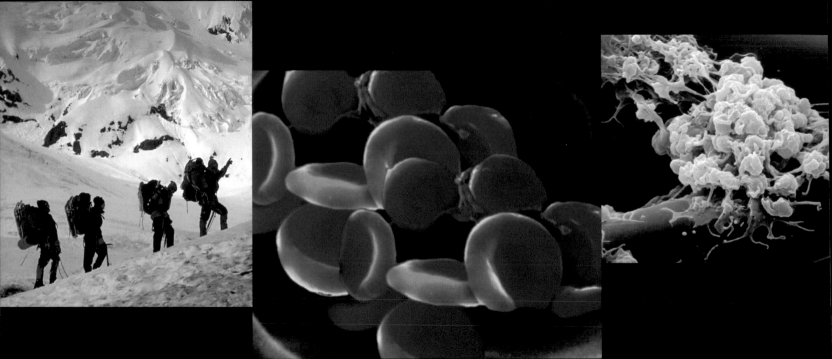

Breath, blood and defense

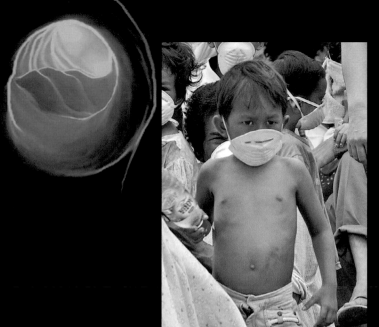

Just as the body requires communication networks, it also needs transport systems to move substances, and some cells, from place to place. The respiratory system imports oxygen and exports carbon dioxide. Working in concert with it and with body defenses is the blood-transporting cardiovascular system.

The breath of life	138
Exchanging carbon dioxide for oxygen	140
The mechanics of breathing	142
Breathing controls	144
Blood: the all-purpose transporter	146
Red blood cells	148
White blood cells	150
Why and how blood clots	152
The cardiovascular system	154
The heart: two pumps in one	156
Heartbeat: the cardiac cycle	158
The heart's pacemaker	160
Dual circuits of blood	162
Blood pipelines	164
Return transit to the heart	166
The cardiovascular system under siege	168
The lymphatic system	170
Hidden enemies	172
Defensive strategies	174
The inflammation response	176
Immune responses	178
Recognizing the invader	180
Antibodies	182
Killer cells and other defenders	184
Medical immunology	186

The breath of life

The body can do without food or water for days at a time, but we perish in minutes without a steady supply of oxygen. Moment to moment, this gas plays a central role in the process that makes a chemical called ATP, the fuel for all living cells. Making ATP is just one facet of a cell's metabolism, the sum of activities that allow a cell to carry out its designated functions. In addition to requiring oxygen (O_2), metabolism generates wastes, including a great deal of the gas carbon dioxide (CO_2). Released into the bloodstream, carbon dioxide is a potential threat, because if it accumulates the blood becomes too acidic—an imbalance that eventually can poison the central nervous system. Meeting our biological need to obtain oxygen from the air and jettison carbon dioxide wastes is the role of the respiratory system. At first glance, the system seems uncomplicated, for it is divided into just two main parts, the airways that provide passages for O_2 and CO_2, and the lungs, where these gases enter and leave the bloodstream. In reality, a variety of structures play roles in supplying us with the breath of life.

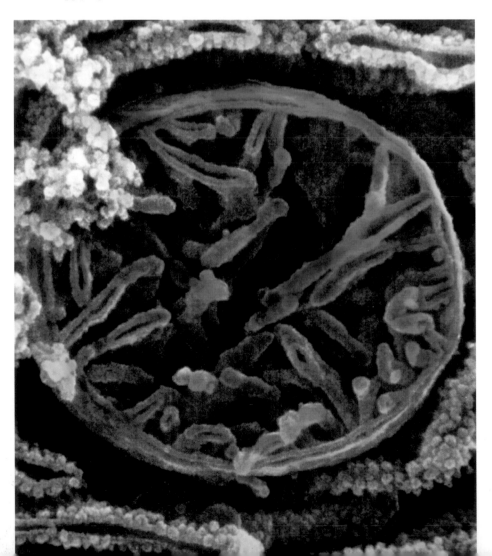

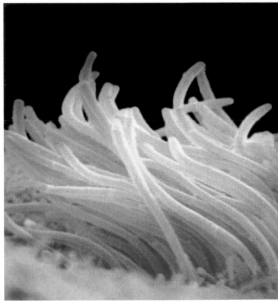

↑ **Hair-like, mucus-coated cilia** lining the trachea sweep dirt and other debris up toward the throat and mouth so it can be expelled.

⇑ **Competitive cycling,** like all activities, depends on the body's ability to extract oxygen from the air and to remove potentially toxic CO_2 produced in contracting muscle fibers and other body cells.

← **Oxygen in the air** we breathe is used in the mitochondria of body cells. A mitochondrion (colored pink) is shown here in cross-section.

THE RESPIRATORY "TREE"

The respiratory system resembles an inverted tree. Its "trunk" begins at the nose, where inhaled air enters during normal breathing. Air moves downward through the throat (pharynx) and then through the 2-inch (5 cm) larynx. At this point a valve-like flap, the epiglottis, shunts incoming air into the windpipe (trachea) or diverts incoming food into the esophagus. Air flows into the windpipe , a flexible tube about 4 inches long (10 cm). At its base, the airway splits into two bronchi that branch repeatedly after they enter the lungs. Deep in the lungs at the tips of the "branches" are clusters of sacs called alveoli, where oxygen moves into the bloodstream and carbon dioxide moves outward. In its journey from the nose to the alveoli, air is warmed and moistened as it passes over mucous membranes, while hairs and cilia clean it of dirt, debris and bacteria.

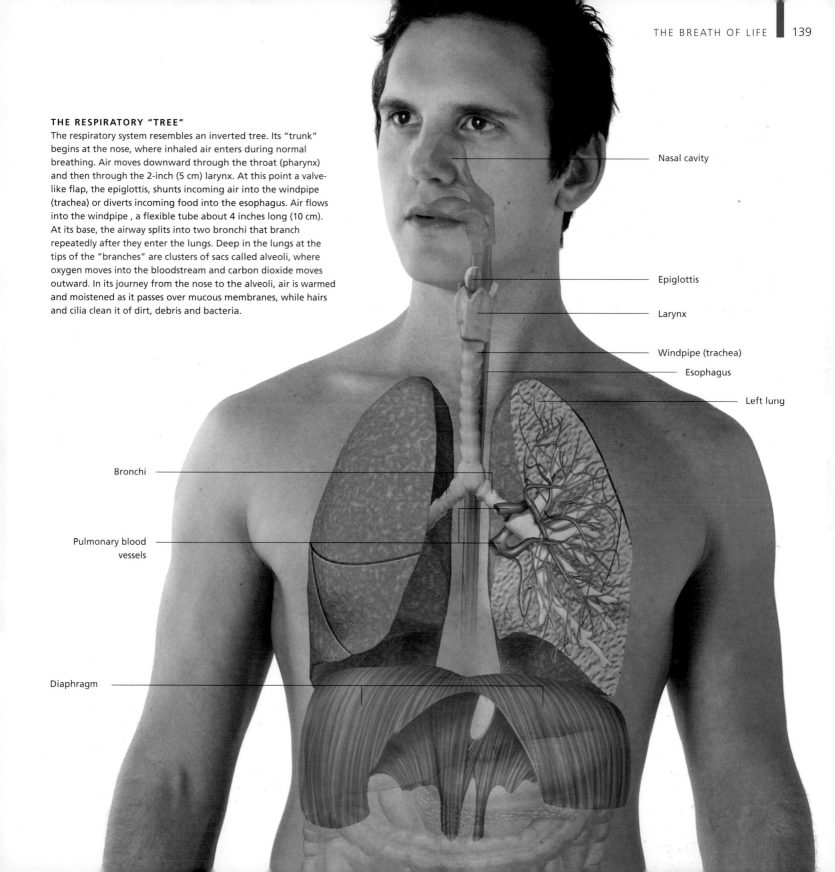

Nasal cavity

Epiglottis

Larynx

Windpipe (trachea)

Esophagus

Left lung

Bronchi

Pulmonary blood vessels

Diaphragm

Exchanging carbon dioxide for oxygen

Together our two lungs have an estimated 300 million of the delicate air sacs called alveoli, which collectively provide a vast surface for gas exchange—moving oxygen into the lungs, and thence into the blood, while simultaneously moving carbon dioxide outward. At each step, this life-sustaining traffic of gases is governed by several factors. For one, at sea level, only about 21 percent of the air we breathe is oxygen. The rest is mostly nitrogen, mixed with a small amount of carbon dioxide and other gases. To gain oxygen, the respiratory system must be able to remove it from air.

Like all gases, air can be moved from place to place when it is "pulled" by pressure changes. Our stretchable lungs exploit this property. When the lungs expand, they become a low-pressure zone, and so air rushes into them. As they recoil a moment later, carbon dioxide is flushed out. Oxygen in inhaled air moves into the blood, and carbon dioxide moves out, by a low-pressure gradient. When blood reaches the lungs from tissues, its load of oxygen is low because cells have been removing it all along the way. As a result, the "pressure" of oxygen in the blood is lower than in the fresh air in the alveoli, and so oxygen moves from alveoli into the blood. On the other hand, the carbon dioxide that cells have off-loaded moves from the bloodstream into the lungs because it is relatively scarce there.

→ **High-mountain trekkers** risk apoxia—oxygen deprivation—if they venture above 10,000 feet (3000 m). At that altitude there is much less oxygen in the air than at sea level.

BREATHING EXTREMES

In extreme environments the human respiratory system encounters major challenges. Although at sea level one molecule in five is oxygen, the higher you go the less oxygen there is because the Earth's gravitational pull is steadily weakening. People with asthma or some other respiratory disease may begin to feel breathing distress even at low elevations. At 10,000 feet (3000 m), most people will feel a bit breathless, especially with physical exertion. As the brain and other tissues begin to run low on oxygen, a person may develop headaches and heart palpitations. By 15,000 feet (about 4500 m), serious symptoms of altitude sickness may set in, including the swelling of lung and brain tissues as small blood vessels begin to leak fluid. Deep water presents other problems. On deep descents divers typically breathe pressurized air from tanks—which, along with providing oxygen, prevents their lungs from collapsing due to the increasing water pressure. Unless a deep-water diver uses a nitrogen-free gas mixture, a too-rapid ascent can trigger "the bends," a sickness caused by nitrogen bubbles in the blood.

↑ **This South American frog** breathes differently from humans. In typical amphibian style, muscles in the floor of its mouth pump air into its mouth and lungs, while carbon dioxide wastes exit through its moist skin.

↓ **Air sacs in the lungs**—alveoli—are shown here as white areas surrounded by many blood capillaries, the small vessels that take up oxygen arriving from the lungs and deliver carbon dioxide to the lungs to be exhaled.

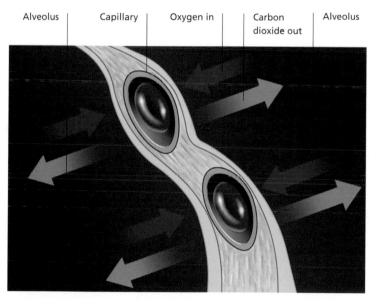

Alveolus	Capillary	Oxygen in	Carbon dioxide out	Alveolus

↑ **The walls of alveoli** in the lungs are thin and wet, so providing a respiratory interface for oxygen and carbon dioxide. These gases can cross into or out of alveoli only if they first dissolve in water.

→ **A gas crossing a respiratory surface** moves from the area of higher pressure to the area where its pressure is lower. The greater the pressure difference, and the larger the surface, the faster the gas will cross.

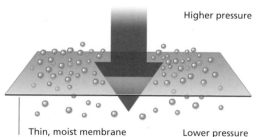

Higher pressure

Thin, moist membrane   Lower pressure

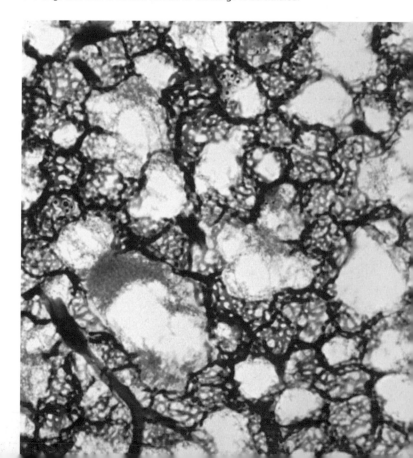

The mechanics of breathing

By the time you reach the age of 75 you will have taken more than half a billion breaths, roughly 12 per minute. In each one, muscle movements expand and then shrink the chest cavity in a rhythmic cycle. As the volume of the chest shifts from larger to smaller, the lungs follow suit, expanding and then recoiling. Each time the lungs expand, the air pressure inside them falls, and when they recoil it increases again. In normal breathing, inhalation draws in about two cups (500 ml) of oxygen-rich air, and when you exhale about the same quantity of air laden with carbon dioxide flows out. Physiologists call this in-and-out process ventilation, and while it usually is automatic we do have some control over some aspects of breathing. If you deliberately suck in a deep breath, you can increase the volume of inhaled air six- or sevenfold, and if you forcefully exhale, you can also empty your lungs of much more. No matter how hard we breathe during strenuous activity, some air always remains in the airways. Overall, however, an adult's lungs take in just over a gallon (4.5 liters) of fresh air each minute.

MOVING AIR IN AND OUT

A variety of muscles are used in breathing. When you inhale, the diaphragm—a broad, rounded sheet of muscle that separates the chest cavity from the abdomen—contracts and flattens. At the same time, external muscles between the ribs draw the rib cage up and outward. Inside this expanded chamber, the air pressure falls and the lungs now stretch and expand. Following the pressure gradient, outside air flows in. When you exhale, the diaphragm and rib cage return to their original positions and the lungs "deflate" as air flows out of them. Air also moves into and out of the lungs in activities unrelated to breathing. Hiccups are short, sudden inward movements of air that happen when the diaphragm contracts in spasms. Before a cough, a person inhales deeply, the passage through the larynx briefly closes, and air begins to rush upward from the lungs. When the pressurized air reaches the larynx, the passage reopens suddenly and the air surges out.

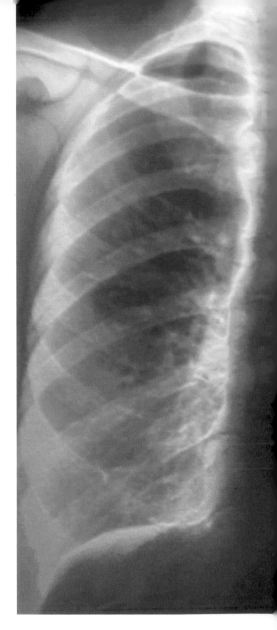

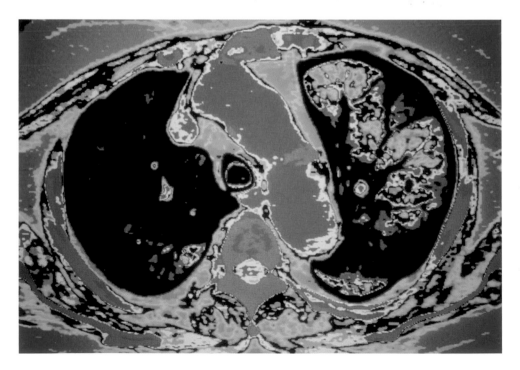

↑ **As a person inhales** (left X-ray), the rib cage moves up, expanding slightly, and the diaphragm moves down. On the right, as a person exhales the rib cage returns to its resting position and the diaphragm relaxes, moving up.

← **In emphysema,** the air sacs (alveoli) enlarge and clump together, reducing the respiratory surface for gas exchange and hampering the lungs' ability to supply body tissues with oxygen and remove carbon dioxide. In this top-down view, the right lung has suffered the most damage, shown as blue-green areas.

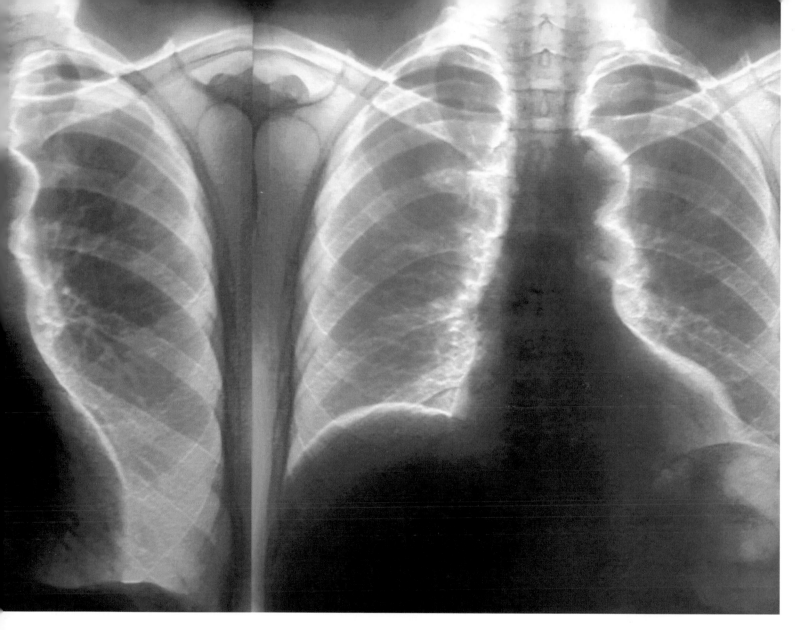

→ **When air passes** through the throat it can also produce sound. The vocal cords are membrane-wrapped ligaments on either side of the larynx. Skeletal muscles open and close the cords. The gap between them is called the glottis. When the cords are closed (left), exhaled air flowing up through the glottis vibrates the cords, and we can produce sounds by controlling the vibrations. You can feel the vibrations when you hum; we make spoken or sung words using our tongue, teeth, lips and palate to modify the sounds. When the vocal cords are open (right), no sound can be made.

VOCAL CORDS CLOSED

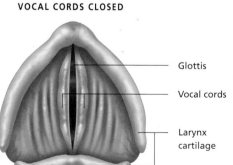

Glottis

Vocal cords

Larynx cartilage

VOCAL CORDS OPEN

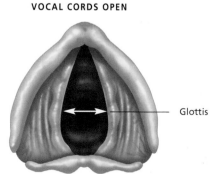

Glottis

Breathing controls

In different ways, the brain and the lungs control the life-sustaining cycle of inhaling and exhaling air. At one level, control centers in the brain stem generate a steady cadence of nerve impulses to the diaphragm and rib muscles the body uses in breathing. These signals maintain the basic mechanical operations that ventilate the lungs. Another set of controls regulates the rate and depth of breathing to meet changing needs, such as during sleep or when you run to catch a bus. These adjustments respond to changes in blood chemistry when the relative amounts of respiratory gases in the blood rise or fall. Surprisingly, the key parameter is not the amount of oxygen in the blood: the body's monitoring mechanisms are much more attuned to the blood level of carbon dioxide. When the level rises above a prescribed value, sensors signal respiratory control centers in the brain to step up the rate and depth of a person's breathing. Huffing and puffing during demanding physical activity is a physiological response that speeds the removal of excess CO_2 from the body.

→ **Divers sometimes hyperventilate,** a risky practice of deep, rapid breathing that delays normal nervous system controls over breathing.

↓ **Chemoreceptors** in the walls of major blood vessels in the neck and near the heart closely track the relative amount of carbon dioxide in the bloodstream.

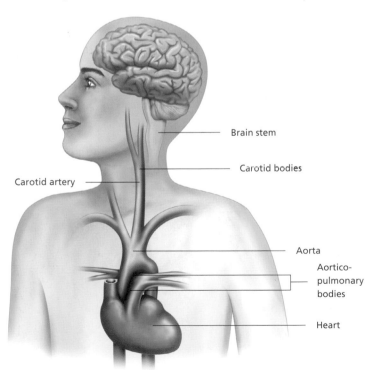

Brain stem

Carotid bodies

Carotid artery

Aorta

Aortico-
pulmonary
bodies

Heart

DANGEROUS REACTIONS

When the supply of oxygen in our blood falls too low, survival is at risk. Oxygen deprivation can be an insidious danger. An example is the effect of carbon monoxide gas (CO), which is produced by burning gasoline, charcoal, wood and other carbon-based fuels. Oxygen normally circulates in red blood cells, where the iron-containing protein hemoglobin binds and holds it until it is delivered to working cells. Carbon monoxide, which is odorless, tasteless and non-irritating, usurps oxygen's binding sites in red blood cells. If enough CO is inhaled in a confined space, the person will die of asphyxiation. The forced deep, rapid breathing called hyperventilation presents a similar hidden danger to people who employ it to hold their breath longer under water. It abnormally flushes so much carbon dioxide out of the body that the CO_2-sensitive controls over breathing may fail to operate. As the blood level of oxygen falls, a swimmer may lose consciousness and drown before the natural controls kick in again. If a person has an iron deficiency, cells and tissues receive less oxygen because there is not enough iron to build the normal supply of hemoglobin in red blood cells. As a result the person feels very tired and sluggish.

↑ **Holding your breath** mobilizes nervous system controls. Eventually carbon dioxide builds up in the blood and triggers an irresistible urge to breathe.

↗ **A whale,** such as this humpback whale, can submerge for extended periods in part because, unlike a human diver, it depends less on oxygen carried in blood and draws on O_2 stored in its lungs and other tissues.

→ **Aerosol inhalers** force a mist—sometimes a steroid drug—into the constricted air passages of asthma sufferers to help reopen them.

Blood: the all-purpose transporter

By weight, blood makes up about 8 percent of the body. It is sometimes considered a fluid tissue because it consists of watery plasma plus several types of cells and cell fragments. Red blood cells—some five billion of them in every milliliter—comprise just under half the volume of whole blood. By contrast, less than 1 percent of blood consists of white blood cells and the fragments called platelets, components with major roles in body defenses and blood clotting, respectively. The white blood cells include lymphocytes, which are important in the immune system. The rest of the blood, about 55 percent, is pale, yellowish plasma. In it are suspended molecules of several proteins that have various functions in the body.

These components are not the only substances blood transports. Red blood cells ferry oxygen throughout body tissues, while plasma carries along with it nutrients from the digestive system, hormones from endocrine glands, and waste products destined to be eliminated in urine. An equally vital blood function is transporting body heat toward or away from internal organs—a role that helps maintain core body temperature within life-supporting limits.

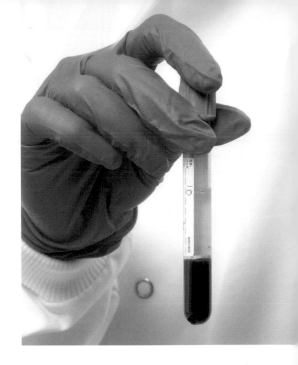

→ **When whole blood** is spun in a centrifuge, it separates into its components—yellow plasma at the top, reddish-brown blood cells and platelets at the bottom, and a thin layer of white blood cells in between.

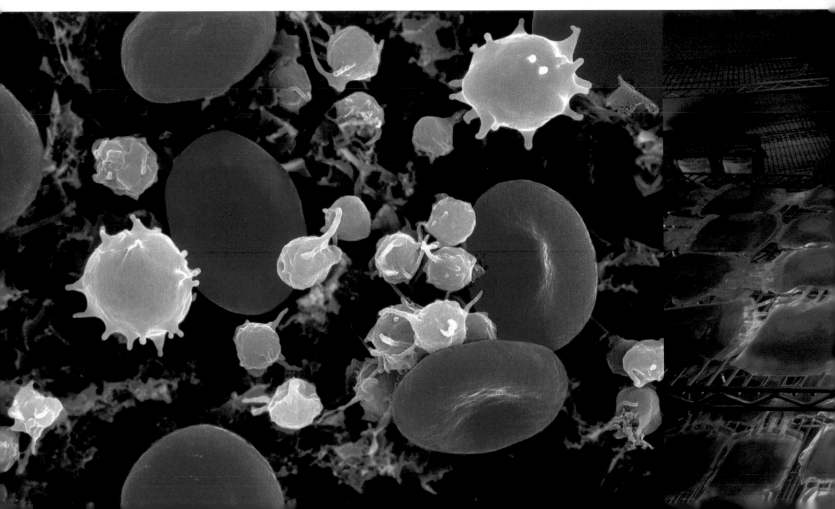

WHAT PLASMA PROTEINS DO

More than half of the blood circulating in the body is plasma. This vital volume of plasma helps determine blood pressure, among other important functions. Keeping an adequate volume of plasma is one role of several proteins that flow along in the blood, suspended in plasma. Their presence influences the movement of water into and out of the bloodstream in ways that prevent too much water from being lost. Albumin, made in the liver, makes up nearly two-thirds of plasma proteins. In addition to its role in the blood's water balance, albumin also transports substances such as therapeutic drugs and chemical wastes. Roughly another 35 percent of plasma proteins are fibrins and globulins. Fibrins are a protein formed in the clotting of blood. Some globulins transport fats and certain vitamins, among other cargoes. A number of different gamma globulins make up the immune system's antibodies. Although substances constantly move into and out of the blood, overall its chemical make-up remains remarkably consistent.

→ **Automated equipment** separates the constituents of donated whole blood. The red blood cells may be used in patients who need repeated transfusions.

→ **Stem cells** in bone marrow give rise to blood cells. Certain stem cells produce the forerunners of red blood cells and others produce platelets. Still others are the source of the five different kinds of white blood cells.

← **Sterile bags of blood plasma** await use in treating bleeding disorders such as hemophilia. The plasma, which contains factors that help clot blood, is frozen and stored until it is needed.

↞ **Red blood cells,** the red discs in this image, are shaped like flattish doughnuts, but with an indentation in the center instead of a hole. The two large, spiky orbs are white blood cells. The smaller bluish balls are platelets.

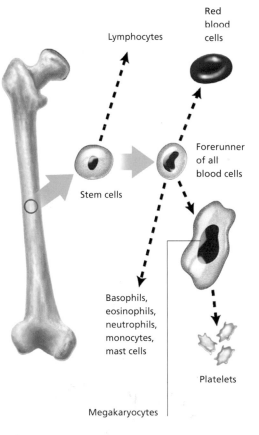

Lymphocytes

Red blood cells

Stem cells

Forerunner of all blood cells

Basophils, eosinophils, neutrophils, monocytes, mast cells

Platelets

Megakaryocytes

Red blood cells

Red blood cells, or erythrocytes, ceaselessly pick up oxygen in the lungs, transport it to tissues and offload it to the living cells there. Generated in red bone marrow, red blood cells acquire their crimson color from the iron in hemoglobin, a large protein that carries the blood's oxygen. Compared to the body's other living cells, red blood cells are mere shells. Before a red blood cell is released from the bone marrow, most of its internal parts, such as its nucleus and other organelles, are ejected like so much excess baggage. When a freshly formed red blood cell enters the bloodstream, it has essentially been reduced to a circulating capsule of hemoglobin devoted almost entirely to transporting oxygen. Although red blood cells also pick up some carbon dioxide and return it to the lungs, about 80 percent of CO_2 travels in the bloodstream to the lungs in other forms.

Like plastic bags of soft dough, hemoglobin-packed red blood cells are pliable. As they crowd through blood vessels the cells can flex and fold, then return to their normal disc shape. Not all shape changes in red blood cells are normal or reversible, however. An example is sickle cell anemia, a genetic disorder common in parts of Africa and in people whose ancestors lived there. A gene mutation produces malformed hemoglobin, and as a result the red blood cells stiffen and take on a sharply curved, "sickle" shape. They don't deliver oxygen to tissues in a normal fashion, so the affected person may gasp for breath and experience great pain. Sickle cell anemia can be treated with blood transfusions, but it often is eventually fatal.

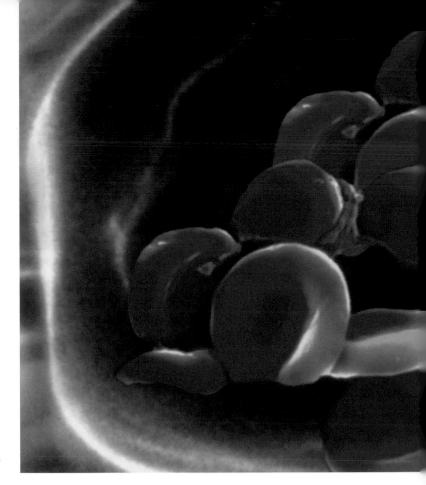

↑ **These red blood cells** are flowing through a fairly small blood vessel. When they reach capillaries, the narrowest vessels, they will have to squeeze through them in single file.

← **DNA from white blood cells** yields a "fingerprint," a pattern of its chemical make-up. Red blood cells, however, do not contain any nuclear DNA.

THE LIFE OF A RED BLOOD CELL

Each hour, about 180 million newly formed red blood cells enter the bloodstream. At the same time, tens of millions of older red blood cells are wearing out and being dismantled. Overall, the cell count—the number of blood cells in a given volume of blood—remains about the same. In an adult, the stem cells that routinely generate new red blood cells are located in the red marrow of bones such as the breastbone and vertebrae. They do their vital work in response to a signal from the hormone erythropoietin, made in the kidneys. Normally, each new red blood cell will live about 4 months. As it wears out or is damaged, a red blood cell is removed from the blood, often in the spleen. Cells called macrophages, or "big eaters," chemically digest the doomed cell and release its components, such as amino acids and iron from hemoglobin. Amino acids return to the circulation where they can be used by other cells, and iron is returned to red bone marrow, where it will be recycled to produce new red blood cells. Other components become part of an orange-brown pigment called bilirubin, which, eliminated in feces, helps give the body's solid waste its typical brown color.

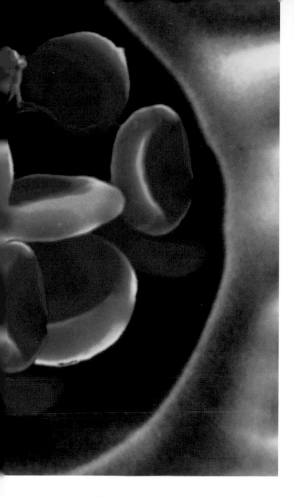

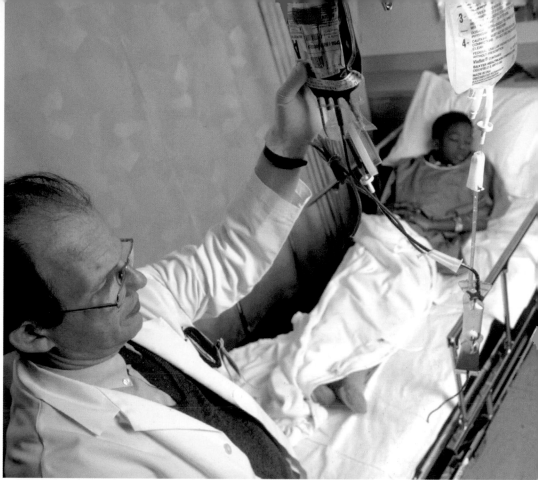

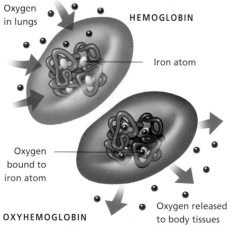

Oxygen
in lungs

HEMOGLOBIN

Iron atom

Oxygen
bound to
iron atom

Oxygen released
to body tissues

OXYHEMOGLOBIN

↑ **A hemoglobin molecule** has four strands, each
folded around a chemical unit containing iron.
When oxygen molecules bind to the iron atoms,
the resulting oxyhemoglobin molecule transports
four molecules of oxygen to body tissues.

↑ **In a transfusion,**
donated blood must
be compatible with the
recipient's blood type.
If it is not, the person's
immune system will
mount a potentially
lethal attack on the
"foreign" cells.

← **Unethical competitors**
in endurance sports
may resort to blood
"doping." Some of the
athlete's blood is drawn
and stored, so stem cells
replace the "lost" red
blood cells. The stored
blood is then returned
to the athlete's body,
so boosting the amount
of oxygen delivered
to the athlete's muscles.

White blood cells

Leukocytes, or white blood cells, make up only about 1 percent of the cells in blood. Like red blood cells, they are produced by specialized stem cells in bone marrow. Lacking hemoglobin, however, white blood cells not only look very different from red blood cells, but function very differently as well. There are a variety of types, ranging from monocytes, which in turn become either macrophages or defensive dendritic cells, to lymphocytes, which are important for immune responses. Some 70 percent of white blood cells are neutrophils, which launch a lethal chemical barrage against harmful bacteria.

Most types of white blood cells can leave the bloodstream and counter threats in the body's tissues directly. Others, like basophils, stay in the blood but release chemicals that signal to other defenders that damage is under way. Although most white blood cells live for only a few hours or days, they develop rapidly in the bone marrow. When the body responds to an infection, the number of white blood cells can double in less than a day and be constantly replenished.

→ **Soft and spongy, bone marrow** (shown here as pink) contains stem cells that give rise to the various kinds of white blood cells. Proteins called growth factors prompt the development of each type.

↘ **A basophil** is a type of white blood cell which releases histamine, a chemical that promotes inflammation. Antihistamine drugs counter this effect.

↓ **Lymphocytes,** the three smaller cells in this image, are mobilized during immune responses. The large, spiky cell is a monocyte, a white blood cell that will eventually develop into a macrophage.

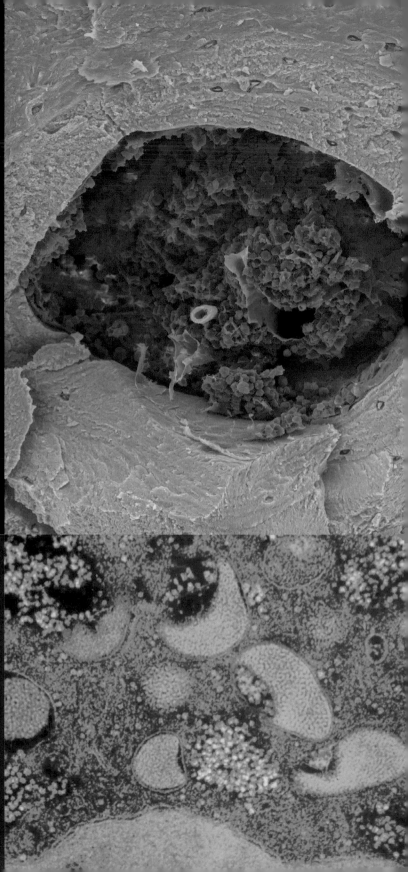

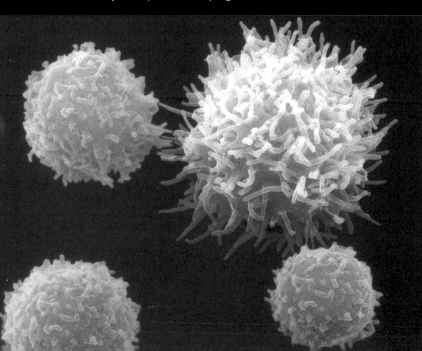

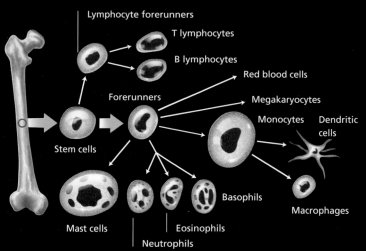

Lymphocyte forerunners

T lymphocytes

B lymphocytes

Red blood cells

Forerunners

Megakaryocytes

Monocytes

Dendritic cells

Stem cells

Mast cells

Neutrophils

Eosinophils

Basophils

Macrophages

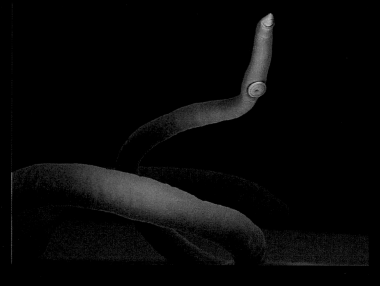

MIGHTY MACROPHAGES

Macrophages, or "big eaters," which develop from monocytes, are large cells that have a ravenous appetite for nearly any substance that is "foreign" or disposable. They are the body's most efficient phagocytes— cells that almost literally "eat" small bits of material as their outer membrane folds around it and brings it into the cell. Shards of cellular debris, dead cells, motes of dust—these and more are engulfed by macrophages in connective tissues, lymph nodes and elsewhere. In the spleen, macrophages help cleanse blood passing through it by removing worn-out red blood cells. In addition to these housekeeping duties, macrophages also consume bacteria, secrete antibiotic chemicals and play other major roles in defense. When they encounter foreign substances associated with an infection, they secrete substances called pyrogens ("fire-starters"), which can trigger a fever. Macrophages are among several types of defensive cells that can attack and kill cancer cells. They also participate in key chemical reactions that set the stage for massive defensive responses mounted by the immune system.

↑ **Parasitic worms** evoke an immune response by eosinophils. This worm is one of several that cause schistosomiasis in millions of humans in tropical areas. Infection can lead to severe, permanent damage to the lungs, liver, bladder and colon.

↖ **Specialized white blood cells,** including B and T lymphocytes, macrophages, neutrophils and others, have their origins in the stem cells in bone marrow, as do red blood cells and platelets.

↓ **Mast cells** do not circulate in blood but instead stay in tissues such as the skin and mucous membranes of the respiratory passages. Like basophils, they release histamine and other substances that feed inflammation. The chemicals are contained in granules that are visible here as blue-green dots.

↙ **When inflammation** signals an invasion, neutrophils quickly leave the bloodstream and follow a chemical trail to the site. Although each neutrophil lives for only about 6 hours, it unleashes chemicals that rapidly kill bacteria.

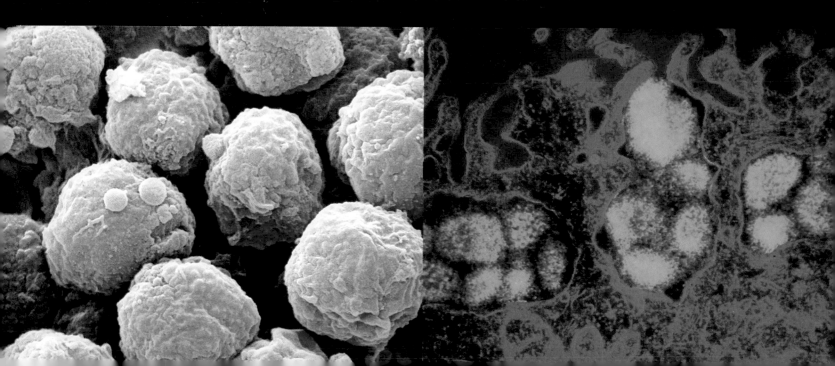

Why and how blood clots

The blood's capacity to clot is called hemostasis. "Stasis" simply means a steady state, and maintaining stability is what blood clotting is designed to accomplish. The bloodstream transports substances such as oxygen and nutrients that we cannot live without. If more than a small amount of blood is lost from the closed system of vessels that thread through the body, the reliable circulation of blood is thrown into disarray and tissues quickly begin to die. Any massive, sudden blood loss is a major medical emergency because the body has no effective means of staunching the flow.

More often, however, daily life brings punctures, cuts, scrapes or other damage that opens up small vessels and allows blood to escape. In these situations, blood clotting is our first line of defense. In a series of steps, it constricts an injured vessel, mobilizes platelets and specialized proteins to plug the breach and to help ward off harmful bacteria, and sets the stage for subsequent long-term repair processes that generate a permanent seal. Although the process may seem elaborate, it is clearly essential: without it, even the smallest cut or tear would be life-threatening—as it is for those who suffer from the genetic "bleeder's disease," hemophilia.

→ **Platelets converge** at the site of blood vessel damage, secreting chemicals that promote blood clotting. Appearing pale green in this microscope photograph, the platelets are clumped atop strands of a filter (blue) that was used to extract them from their fluid surroundings.

↓ **A net of fibrin** traps red blood cells to make blood clot, a key step in stopping bleeding. The fibrin strands (colored pale green in this image) form through a series of chemical reactions that begin as soon as a blood vessel is damaged.

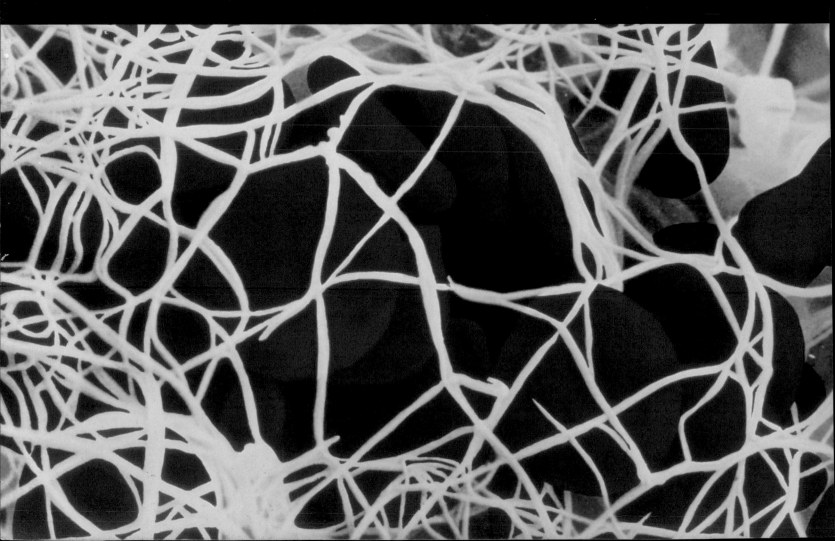

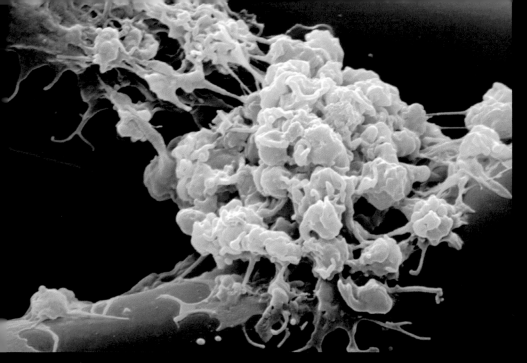

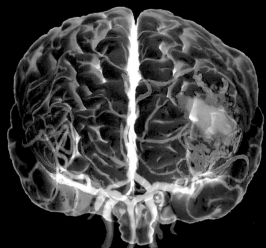

SEALING A RUPTURED BLOOD VESSEL

As soon as a small blood vessel ruptures, its walls constrict and so reduce the flow of blood. When platelets begin to arrive, they expand, become sticky and release chemicals that draw in reinforcements. As they adhere to collagen fibers exposed by the injury, the platelets plug the rupture. If the damage is minor, this plug may be all that is needed to seal the break while it heals. Otherwise, a clot forms. Chemicals from damaged tissues and from the blood itself can set this process in motion. An injured blood vessel or the tissues around it may release substances that activate a clotting factor. This protein triggers reactions that form an enzyme called thrombin. Thrombin in turn causes molecules of the plasma protein fibrinogen to coalesce into strands called fibrin. As they tangle like cooked spaghetti, the fibrin strands form a net that snares red blood cells. The blood's clotting signal comprises substances, including vitamin K, that also launch reactions to trigger the formation of thrombin and the fibrin net.

↑ **Unchecked internal bleeding in the brain** is one sign of a cerebrovascular accident (CVA), which is also called a stroke. In this image the hemorrhage in the brain's left hemisphere (center right) has damaged the brain in ways that have caused partial paralysis on the right side of the body, as well as impaired speech.

THE BLOOD CLOTTING PROCESS

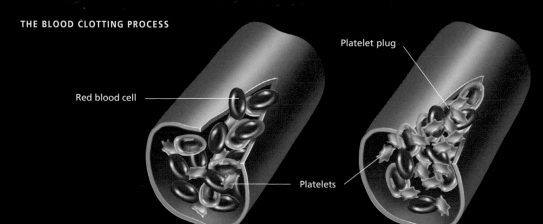

Platelet plug

Red blood cell

Platelets

Fibrin net

1. The clotting process begins when platelets become sticky.

2. Platelets form a plug to stop blood loss during healing.

3. Clotting factors trigger the formation of a fibrin net.

The cardiovascular system

The ancient Greek naturalist Galen, pondering the flow of blood through the human body, proposed that blood moved like a tidal flux, washing out from the heart and then returning like waves to the shore. Today we know that the heart and blood vessels jointly make up the body's cardiovascular system, which circulates blood throughout the body in a double loop day in and day out for as long as we live.

The heart pumps blood under pressure in two coordinated pathways—one carrying blood to the lungs where oxygen enters it and carbon dioxide leaves it, and the other transporting blood to the rest of the body. Like many other animals, humans have large, complex bodies, and a powerful pump connected to a vast distribution system is the only efficient means of delivering blood and its cargoes to the trillions of body cells. The human cardiovascular system transports oxygen, nutrients, secretions such as hormones and other needed substances, and carries away wastes and excess metabolic heat.

THE VITAL ROLE OF CAPILLARIES

The circulatory system includes an estimated 60,000 miles (nearly 97,000 km) of vessels, most of which are simply transport tubes for blood. The many substances that move into and out of the blood do so only in the tiniest blood vessels— capillaries, whose thin walls bring the blood close to body tissues. This proximity is crucial, because oxygen, nutrients and wastes all make their way between cells by diffusion, a slow process relative to the moment-to-moment needs of living cells in a large, active animal such as a human. Diffusion can take place fast enough to sustain metabolism only if the cardiovascular system brings circulating blood to within a minuscule distance of each cell.

↑ **Many thousands of blood vessels** like this supply the spinal cord with nutrients and other substances, and remove wastes produced by neurons and other cord cells.

↙ **Unlike humans, insects** like this luna moth have a simple tube-like heart and a single blood vessel. Insect "blood" is a fluid called hemolymph that does not transport oxygen.

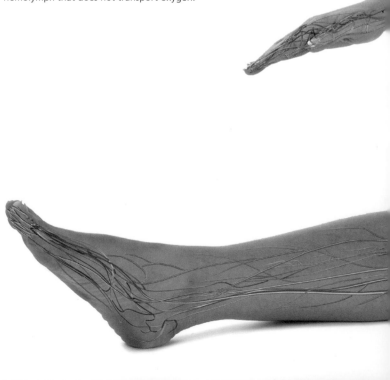

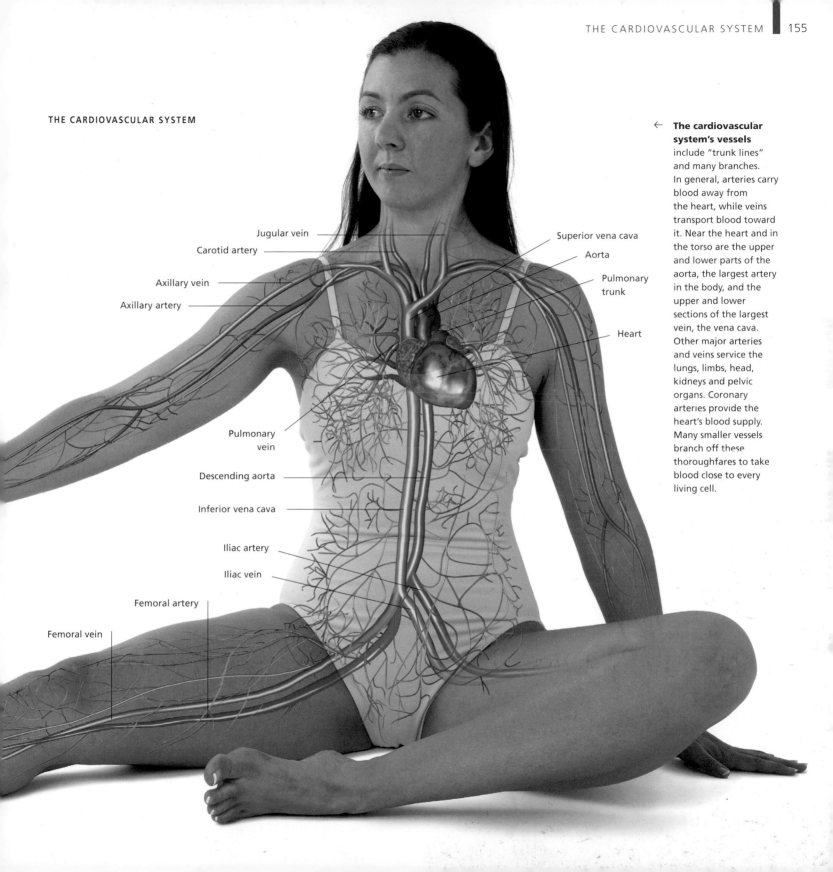

THE CARDIOVASCULAR SYSTEM

Jugular vein

Carotid artery

Axillary vein

Axillary artery

Pulmonary vein

Descending aorta

Inferior vena cava

Iliac artery

Iliac vein

Femoral artery

Femoral vein

Superior vena cava

Aorta

Pulmonary trunk

Heart

← **The cardiovascular system's vessels** include "trunk lines" and many branches. In general, arteries carry blood away from the heart, while veins transport blood toward it. Near the heart and in the torso are the upper and lower parts of the aorta, the largest artery in the body, and the upper and lower sections of the largest vein, the vena cava. Other major arteries and veins service the lungs, limbs, head, kidneys and pelvic organs. Coronary arteries provide the heart's blood supply. Many smaller vessels branch off these thoroughfares to take blood close to every living cell.

The heart: two pumps in one

The centerpiece of the cardiovascular system is the heart, a hard-working organ that is about the size of your fist and weighs about 1 pound (450 g). Located in the center of the chest, the heart rests snugly between the lungs and is supported from below by the diaphragm. A sturdy, layered sac called the pericardium wraps around the heart and a film of fluid between two of the layers prevents friction as the heart muscle contracts and relaxes.

In effect the heart is two pumps in one. It has two halves, right and left. Each half is divided into two hollow chambers, an atrium and a ventricle, and the two ventricles each pump blood into vessels with different destinations. The right side receives blood from tissues and pumps it to the lungs, and the left side receives blood from the lungs and pumps it out to tissues. Valves between the heart chambers ensure that blood flows in the proper direction. While the heart's atria have relatively thin walls, the heart's "pumping stations"—the ventricles—have thick, muscular walls—a structure that befits their round-the-clock function of circulating the blood.

↑ **The "heart strings,"** or chordae tendineae, are stabilizers built of collagen. They hold the valve between the atrium and the ventricle closed while blood is propelled into an artery.

STRUCTURE OF THE HEART

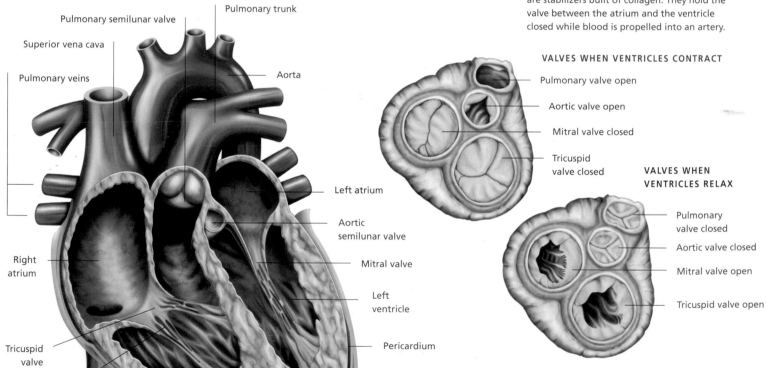

- Pulmonary semilunar valve
- Pulmonary trunk
- Superior vena cava
- Pulmonary veins
- Aorta
- Left atrium
- Aortic semilunar valve
- Right atrium
- Mitral valve
- Left ventricle
- Tricuspid valve
- Pericardium
- Chordae tendineae
- Inferior vena cava
- Right ventricle
- Myocardium
- Descending aorta

VALVES WHEN VENTRICLES CONTRACT

- Pulmonary valve open
- Aortic valve open
- Mitral valve closed
- Tricuspid valve closed

VALVES WHEN VENTRICLES RELAX

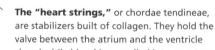

- Pulmonary valve closed
- Aortic valve closed
- Mitral valve open
- Tricuspid valve open

← **The heart's architecture** suits its operation as a double pump. Internally, the right and left sides are separated by a thick wall that merges with the myocardium—the heart's strong, contracting muscle layer. On each side, the atria serve as "reception rooms," where blood enters and is briefly stored before it moves into the ventricles below. Fibers of cardiac muscle are arrayed in ribbon-like bundles that spiral around the heart. This allows the contracting ventricles to forcefully squeeze blood outward. Valves (above) open or close to control the flow of blood through the heart.

HEART VALVES

Two kinds of valves direct the one-way flow of blood through the heart and then out into arteries. An atrioventricular (AV) valve at the base of each atrium prevents incoming blood from washing backward. In the right atrium the AV valve has three pliable flaps and so often is called the tricuspid ("three cusped") valve. The mitral valve in the left atrium has only two flaps. In between each beat, the heart relaxes briefly as the atria fill with blood. Then the AV valves open and let the blood flow into the ventricles. A moment later when the ventricles

contract, their squeezing motion forces blood upward against the underside of the valve flaps and they close. Each ventricle also opens to a large artery that receives the blood it pumps. The entrance to the artery is guarded by a semilunar valve, which has three pouched flaps shaped like half-moons. Blood spurting into the artery collapses the flaps open against the artery wall. When the flow slackens and blood begins to wash back toward the ventricle, it expands the valve pouches so the semilunar valve closes and blood cannot re-enter the ventricle.

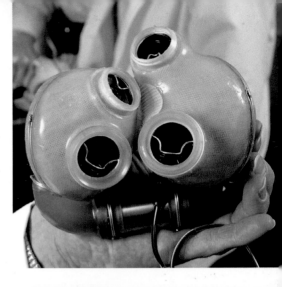

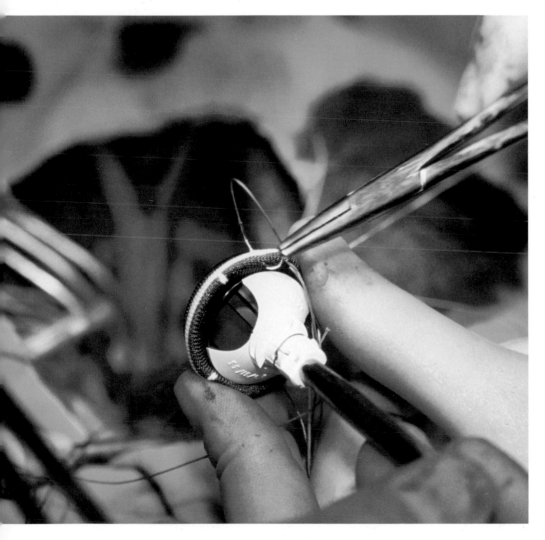

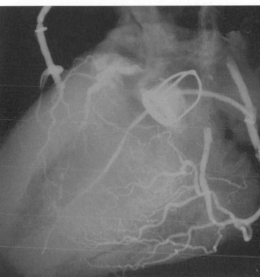

↑ **An artificial valve** between the left ventricle and the aorta can be seen in this X-ray. Replacing a faulty semilunar valve, the new valve will prevent blood from continually washing back into the ventricle.

⇑ **An artificial heart** can be an emergency life-saver. If a person's own heart is failing, the device can be implanted temporarily to allow time for a transplantable heart to be found.

← **When a heart valve** leaks or becomes too stiff to function properly, the heart has to work overtime to pump blood. Seriously defective valves can be replaced by artificial valves, like the one in the hands of surgeon here, or by valves harvested from cadavers or even pigs.

Heartbeat: the cardiac cycle

If you live to the age of 80, your heart will have produced nearly 300 billion beats, pumping blood with each one. In a heartbeat, the heart's chambers—the two atria above and the two ventricles below—alternately contract and relax in a sequence called the cardiac cycle. The electrical signals for contraction come from the heart muscle itself. The contraction phase is called systole, and the relaxation phase is diastole. The heart's atria, its two smaller chambers, contract first, pushing blood into the larger ventricles. Then the ventricles contract, pumping blood out of the heart. Ever so briefly, the heart rests, then the cycle begins anew. During the cycle, movements of the heart and blood generate the "lub-dup" sound heard through a stethoscope. The entire sequence takes just under a second, but in that time a significant portion of a person's blood is moved through the heart and into arteries.

The blood is sent on its way under pressure. Systolic pressure—the top number in a value such as 120/80—is generated when the ventricles are contracting at their peak, while the bottom number, diastolic pressure, tracks the low point when the heart is relaxed. Blood pressure can be a window into a person's health. High blood pressure, or hypertension, may indicate that arteries are clogged by fat deposits or that the kidneys are not removing excess water. By contrast, low blood pressure that is still within normal limits—for example, 115/60—is associated with an efficiently functioning cardiovascular system. Studies suggest that people with low-normal blood pressure enjoy a longer, healthier life.

↑ **Young children** often have heart murmurs, even though their hearts are healthy. The "abnormal" sound can result when the walls of a child's heart vibrate as blood moves through. Sometimes a murmur may indicate a defective heart valve.

THE CARDIAC CYCLE

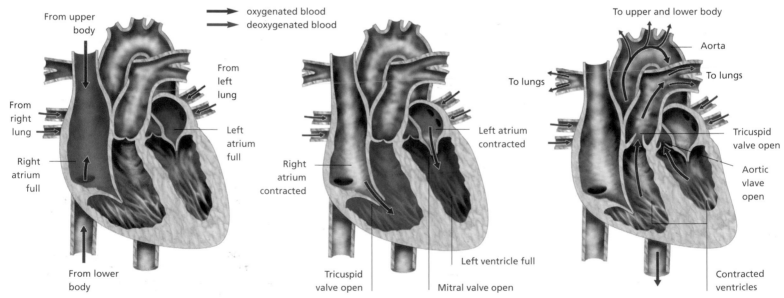

1. At the start of the resting phase of the cardiac cycle, blood flows into the atria. The AV valves stay closed until the atria are nearly full. At that point, some blood begins to dribble into the ventricles.

2. Nerve impulses signal the atria to contract and eject blood into the ventricles through the open AV valves. These then close as pressure builds inside the ventricles, which are still in their resting phase.

3. When nerve impulses reach the now-full ventricles, they contract and pump blood through the semilunar valves into arteries: from the right ventricle to the lungs and from the left to the aorta.

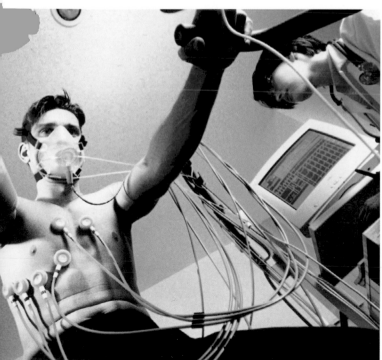

↑ **Swimming and other forms of aerobic activity** exercise the heart muscle so that over time it gains strength and pumps blood more efficiently. During exercise a healthy heart may pump up to 30 quarts (34 liters) of blood per minute.

← **A cardiac stress test** monitors heart health. The person being tested exercises with increasing intensity while attached to a machine that records the pattern of the heart's electrical activity. The recording, called an electrocardiogram (ECG), can reveal signs of heart disease.

HEART RATE

Not everyone's heart beats at the "normal" rate of 70 to 80 beats a minute. In a condition called bradycardia, the resting heart rate is less than 60 beats per minute. Certain drugs and other factors can trigger bradycardia, but it is also common in endurance athletes, such as swimmers and cyclists. It develops when the heart muscle enlarges due to regular aerobic training and therefore can pump more blood during each "turn" of the cardiac cycle.

The heart's electrical activity also may become irregular, a phenomenon called arrhythmia. Some arrhythmias are no cause for concern, but others are red flags. In tachycardia, the heart beats 100 or more times per minute. Triggers include stress, fever and heart disease. Long-term, tachycardia is a generally a sign of illness. More serious is a rapid, chaotic heart rhythm called fibrillation. When one or both atria are affected, they may not supply adequate blood to the ventricles. Fibrillation of the ventricles is a medical emergency that is often associated with a massive heart attack or strong electrical shock. The ventricles stop pumping altogether, halting blood flow to the brain.

The heart's pacemaker

Unlike other muscles, cardiac muscle contracts—and the heart beats—without commands from the nervous system. This property comes from the heart's pacemaker, a small cluster of cells that is called the sinoatrial (SA) node. Located near the top of the right atrium, the pacemaker cells are genetically programmed to produce electrical signals 70 to 80 times a minute, or more if the situation warrants. The pacemaker continues to stimulate a heartbeat even if all nerves to the heart are cut! Its signals reach the cardiac muscles of the atria first, prompting the atria to contract and pump blood into the ventricles. Then the impulses travel on to a second cluster of cells near the base of the right atrium. This cluster, called the atrioventricular (AV) node, transfers the commands for contraction to the ventricles. From there, the commands spread through the ventricles along a branching system of conducting fibers called Purkinje cells. It takes only 30 one-thousandths of a second for Purkinje cells to transmit the impulses to all the muscle fibers in the ventricles, so the heart's chambers can contract in the coordinated fashion that keeps blood flowing steadily through the cardiovascular system.

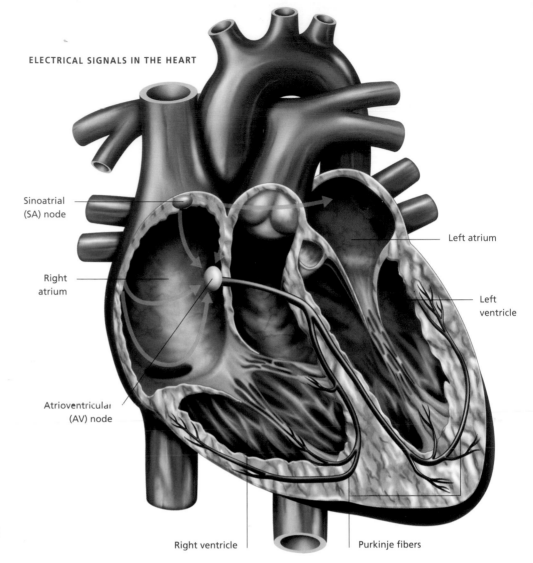

ELECTRICAL SIGNALS IN THE HEART

Sinoatrial (SA) node

Right atrium

Atrioventricular (AV) node

Left atrium

Left ventricle

Right ventricle

Purkinje fibers

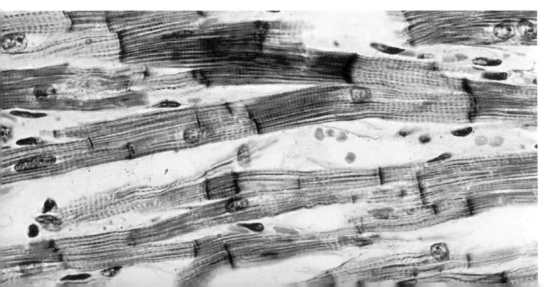

↑ **Signals from the SA node** travel to both atria and also to the AV node. From there a network of Purkinje fibers extends down through the thick wall that separates the two ventricles, then branches upward, distributing the pacemaker's signals throughout the heart.

← **Muscle fibers in the heart** are linked by specialized junctions. Visible as dark purple lines in this microscope photograph, the junctions, called intercalated discs, knit the fibers tightly together in a fashion that allows nerve impulses to travel easily from fiber to fiber.

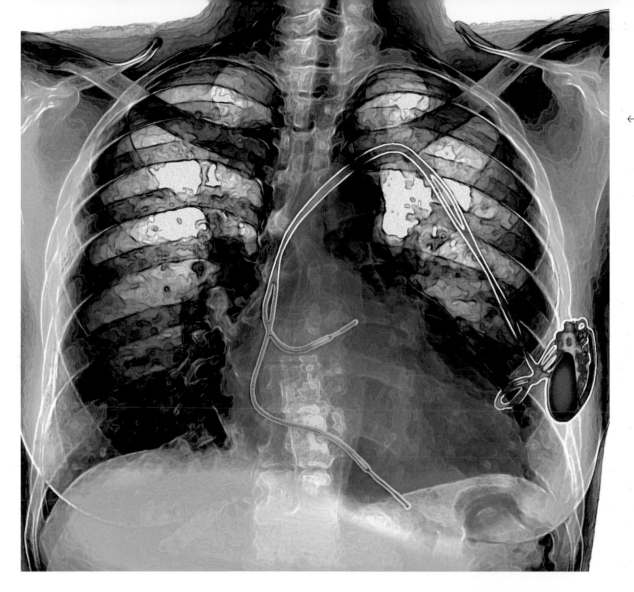

A **battery-powered pacemaker** that has been surgically implanted takes over the duties of a malfunctioning SA node, emitting electrical signals that trigger a normal rhythm of heart contractions. Impulses from the device travel along one or more wires to the heart, which is shown as a pale orange mass in this colored X-ray. The wires may be threaded through large blood vessels that service the heart.

DISRUPTED HEART RHYTHMS

A variety of factors can disrupt operations of the heart's pacemaker or its system for conducting electrical impulses. Stress, anxiety, or too much alcohol, caffeine or nicotine (from tobacco) can overexcite part of the heart muscle, and the heart chamber may contract before a signal arrives from the pacemaker. When it does arrive, the chambers have been filling with blood for longer than normal, and when they contract again the person feels an odd "knock" in the chest.

A disorder known as a "heart block" develops when the conducting fibers between the atria and ventricles are damaged. As a result, the ventricles contract at only 30 beats per minute. An artificial pacemaker can restore the normal rhythm. Otherwise, such a low heart rate makes all usual activities, as well as consciousness, impossible. If the SA node itself is faulty, the AV node may assume the role of setting the heart rate. Its maximum pace is only about 50 beats per minute, so an affected person may feel lethargic due to reduced blood flow to body tissues. The pacemaker and AV node also may become scarred in normal aging, which is why elderly people are more likely to develop abnormal rhythms.

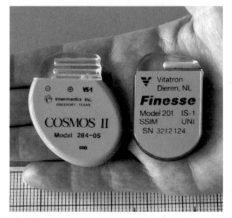

Artificial pacemakers are implanted under the skin. They are powered by batteries that must be replaced every few years. A person who has a pacemaker is advised to avoid some kinds of electromagnetic signals, such as powerful radio waves, which can interfere with the electrical signals that the device generates.

Dual circuits of blood

Blood courses through your body in two loops, one beginning in the right side of the heart and the other beginning in the left side. The loop called the pulmonary circuit begins as veins deliver blood from tissues to the heart's right atrium. Cells have been exchanging gases with that blood all along its route through the body, so it contains little oxygen and a great deal of waste carbon dioxide. After it moves into the right ventricle, it is pumped into arteries leading to the lungs. There, the oxygen is replenished and carbon dioxide is exhaled.

From the lungs, oxygenated blood flows into the second circulation loop, which begins in the left side of the heart. Entering the left atrium, blood is sent on to the powerful left ventricle and pumped forcefully into the aorta. This is the beginning of the systemic circuit, which transports oxygenated blood to the rest of the body. Almost immediately, some of the flow is diverted into coronary arteries that keep the heart supplied with oxygen-rich blood. The brain routinely receives about 13 percent of the heart's output of blood, and elsewhere blood is distributed according to the demands of the moment—with more going to the digestive tract after a meal, for example, or to skeletal muscles when you are physically active.

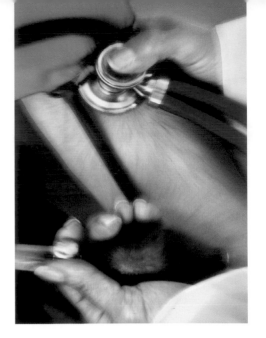

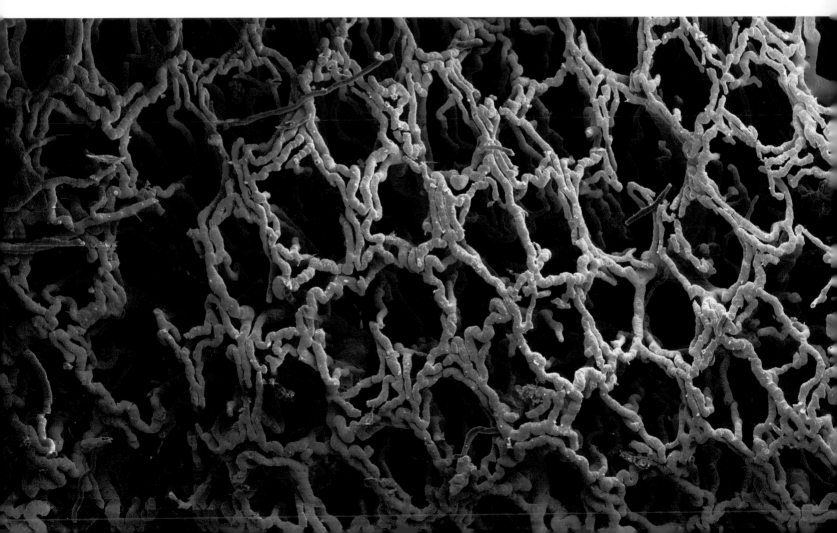

DELIVERIES AND DETOURS

Circulating blood continually picks up substances from cells and delivers others to them. If this were not so, our cells would quickly be starved of nutrients and oxygen, as well as smother in their own metabolic wastes. These vital exchanges start in capillaries, the smallest blood vessels. In the liver, intestines and many other organs, capillaries interweave in elaborate networks called capillary beds. Here blood flowing in the general circulation is diverted from larger outflow vessels, the arteries and arterioles, into countless smaller channels. Its flow slows to a trickle because the red blood cells can only move through the narrow capillaries one by one. Along the way there is time for body cells to take up needed substances as well as for wastes or secretions such as hormones to cross into the bloodstream. At the other end of a capillary bed, the capillaries merge into the venous system. Small venules, then larger veins transport blood back to the heart.

← **Blood pressure is measured** in part of the systemic circulation using a pressure gauge that is attached to a hollow cuff wrapped around the upper arm.

THE BLOOD'S TWO CIRCUITS

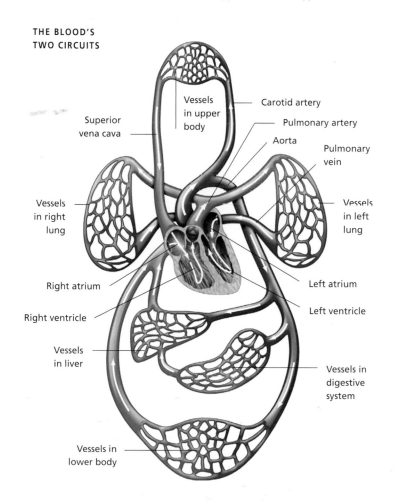

Carotid artery
Vessels in upper body
Superior vena cava
Pulmonary artery
Aorta
Pulmonary vein
Vessels in right lung
Vessels in left lung
Right atrium
Left atrium
Right ventricle
Left ventricle
Vessels in liver
Vessels in digestive system
Vessels in lower body

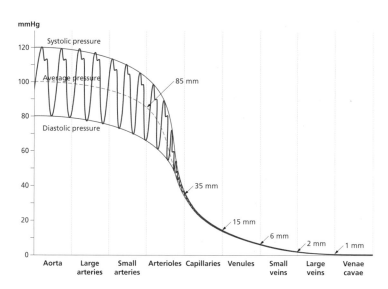

mmHg
Systolic pressure
Average pressure
Diastolic pressure
85 mm
35 mm
15 mm
6 mm
2 mm
1 mm

Aorta | Large arteries | Small arteries | Arterioles | Capillaries | Venules | Small veins | Large veins | Venae cavae

↑ **As blood circulates** its pressure drops. Pressure, measured in milligrams of mercury (mmHg), is highest in the aorta, which receives blood straight from the left ventricle, but then falls off steeply. By the time blood reaches veins that return it to the heart, its pressure is a fraction of the starting value.

← **Both circuits of the cardiovascular system** carry oxygen-rich (red) and oxygen-depleted (blue) blood, but in the pulmonary loop oxygen-poor blood is pumped to the lungs to be refreshed. Although we visualize the pulmonary and systemic loops as separate, they operate as part of a whole that ensures a seamless flow of blood throughout the body.

←← **A network of small blood vessels** in the intestines is shown in this resin cast. This elaborate array picks up nutrients absorbed from food and also services the intestinal cells themselves.

↓ **At rest,** the brain, kidneys, digestive tract and skeletal muscles collectively receive about 70 percent of circulating blood. About 3 percent goes to the heart, with the rest allocated to the bones, liver and other parts of the body.

Blood pipelines

Like a city's water system, the cardiovascular network of blood vessels includes pipelines of various sizes with different functions. By far the largest are the massively built arteries, such as the aorta and femoral arteries to the thighs. An artery is somewhat elastic but also strong enough to withstand and maintain the pressure of the blood pumped out by the heart's ventricles. Sustaining this pressure, arteries keep large amounts of blood coursing away from the heart.

Arteries ultimately branch into narrower, thinner-walled arterioles. Less sturdily built than arteries, their role is to distribute blood to tissues and organs in various body regions. Blood pressure drops and blood flow slows inside arterioles, and they can adjust the movement of blood into different tissues and organs to meet varying demands. Where arterioles end, blood enters millions of tiny, thin-walled capillaries, and substances move into and out of the cells in tissues. From there, venules and veins transport blood back to the heart.

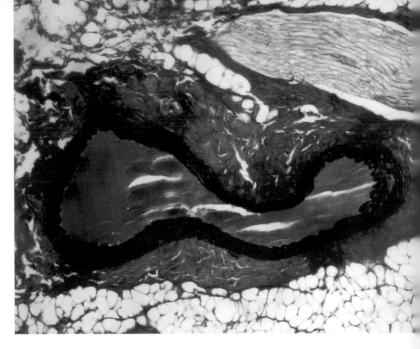

HOW ARTERIOLES MANAGE BLOOD FLOW

Because they have less bulky walls than arteries, arterioles can contract or expand more easily. This mechanical versatility allows them to adjust the movement of blood into and through different tissues. If a person's blood pressure falls below a set point, nerve impulses from the brain signal the muscle in arteriole walls to contract. As the vessel narrows, blood pressure rises. Similarly, blood pressure can be lowered by neural signals that increase the diameter of arterioles. Shifts in blood chemistry that correlate with physical exertion, digesting a meal, or some other physiological event also trigger signals that dilate arterioles in one site and constrict them elsewhere. This mechanism temporarily increases the amount of blood flowing to regions where cells and tissues are especially active.

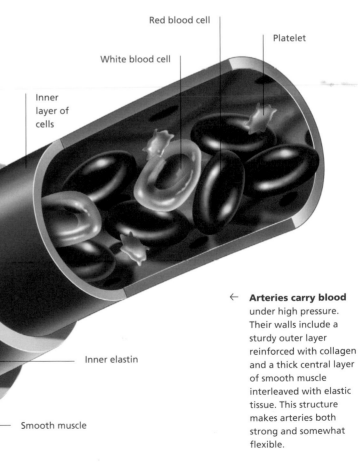

Red blood cell

Platelet

White blood cell

Inner layer of cells

Outer elastin

Outer layer reinforced with collagen

Inner elastin

Smooth muscle

← **Arteries carry blood** under high pressure. Their walls include a sturdy outer layer reinforced with collagen and a thick central layer of smooth muscle interleaved with elastic tissue. This structure makes arteries both strong and somewhat flexible.

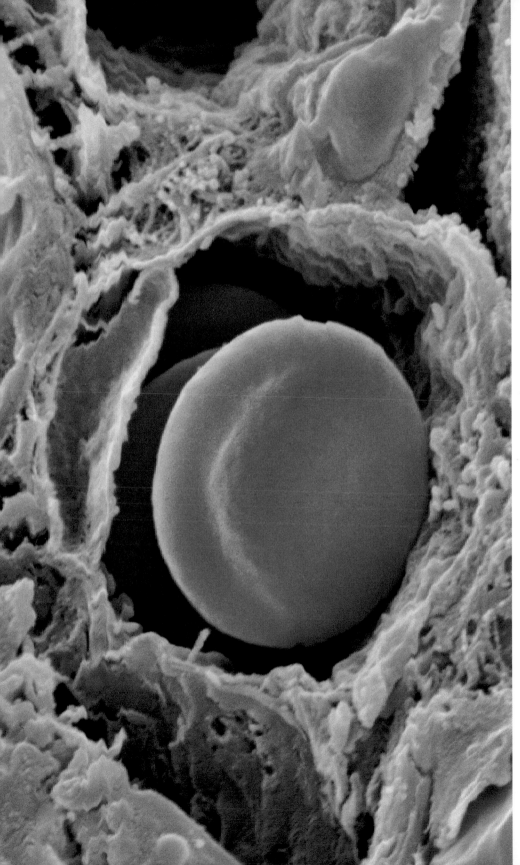

← **Millions of capillaries** permeate body tissues, moving substances into and out of cells. This image shows how red blood cells must squeeze single file through these tiny, narrow, tunnel-like vessels.

⇇ **Arterioles have** stretchy, flexible walls. In this cross-section of an arteriole in the intestines, its thin, elastic inner wall is crimson, while its outer wall of smooth muscle is dark pink.

↓ **With every heartbeat,** pressurized blood is forced into arteries. We can feel the surges—and so monitor our pulse rate—where arteries are near the surface, such as in the neck or at the wrist.

Return transit to the heart

At any given moment, your veins may contain more than 60 percent of the total volume of blood in your body. As the main components of the venous system, their task is to transport blood from tissues back to the heart. Much like the various drainpipes in a house, the system's smallest vessels, called venules, receive blood from capillaries, then merge with veins, which become progressively larger closer to the heart. Blood in the venous system is pumped to the lungs by the right side of the heart—the beginning of the cardiovascular system's pulmonary circuit. In the lungs it offloads carbon dioxide, takes on oxygen, and is propelled back to the heart and into the systemic circuit once again. Like arteries, veins are relatively large-capacity vessels, but they do not transport blood under high pressure. In fact, were it not for features such as one-way valves in the veins of the lower body, gravity might cause the blood in them to flow backward.

SUSTAINING THE FLOW

Peer through a microscope at a cross-section of a vein, and you might be struck by its thin walls and large lumen—the open space inside it. These features combine to give veins two of their most distinctive properties, a capacity to hold a great deal of blood and to keep it moving. In vessels with a narrow lumen, friction between blood and the vessel walls slows the flow. As blood gathers from venules into veins, the relatively wide-open interior allows blood in veins to move fast. As a result, even though it has made a long, circuitous trip through the arterial system, blood in veins reaches the heart traveling at virtually the same velocity as when it was pumped out. The valves in veins of the limbs help sustain this flow even when a person is standing or sitting for long periods.

↓ **Near the surface of the skin,** as at the inner wrist, veins often appear blue. This color is due in part to the presence in skin of the yellowish-brown pigment melanin. When venous blood is drawn for medical testing, we can see its true, dark red color—a hue that blood takes on as it loses oxygen.

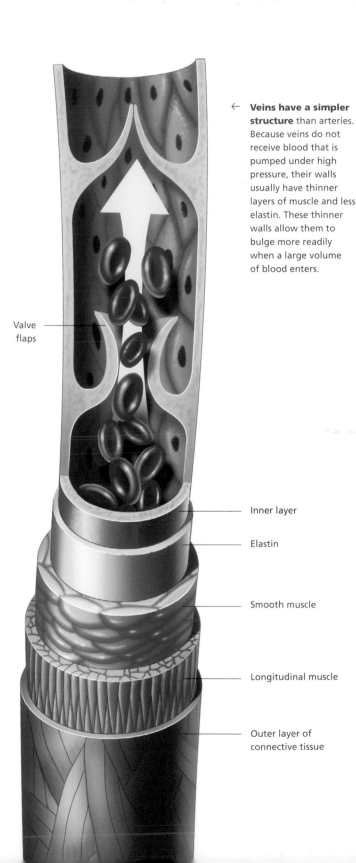

← **Veins have a simpler structure** than arteries. Because veins do not receive blood that is pumped under high pressure, their walls usually have thinner layers of muscle and less elastin. These thinner walls allow them to bulge more readily when a large volume of blood enters.

Valve flaps

Inner layer

Elastin

Smooth muscle

Longitudinal muscle

Outer layer of connective tissue

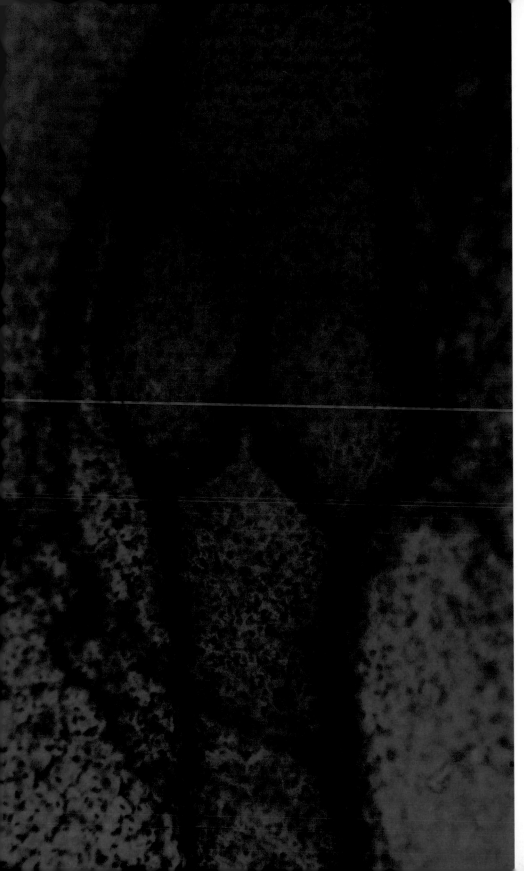

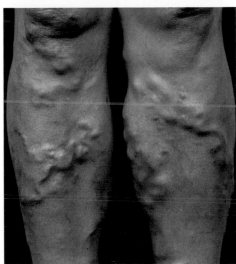

↑ **Being active** helps return blood to the heart. When skeletal muscles in the arms and legs contract, they press on veins and help force blood upward through the valves.

↑ **Varicose veins** usually develop in the legs. Some people naturally have weak valves in their veins, but more often valves are weakened by conditions like obesity and pregnancy, which add weight in the torso and increase downward pressure on leg veins.

← **Valves in veins** allow blood to flow in just one direction, toward the heart. Veins in the arms and legs have valves, and so do some in the trunk. Blood pushed through the valve from below fills the cuplike pockets and keeps the valve closed.

The cardiovascular system under siege

Cardiovascular disorders damage blood vessels and force the heart to work overtime. One of the most stealthy is atherosclerosis, in which fatty plaques slowly develop in artery walls. Years in the making, eventually the bulging plaque obstructs blood flow and may lead to a heart attack. Excess cholesterol in the blood is a common culprit in atherosclerosis, but it may be only one contributing factor. Studies suggest that it begins when some kind injury, perhaps from high blood pressure, microbial attack or some other source, triggers inflammation. This response leads to other changes that result in a plaque.

Chronic high blood pressure, or hypertension, develops as smaller arteries become constricted, hampering blood flow. Hypertension is a common cause of strokes. It can also contribute to heart failure because the heart must constantly work harder to pump blood through stiff, narrowed vessels. The heart is so weakened that it can no longer pump blood efficiently, and people with heart failure often find it challenging to walk across a room. Genetic factors and aging play a role in most cardiovascular diseases, but so do lifestyle choices like smoking, being sedentary, carrying excess weight and having an unhealthy diet.

BLOOD PRESSURE VALUES		
Blood pressure	Systolic (mmHg)	Diastolic (mmHg)
Normal	10–119	60–79
Low (hypotension)	Less than 100	Less than 60
Prehypertension	120–139	80–139
High (hypertension)	140 and above	90 and above

↓ **Smoking tobacco is just as dangerous** for the cardiovascular system as it is for the lungs. Among other ill effects, the nicotine in tobacco stimulates the constriction of arteries and arterioles, including those that service the heart itself.

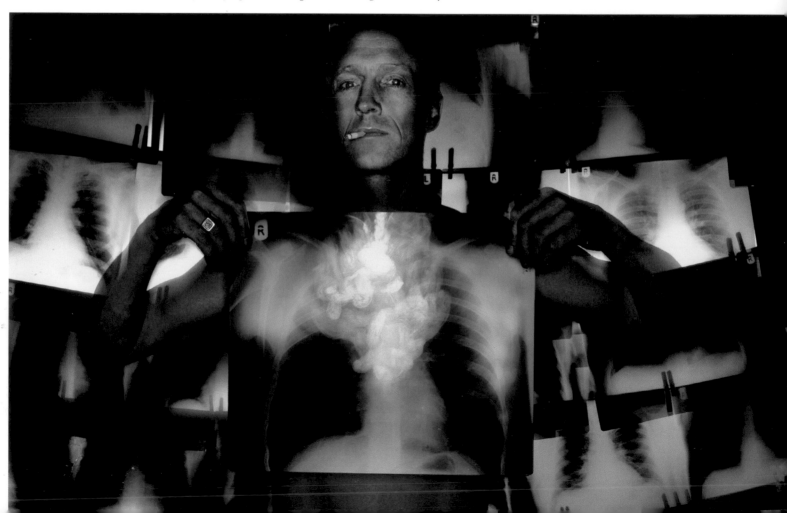

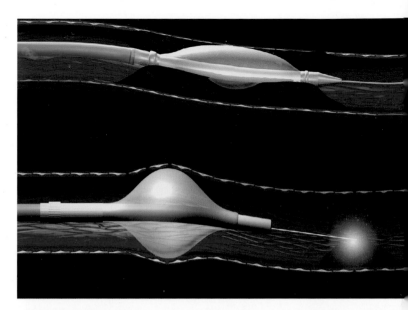

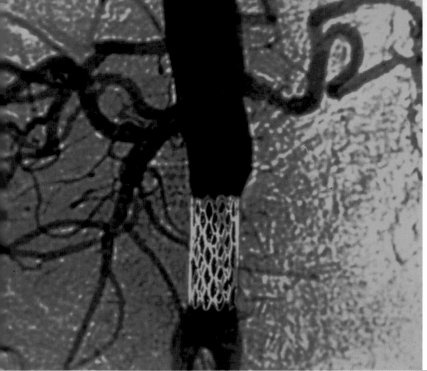

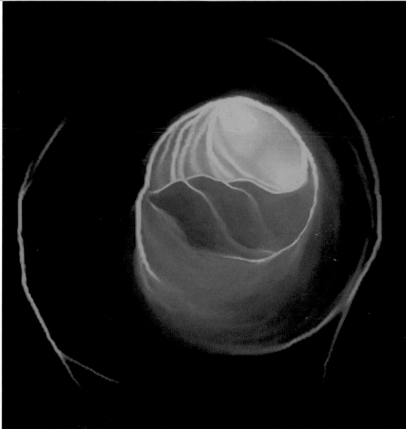

↑ **An implanted stent** holds a clogged artery open in a large abdominal vessel. The hollow, wire-mesh device is threaded into the affected artery during a procedure called angioplasty.

↗ **Angioplasty offers options** for restoring blood flow. In the most common approach (top) a balloon is threaded through a catheter inserted in the artery and inflated at the blocked point, widening the artery. A variation on this method (bottom) uses a laser to destroy the blockage ahead of the catheter.

→ **Plaques of fat and cholesterol** that look like miniature mountains close off part of a carotid artery, a blockage that will restrict the flow of blood. Located in the neck, this artery provides the main blood supply for the brain.

HEART ATTACK

Often a diseased cardiovascular system is first discovered when a person has a heart attack. The problem begins when some part of the heart muscle receives too little oxygen for an extended period, typically when a coronary artery is blocked by a plaque or blood clot. The lack of oxygen does not directly kill the affected heart muscle, however. Instead, it launches a sequence of events in which immune system responses to the heart's distress cause the lethal damage. If too much of the heart muscle is affected, the attack is likely to be fatal.

People in middle age and beyond are the most vulnerable to heart attack, but we now know that the symptoms may be different in men and women. Men generally have the "traditional" heart attack symptoms, such as shortness of breath, crushing chest pain or pain radiating down the left arm. Women may have these symptoms, or they may mainly feel sudden, deep fatigue and pain between the shoulder blades.

The lymphatic system

Extensive as it is, the cardiovascular system depends heavily on another, less familiar network of organs and vessels. Each day this network—the lymphatic system—retrieves and returns to the bloodstream about a gallon and a half (6.8 liters) of water and other substances that have seeped out of blood vessels and into the spaces in tissues. If many of these substances were not returned to the blood, its volume would dwindle, the heart would no longer have sufficient blood to pump and our blood vessels would collapse. Lymphatic vessels also collect unwanted and foreign material, and certain of them take up fats that have been absorbed from the digestive tract and transfer them to circulating blood.

Like the cardiovascular system, the lymphatic system consists of small capillaries that connect to progressively larger tubes. Unlike blood capillaries, however, lymphatic capillaries have closed ends, forming arrays that resemble the root system of a plant. Moreover, their walls are built so that water and other substances can easily slip into them. This cargo is not pumped through the system. Instead, once it reaches a larger lymphatic vessel, one-way valves keep the lymph, as it is then called, moving toward veins that empty into the heart. En route, lymph is filtered through bean-like lymph nodes, where macrophages and other defensive cells wait to intercept and destroy bacteria and other unwanted material. In fact, many elements of the lymphatic system—the lymph nodes, spleen, thymus, tonsils and patches of tissues elsewhere—are essential to body defenses.

LYMPH AND LYMPH NODES

Much thinner than blood and colorless, lymph is a mixture of water, lymphocytes, tissue "garbage," microbes and even cancer cells. It also contains proteins that have escaped from blood plasma as well as other essential substances. For instance, hormones from endocrine organs circulate in lymph vessels, making the lymph an auxiliary delivery system for hormone signals. Droplets of fat from digested food enter lymph vessels in the lining of the small intestine. They form part of the lymph until they are "off-loaded" into the bloodstream in neck veins. Lymph nodes are arrayed along lymph vessels at various points around the body, especially in the neck, groin and armpits. Because the nodes and the white blood cells in them trap invading or unwanted cells, they can provide vital health information, including evidence that a cancer has spread. When the body fights an infection, lymph nodes can swell with bacteria or other debris.

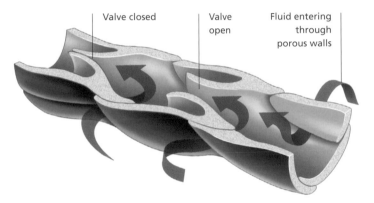

Valve closed   Valve open   Fluid entering through porous walls

↑ **A lymph vessel** is a porous one-way tube, with valves that help keep lymph moving in the proper direction through the system.

← **The spleen** contains soft red and white tissues, referred to as "pulp." Red pulp breaks down aging or defective red blood cells. In the white pulp, macrophages destroy other unwanted material, such as fragments of bacteria.

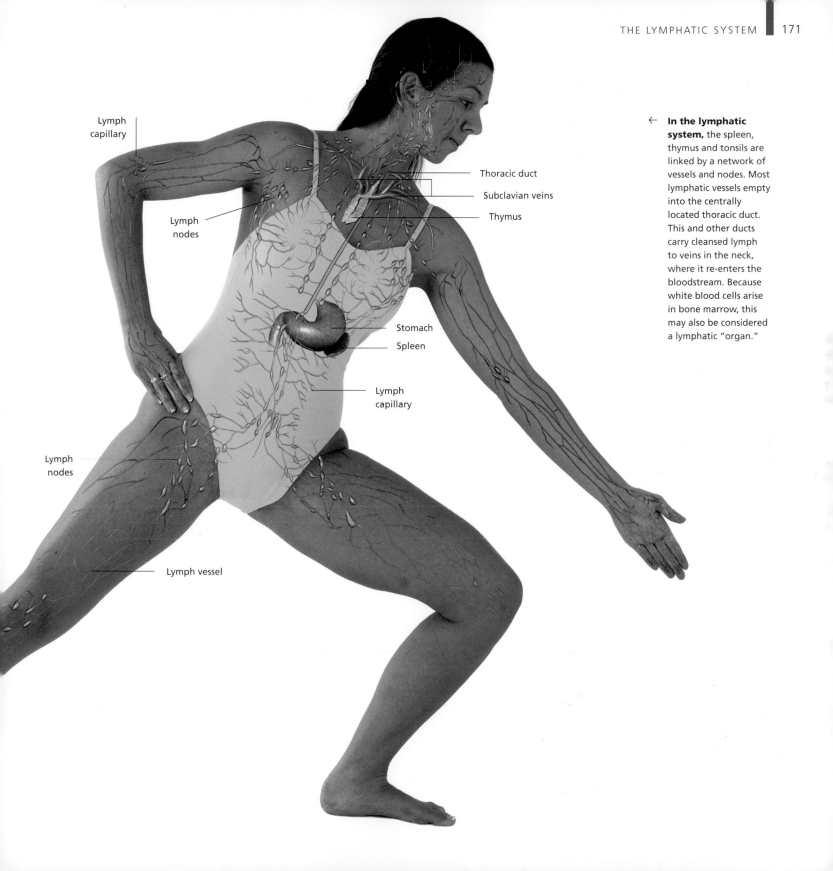

Lymph
capillary

Thoracic duct

Subclavian veins

Lymph
nodes

Thymus

Stomach

Spleen

Lymph
capillary

Lymph
nodes

Lymph vessel

← **In the lymphatic
system,** the spleen,
thymus and tonsils are
linked by a network of
vessels and nodes. Most
lymphatic vessels empty
into the centrally
located thoracic duct.
This and other ducts
carry cleansed lymph
to veins in the neck,
where it re-enters the
bloodstream. Because
white blood cells arise
in bone marrow, this
may also be considered
a lymphatic "organ."

Hidden enemies

We inhabit a world of small but powerful enemies, many of them invisible to the unaided eye. They include bacteria, viruses, disease-causing fungi, and parasitic worms and mites. Some 1000 species of bacteria inhabit the surface and crevices of our skin, mouth, intestines and other parts of the body, most usually posing no threat. Other bacteria, however, produce toxins that make them agents of disease and sometimes death. Smaller, and in some ways more insidious, are viruses. Consisting of protein-coated DNA or RNA, a virus cannot be touched by antibiotics, because it lacks the working parts of a cell. Similarly we have no biological defenses against the mysterious infectious proteins called prions. Transmitted in infected food, prions devastate the brain, as in Creuzfeldt-Jakob disease, the human version of mad cow disease. Common fungal infections, such as athlete's foot and yeast infections, often are merely annoying, but some fungi, such as the one responsible for retina-destroying histoplasmosis, can cause permanent damage. Millions of people also spend much of their lives afflicted with larger but equally destructive parasites that survive by using human tissues and organs as sources of food.

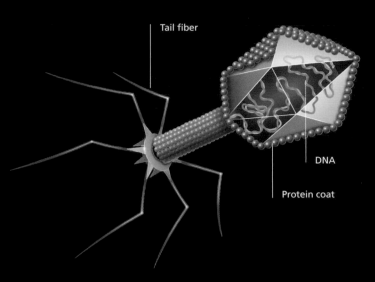

Tail fiber

DNA

Protein coat

↑ **Looking like tiny lunar landers,** some viruses have spider-like tail fibers and other parts. A protein coat wraps around their genetic material, DNA, which commandeers infected cells and reprograms their metabolic machinery for one function—to make more viruses.

↓ **Tuberculosis bacteria** (*Mycobacterium tuberculosis*), visible here as blue-green rods, infect the lungs—and sometimes other parts of the body.

↓ **Helicobacter pylori bacteria** (pink) cause most stomach ulcers, once thought to be the result of stress.

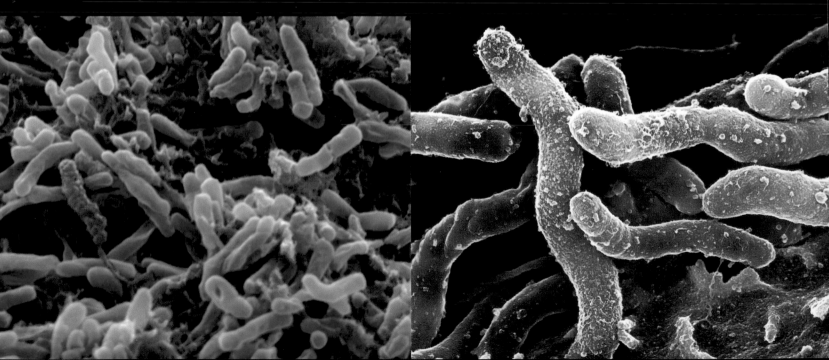

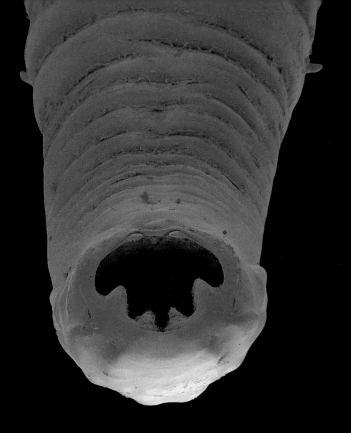

ONGOING BATTLES

While we often think of microbes as threats, some bacteria and fungi have long served us in the making of fermented cheeses, yeast breads, beer, wine and many other products. Massive vats of natural and genetically engineered fungi and bacteria routinely produce therapeutic drugs, including antibiotics. Ironically, the more antibiotics we make and use, the more we see new strains of bacteria that antibiotics cannot kill. Bacteria can quickly develop genetic mutations that change their characteristics, which can give rise to antibiotic resistance. The widespread, misguided use of antibiotics, such as fruitlessly using them to treat a viral infection or not taking the full prescribed course, simply increases the likelihood that resistant strains will become established. Clinicians regularly see drug-resistant tuberculosis and syphilis, as well as resistant infections of the middle ear, urinary tract and surgical wounds, among others. Now and in the future, one of the greatest medical challenges will be finding ways to stay one step ahead of our microbial enemies.

← **A hookworm's head** attaches to the small intestine, where adult worms suck blood and breed. Hookworms infest millions of humans.

↙ **The Ebola virus** causes high fever and severe hemorrhaging. Often fatal, Ebola fever has been limited mostly to central Africa.

↓ **The herpes simplex virus** causes blisters on skin and mucous membranes. Type 1 triggers "cold sores" and type 2 the lesions of genital herpes.

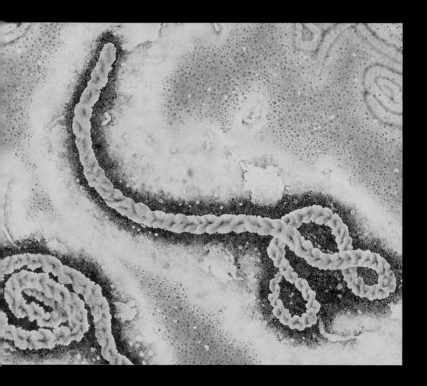

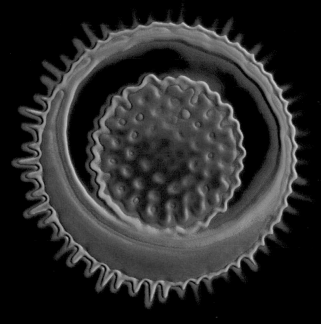

Defensive strategies

The body has three major lines of defense against harmful organisms. Some we bring with us when we enter the world, while others develop as we go through life. The first set of countermeasures consists of physical barriers. Paved with a thick top layer of dead epithelial cells, the skin's surface also swarms with harmless bacteria. Both these features make it difficult for potential pathogenic types to even gain a foothold. Less exposed linings also bar the easy passage of microbes. For example, mucous membranes in the reproductive tract have resident populations of defensive white blood cells. Acidic stomach fluid creates a chemical environment intolerable to most microorganisms in food. Our acidic urine is equally inhospitable—and it also flushes intruders out. Mucus bathing the lining of the airways contains a natural antibiotic called lysozyme.

A second set of countermeasures is mobilized when a body tissue is damaged, regardless of the cause. Strategies such as inflammation rally white blood cells and proteins in blood plasma.

The body's third and most versatile defense is the immune system. Based in organs of the lymphatic system, this response to harm comes from warrior cells and proteins that can single out specific threats and mount a vigorous counterattack. At birth, the immune system is a mere shadow of what it will become. Only over time does it acquire the ability to recognize, defend against and "remember" the body's foes.

→ **Combating a measles infection,** a defensive cell called a macrophage (colored red) engulfs a virus particle. Other white blood cells (blue) have also assembled, attracted by bits of virus protein that the macrophage has displayed on its surface.

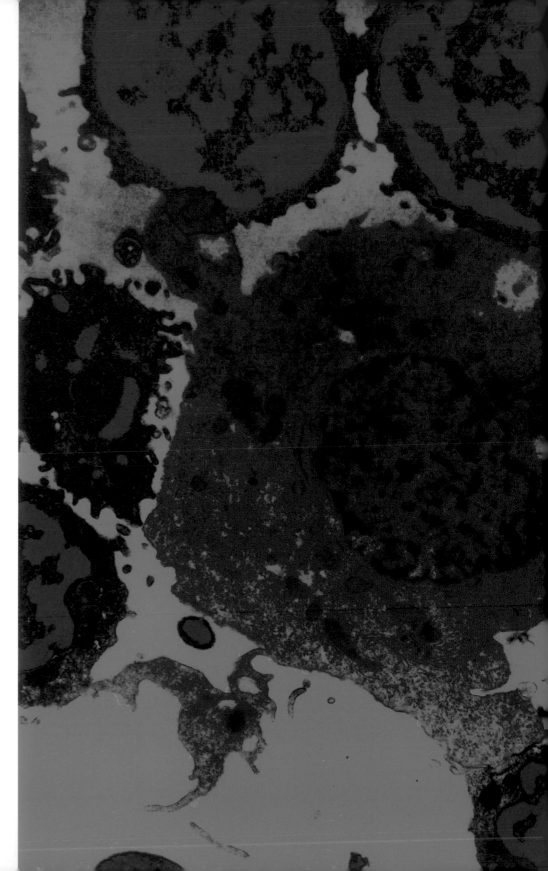

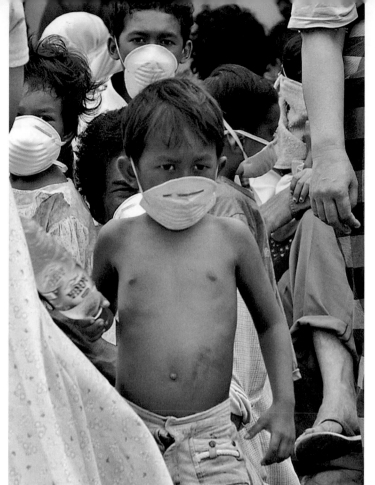

THE BODY'S THREE LINES OF DEFENSE
1. Surface barriers
Unbroken skin, mucous membranes
Harmless bacteria on skin and elsewhere
Antimicrobial substances in saliva, tears and stomach fluid
Removal of microbes by tears, saliva, urine, coughs and sneezes
2. General defenses
Inflammation: neutrophils, basophils, eosinophils, macrophages, mast cells (histamine)
Natural killer cells
Pathogen trapping in lymph nodes
Complement proteins
3. Immune responses
Lymphocytes: T cells, B cells, macrophages
Defensive proteins and other chemicals: antibodies, interleukins, interferons, complement proteins

↑ **In the wake of the 2004 tsunami** in Indonesia, refugees cope as best they can. Water and soil contaminated by sewage and decaying bodies and livestock added to the threat of dysentery, typhus and other diseases to the misery of millions.

→ **Tears can be defensive weapons.** Like mucus in the airways, they contain lysozyme and other antimicrobial substances that moisten and help protect the exposed surface of the eyeball.

PROTEIN DEFENDERS

The body makes tens of thousands of proteins, using them for purposes ranging from structural building blocks to hormones to catalysts that speed chemical reactions. About 20 proteins jointly form a defensive unit that is called the "complement" system because it enhances other defenses. Complement proteins circulate constantly in the bloodstream, ready to respond to chemical clues left by intruders, including bacteria, some types of viruses and various kinds of parasites. At first only a few complement proteins may be activated, but the blood soon teems with them as the response develops. If bacteria are the quarry, the proteins become organized into tiny structures that puncture the surface of the cell so the bacterium bursts open and quickly dies. Complement proteins also serve as signals that an invasion is under way. As they converge on a pathogen, they form a chemical trail that white blood cells can follow to damaged tissue.

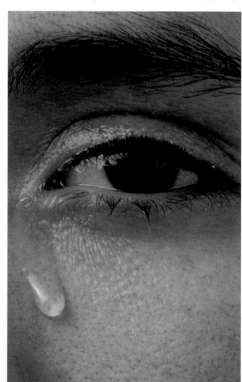

The inflammation response

When an infection or wound kills or damages cells, the body mounts the response we call inflammation. A variety of combatants, including many of the body's white blood cells, are involved at various stages, fighting the invasion physically, chemically or both. Early on, mast cells produce histamine. Although histamine is the culprit in troublesome allergic reactions, in this setting it is part of a potent defensive response. As histamine dilates blood vessels in the injured area, more blood flows into it, bringing other defenders. Histamine also makes it easier for blood cells to pass through the walls of blood capillaries.

In a matter of hours, neutrophils circulating in the bloodstream arrive at the besieged site, where they squeeze out of the leaky vessels and begin destroying invaders by a "cell-eating" process. Proteins that operate in blood clotting and in the complement system can also slip out of capillaries and into the tissue. In time macrophages arrive. The body's most long-lived "cell-eaters," they may help combat an infection for many weeks or months. Macrophages, neutrophils and other white blood cells also flood the blood with signaling chemicals, which quickly "advertise" that a damaging attack has begun so reinforcements can join the battle. These defenses include prostaglandins, which set in motion steps that produce a fever and lethargy. As a feverish person rests, energy reserves can be used in the battle against illness.

WHEN INFLAMMATION GOES AWRY

Inflammation should stop when the damage is repaired, but if healthy tissue becomes inflamed or the process continues too long, it becomes a liability and can even pose a major threat to health. Headaches and swollen insect bites are mild inflammatory discomforts, but there are numerous diseases, including coronary artery disease, rheumatoid arthritis and some bowel disorders, that involve inflammation gone seriously awry.

Therapeutic drugs that subdue inflammation work in various ways. Aspirin reduces the release of histamine, as do the antihistamines in cold and allergy medicines. Corticosteroid drugs can halt nearly every part of the inflammation process. Unfortunately, they also are toxic to lymphocytes, the white blood cells responsible for targeted immune responses. As a result, someone who uses corticosteroids heavily—for example, to prevent asthma attacks or deal with persistent back pain—may be at higher risk for infections. For this reason, patients who must regularly take large doses are carefully monitored. The dangers of steroid overuse helped spur the development of nonsteroidal anti-inflammatory compounds (NSAIDs). They reduce the release of prostaglandins, which can promote inflammation, cause muscle spasms and trigger pain signals to the brain.

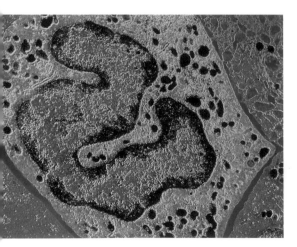

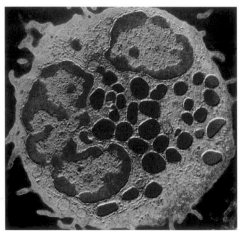

↑ **Fever is a defensive maneuver.** In a chemical chain reaction initiated by white blood cells, the body's "thermostat" in the brain is reset upward. A high fever is dangerous, but one of about 100°F (39°C) can make body tissues too warm for microbes to multiply.

← **Basophils are the rarest** white blood cells. The red spheres in this image are granules filled with chemicals, including histamine and the anti-clotting agent heparin. Lobes of the basophil's nucleus appear as orange and green areas.

← **Eosinophils are white blood cells** that attack parasites such as worms. The one pictured here has a U-shaped nucleus with two lobes (blue and dark pink), a typical eosinophil trait.

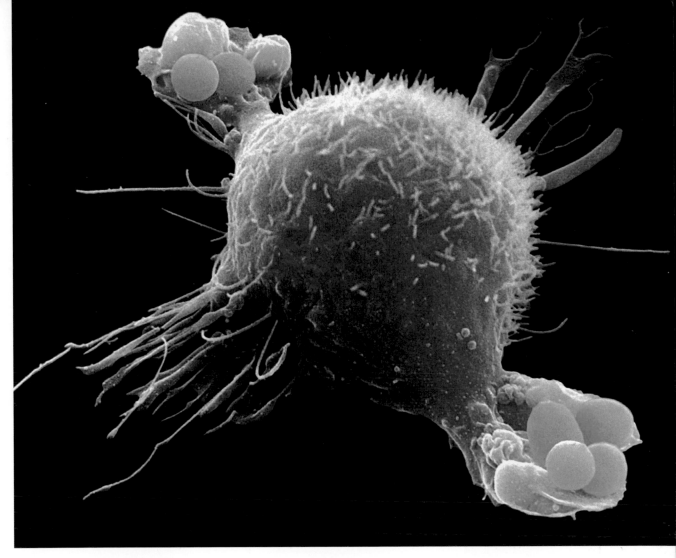

→ **When a macrophage** (blue) detects bacteria (green), it extends pseudopods, or "false feet," which wrap around the offenders and pull them inside it to be destroyed.

↓ **Defensive chemicals and white blood cells** exit dilated blood capillaries during an inflammatory response. As the response unfolds, inflamed tissue becomes red, swollen, warm to the touch and painful.

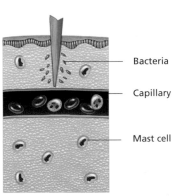

Bacteria

Capillary

Mast cell

1. When tissue is damaged, mast cells are activated.

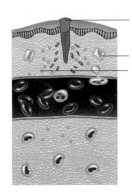

Debris

Damaged cell

Histamine

2. Blood vessels dilate and blood flow increases.

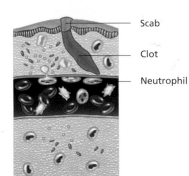

Scab

Clot

Neutrophil

3. A clot forms. Neutrophils migrate into damaged tissue from capillaries.

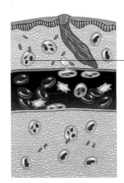

Neutrophil engulfing bacterium

4. Neutrophils engulf and destroy bacteria and dead cells.

Immune responses

Instead of interacting organs and tissues, the immune system consists of interacting white blood cells . Its foremost components are lymphocytes called T cells and B cells, which begin their defensive odyssey in bone marrow. The B cells remain there to develop while the T cells migrate to the thymus, where they take on their final form. When both types of lymphocytes are mature, they enter the lymph nodes, spleen and other lymphatic tissues. There, they are in position to intercept bacteria, virus particles or anything else they recognize as foreign or abnormal, such as a cell that is infected or cancerous.

Lymphocytes are specialized for particular tasks and they can combat millions of the body's biological enemies. This versatile response has three key features—it targets only particular invaders, it operates throughout the body, and the immune system "remembers" its initial battle with a pathogen, so it can mount a faster, stronger response if the intruder returns.

SELF AND NON-SELF

To identify and combat potential threats from foreign material entering the body, the immune system must be able to tell the difference between "self" and "non-self." T and B cells, and auxiliary cells such as macrophages, use chemical differences to distinguish potential adversaries from the body's own cells. This feat is possible because body cells are marked on their surface with "self labels" called MHC proteins, which are encoded by each person's genes. Only identical twins have the same ones. Cells or substances from outside the body have different, and usually incompatible, chemical markers. Autoimmune disorders develop when the immune system mistakenly identifies normal body cells as foreign and mobilizes its weaponry to destroy them.

→ **Spiky ragweed pollen** triggers the immune response we know as an allergy. When antibodies in the nasal passages respond to it, cells there release histamines that trigger swelling and other symptoms.

← **A sneeze** may be just one sign of "the flu." Sneezing is a general defense geared to rid the nasal passages of foreign material. If this man has been infected by an influenza virus, his immune system will also mount a specific counterattack.

← **Genetically engineered piglets** could one day provide transplantable organs. Several biotechnology companies are working to create "donor" animals whose cells lack markers that would normally trigger a dangerous immune response, and rejection of the organ, in a human recipient.

↓ **Dust mites live by the millions** in bed sheets, carpets and upholstery. Microscopic relatives of spiders, the mites feed on skin flakes. In susceptible people their wastes can irritate the skin and respiratory passages, causing rashes and sinus-clogging allergic reactions.

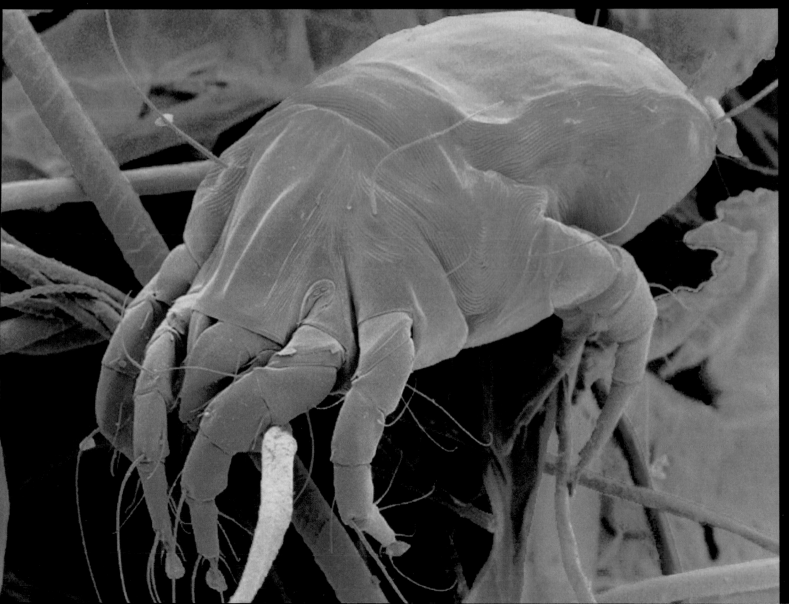

Recognizing the invader

Passing hours and days give pathogens the opportunity to gain a foothold in the body. Yet if the immune system has never before dealt with a particular microbe, it must first organize the appropriate countermeasures. For T cells, this process begins only when macrophages and other patrolling sentinels detect the invader, perhaps a disease-causing bacterium, and then dismantle it and outwardly display some part of it as an antigen—a chemical identity tag—that lymphocytes can "read." This essential first step sets the stage for one of the body's T cells to recognize the antigen as its target. B cells, which make the defensive proteins called antibodies, locate antigens without help. Either way, after the first defender is sparked into action, the immune system produces an army of identical warrior cells to destroy its newly met enemy.

In addition to eliminating the present danger, all this preparation may also pay off many times over. When the body mounts an immune response, some defender cells are held in reserve as "memory cells." Already primed to fight the pathogen, they are like trained, battle-ready battalions that can mount a faster, stronger response if it returns. Memory T or B cells may circulate in the bloodstream for weeks, months or years, conferring immunity against the invader. Vaccination is a powerful public health tool because it spurs the immune system to create memory cells that can counteract disease-causing pathogens as soon as they strike.

CELLS THAT SPECIALIZE

Each T cell or B cell is a specialist. Its antenna-like receptors are shaped so that it can link up with only a single kind of antigen, the chemical identification tags that mark foreign cells as alien. Like doors in a vast castle that require different keys, each of the body's T cells and B cells has receptors shaped so that only one "key"—one type of antigen—will fit with it. This engineering is one of the body's marvels.

Although a given B cell or T cell can mount a defense against only one biological foe, every meeting unleashes the full force of the immune system. The gene-guided system for creating T cells and B cells is so versatile that each of us has hundreds of millions of them, each with a different receptor. As we go through life, they collectively allow the immune system to combat hundreds of millions of different pathogens.

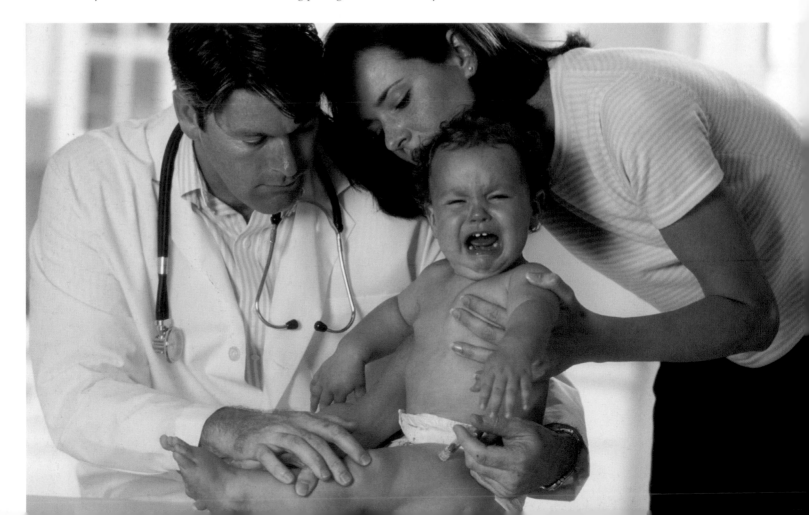

Smallpox was a common and dreaded disease in the eighteenth century. Observing that survivors never caught the disease again, many people inoculated themselves with powdered smallpox scabs, gambling that this would trigger a protective infection. Many died as a result. Seeking a safer alternative, in 1796 English physician Edward Jenner injected an 8-year old boy with pus from a cowpox, a related but milder illness. After the infection subsided he exposed the boy to smallpox. As Jenner hoped, his young patient remained free of disease.

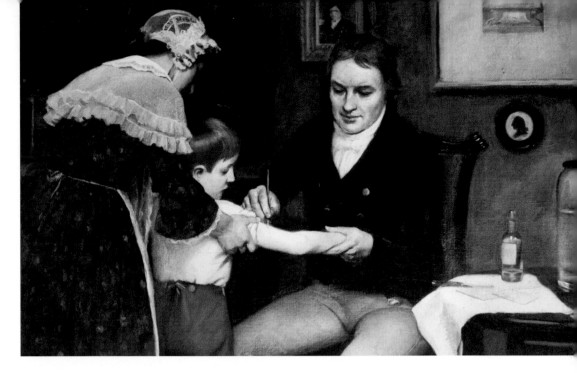

In 1947 New Yorkers lined up to be vaccinated against smallpox. At the time, city officials warned the public of a potential outbreak of the disfiguring and potentially lethal disease. In subsequent decades vaccinations on a global scale all but eradicated smallpox around the world.

Vaccinations protect children from many childhood ills that were once common. In most countries, scheduled inoculations confer long-term immunity against diseases such as polio, diphtheria, whooping cough, tetanus, measles and mumps. "Booster shots" given later extend the protection period, usually into adult life.

Antibodies

Our immune system depends heavily on the large defensive proteins called antibodies. Shaped roughly like the letter Y, antibodies target threats outside of body cells, such as toxins, bacteria or viruses that are circulating in the bloodstream or present in tissue fluid. Antibodies first appear at a B cell's surface. Protruding from the cell, they are the tools by which a B cell identifies its target. Each B cell is equipped with antibodies attuned to a different potential challenger. When a harmful cell or substance actually enters the body, if there is a B cell with the necessary receptor it will engage the invader. This modest event is a springboard, for soon the "first responder" multiplies and produces a vast army of B cells identical to itself. Called plasma cells, these second-generation B cells are high-speed antibody factories, each producing more than 100,000 antibodies per minute. Released into the blood, saliva and elsewhere, the growing flood of antibodies can attach to the offending cells or molecules and label them for destruction by other elements of the body's defensive forces.

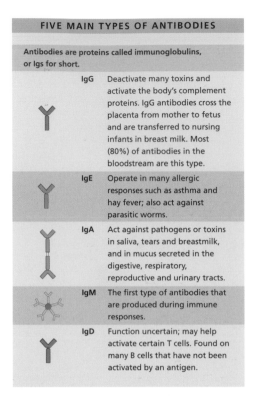

FIVE MAIN TYPES OF ANTIBODIES

Antibodies are proteins called immunoglobulins, or Igs for short.

	IgG	Deactivate many toxins and activate the body's complement proteins. IgG antibodies cross the placenta from mother to fetus and are transferred to nursing infants in breast milk. Most (80%) of antibodies in the bloodstream are this type.
	IgE	Operate in many allergic responses such as asthma and hay fever; also act against parasitic worms.
	IgA	Act against pathogens or toxins in saliva, tears and breastmilk, and in mucus secreted in the digestive, respiratory, reproductive and urinary tracts.
	IgM	The first type of antibodies that are produced during immune responses.
	IgD	Function uncertain; may help activate certain T cells. Found on many B cells that have not been activated by an antigen.

↑ **Hoping to fend off** the SARS virus in 2003, Chinese schoolchildren wore protective masks. SARS is a severe respiratory illness that has only recently been documented in humans. It is particularly dangerous because most people do not have antibodies for it and no vaccine has yet been developed.

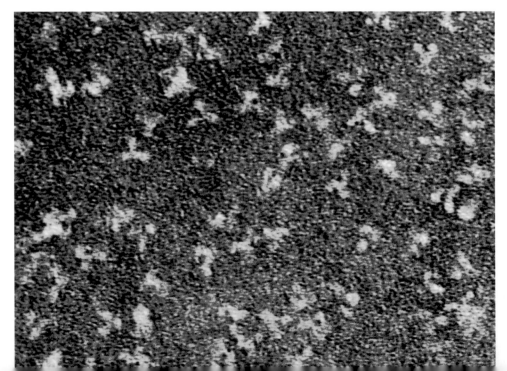

← **Human antibodies** appear yellow in this microscope image. Antibodies are proteins folded into a rough Y shape. About 80 percent of the antibodies in human blood are in the IgG group, like those pictured here.

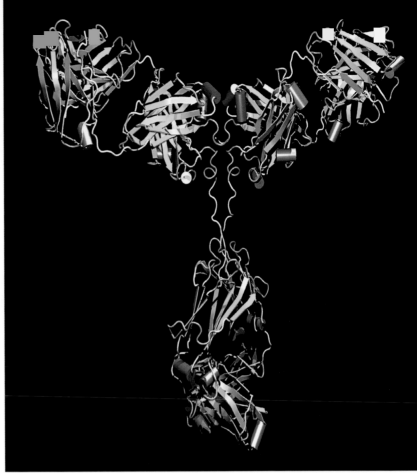

↑ **This ribbon model of an antibody** shows the complex sructure of these proteins. Dips and bumps in an antibody's "arms" fit exactly into dips and bumps in its corresponding antigen. When the two link up successfully, the body can mount its defense response to the antigen.

← **Autoimmune** conditions such as systemic lupus and rheumatoid arthritis are most common in women who have borne children. Researchers are exploring the possibility of a link.

EXERCISE AND IMMUNITY

Anyone who has had common respiratory ailments such as colds and flu will probably have antibodies that can help limit the extent of future bouts of such illnesses. For instance, IgA antibodies combat a variety of infectious agents, including influenza viruses, and research suggests that moderate exercise can help keep them and other immune system weapons at the ready. Those who exercise regularly —taking a brisk walk, riding a bicycle or doing other kinds of aerobic exercise—but not to the point of exhaustion, tend to have more circulating IgA and IgG antibodies, and fewer respiratory infections, than do elite athletes who have recently completed a strenuous competition or training regimen. The "average" exercisers also had more of the T cells that combat entrenched pathogens. One reason for both these findings may be that extreme physical exertion triggers the release of larger than normal amounts of the adrenal hormone cortisol, which suppresses inflammation and other natural defenses. Staying fit without overstressing other body systems may be one key to maximizing immunity.

Killer cells and other defenders

Antibodies target adversaries in blood or lymph, but they are powerless against infected or abnormal body cells. The immune system's T cells, produced by stem cells in the bone marrow, assume this vital role. Some T cells become a general attack force of "natural killer" (NK) cells. Others move on to the thymus, a gland just above the heart, where they become specialized into helper T cells and killer T cells. These later leave the thymus, each carrying two types of receptors—one that allows it to identify a body cell and one that can attach to a specific alien antigen. If a T cell encounters its antigen during a person's life, its defensive task begins.

True to their name, killer T cells release lethal substances that destroy virus-infected body cells, cancer cells or transplanted cells. Only about 20 percent to 40 percent of T cells are killers. The rest are helper T cells that serve as the immune system's "general managers." They release chemicals that organize and boost immune responses, and are so important that without their signals the immune system essentially stops operating. Substances known as interleukins and interferons also aid immunity. Interleukins spur the development of armies of T cells and B cells after a threat has been detected. Interferons, released by dying virus-infected cells, help protect healthy cells by interfering with the ability of viruses to multiply inside them.

ORGAN TRANSPLANTS

Transplants can extend the lives of people with severely diseased organs. Unless the transplanted tissue comes from an identical twin, however, there is a major risk of rejection because defensive cells perceive donated tissues as foreign and launch a powerful immune response against them. To minimize the chances of rejection, the gene-coded "self" markers on donor tissues are matched as closely as possible to those of the patient. The blood types must also be compatible. If at least 75 to 80 percent of self markers match, a transplant has a reasonable chance of success. Even so, recipients must often take drugs that suppress the immune system for the rest of their lives. As well as having unpleasant side effects, the drugs carry a much-increased risk of infections, among other dangers.

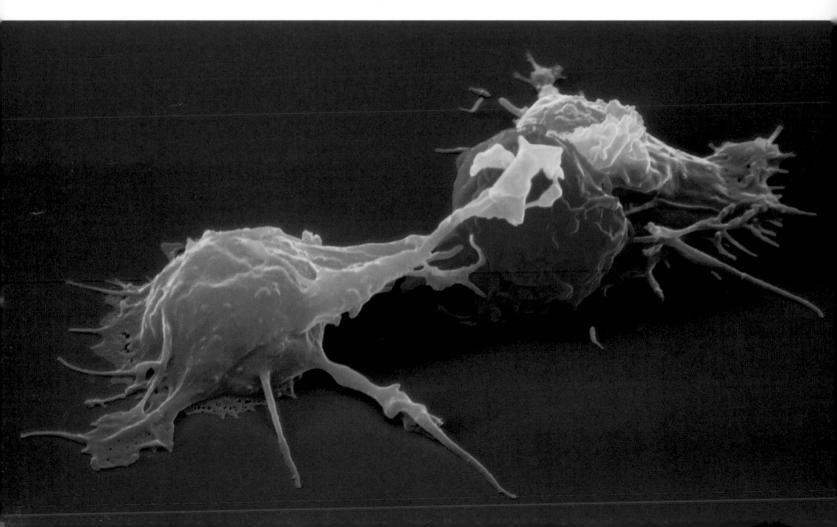

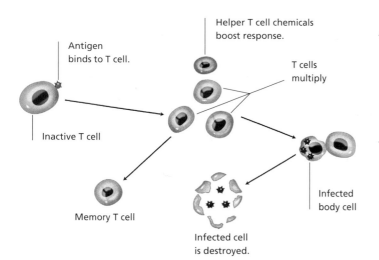

Antigen binds to T cell.

Inactive T cell

Helper T cell chemicals boost response.

T cells multiply

Memory T cell

Infected cell is destroyed.

Infected body cell

← **After recognizing an infected cell,** a killer T cell may bind tightly to it and release chemicals that cause the abnormal cell to burst open.

→ **Test tubes** containing human cells that will be used to make an interferon are frozen before their transfer to storage flasks of liquid nitrogen. Later they will be thawed and stimulated to multiply.

THE IMMUNE SYSTEM'S DEFENDERS	
Killer T cells	Destroy cells infected by specific viruses, cancer cells and cells of transplanted tissues.
Helper T cells	Stimulate the production of large numbers of killer T cells, as well as of B cells.
Macrophages	Activate T cells by processing and presenting foreign antigens to them; engulf foreign cells and promote the inflammatory response.
Interferons	Combat viral infections by entering healthy cells and preventing the virus from multiplying there.
Interleukins	Released by T cells and macrophages; provide chemical signals that stimulate the proliferation of T cells and B cells and promote the inflammatory response.

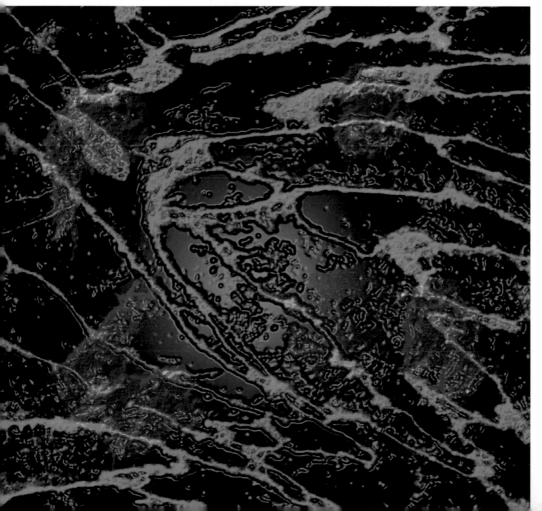

← **T cells** (colored red) leave the bloodstream and migrate through a blood vessel wall (green) to reach and destroy invading organisms.

← **Natural killer (NK) cells** converge on a cancer cell (red). The NK cells make contact with the cancer cell by extending long projections, bind it firmly and then make their lethal attack.

Medical immunology

Our rapidly growing understanding of immunity is reflected in an ever-increasing array of practical applications. For example, among other uses, the body's three kinds of interferons are employed to treat the deadly skin cancer malignant melanoma, some kinds of herpes infections and hepatitis C, a viral disease that can lead to liver cancer. Commercially prepared antibodies are now commonly used to diagnose diseases such as prostate cancer and rabies, in treatments for a common type of leukemia and in home pregnancy tests. Experimental treatments are exploring the power of special vaccines to promote strong immune responses against cancer tumors. Medical researchers are testing whether interleukins released by helper T cells, macrophages and other defenders can bolster the impact of cancer vaccines and other therapeutic drugs. Although many have serious side effects, some of these new approaches promise to improve and extend the lives of millions.

MONOCLONAL ANTIBODIES
Produced commercially, monoclonal antibodies are made by B cells that are descended from a single parent cell. Like natural antibodies, each kind of monoclonal antibody is sensitive to only one kind of antigen—hence they can be used to quickly identify cells or substances. This feature makes them ideal as an emergency treatment for exposure to especially dangerous threats such as rabies, some kinds of snake venom and hepatitis B. Although they only stimulate a short-term immune response, they buy time for the person's immune system to muster its own defenses. Some researchers also are seeking reliable ways of using monoclonal antibodies to carry a lethal dose of radiation or an anticancer drug directly to tumor cells, bypassing healthy body cells.

→ **Monoclonal antibodies** are major tools in healthcare and are now produced commercially for use in research and medicine. Shown here is a start-up flask of cell culture used in the process.

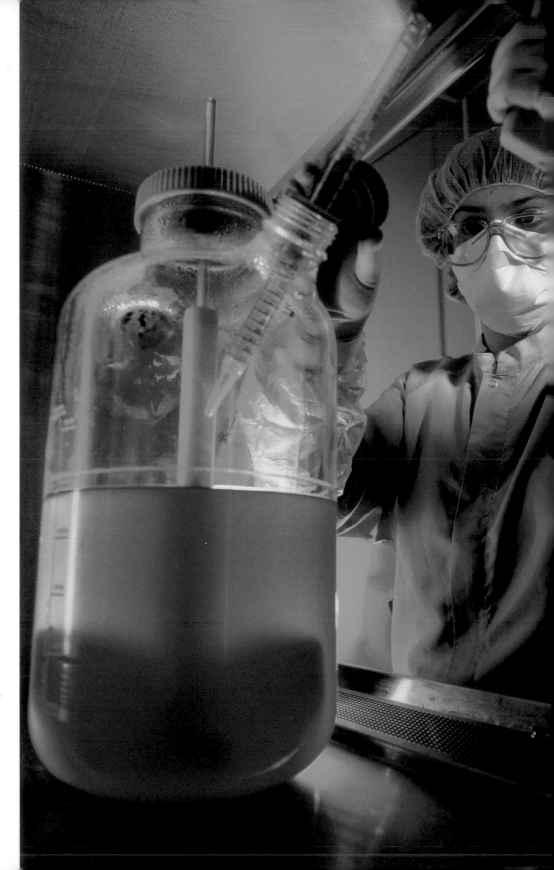

HOW MONOCLONAL ANTIBODIES ARE PRODUCED

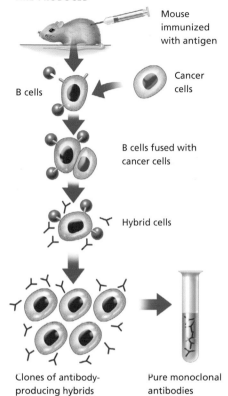

Mouse immunized with antigen

Cancer cells

B cells

B cells fused with cancer cells

Hybrid cells

Clones of antibody-producing hybrids

Pure monoclonal antibodies

← **To prepare monoclonal antibodies,** B cells that make the desired natural antibodies are fused with modified cancer cells. The resulting cells multiply rapidly.

→ **Cellist Jacqueline du Pré** suffered from multiple sclerosis (MS), in which the immune system destroys motor neurons. Today beta interferon is used to treat one form of MS.

↘ **Advanced ovarian cancer** (the large green mass) is one target of experimental immune therapies.

↘ **A pregnancy test strip** changes color if monoclonal antibodies in the kit bind with a hormone produced after an embryo begins to develop.

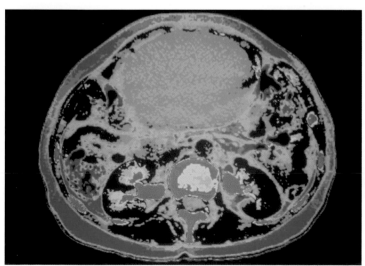

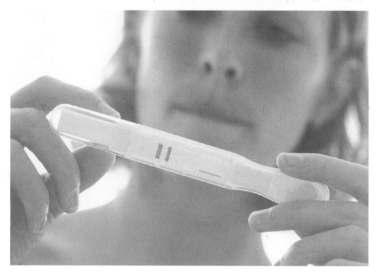

The immune system under siege

Like other body systems, lymphatic organs and cells of the immune system are vulnerable to pathogens, or may malfunction due to a genetic defect or some other cause. Allergies are signs that the immune system is hypersensitive to substances that are normally harmless, such as plant pollen or peanuts. Any disorder that seriously impairs the body's T cells, B cells, macrophages or some other key defender is an immunodeficiency.

The most notorious acquired immunodeficiency disease is AIDS, caused by the human immunodeficiency virus (HIV). Sooner or later AIDS wipes out the body's supply of helper T cells and immune responses grind to a halt. The lymph node cancer called Hodgkin's disease also causes immunodeficiency, drastically reducing the number of defensive cells. Various forms of severe combined immunodeficiency (SCID) are inherited. In affected children a chemical imbalance kills their T cells, or else their bodies make far fewer T and B cells than normal. SCID is a death sentence unless the child receives treatment, such as a bone marrow transplant, or is physically sheltered from microbes. Because faulty immunity is so devastating, research on immune system disorders is one of the most active arenas in the study of human biology.

TURNING AGAINST "SELF"

In autoimmunity, a person's immune system mistakes normal body cells for foreign intruders. As a result, B cells may make antibodies against the body's own cells, and helper T cells may maximize the attack. Affected organs eventually may be severely damaged. Scores of autoimmune diseases afflict 5 to 7 percent of people worldwide, about two-thirds of them women. Common ones include rheumatoid arthritis, systemic lupus, Graves' disease and type 1 (insulin-dependent) diabetes. Autoimmune disorders sometimes run in families, but there are several other triggers. For example, developing T cells and B cells "learn" to recognize self markers on cells they encounter. If an injury or disease later exposes cells with an unfamiliar marker, the immune system may be unleashed. Outside factors such as a viral infection may also make self markers appear alien, while the markers of some microbes are almost identical to those on certain body cells.

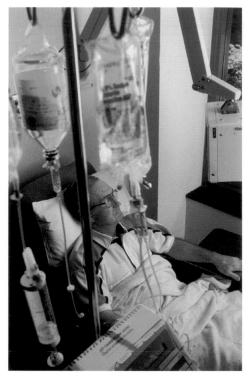

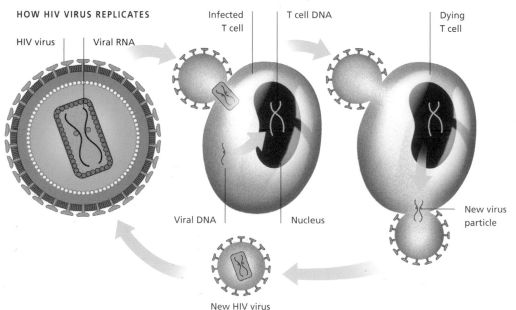

HOW HIV VIRUS REPLICATES

HIV virus | Viral RNA

Infected T cell | T cell DNA

Dying T cell

Viral DNA | Nucleus

New virus particle

New HIV virus

↑ **Chemotherapy often suppresses** a patient's immune system because the highly toxic drugs kill many normal body cells. Some experimental therapies aim to limit the destruction to cancer cells—for instance, by coating tiny metal beads with the drug and using a magnetic device to guide them to a tumor.

→ **A single HIV-infected cell** may produce HIV particles by the thousands. This artist's rendition shows the emerging virus particles in green. Although HIV typically devastates the body's helper T cells, it can also infect other defensive cells, including macrophages.

← **The HIV virus replicates rapidly.** Its genetic instructions (left), which are carried in ribonucleic acid, or RNA, are inserted into a T cell (center) along with reverse transcriptase, an enzyme that converts the RNA message into DNA. The DNA then causes the cell to make new HIV particles that erupt from the dying cell (right and bottom) and infect other cells.

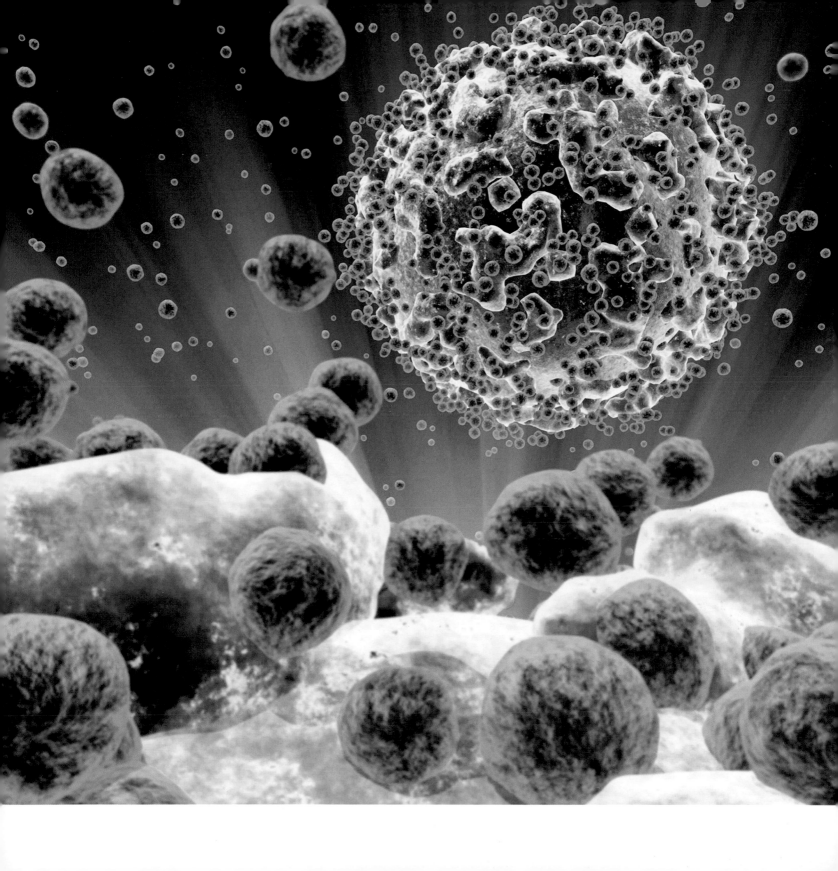

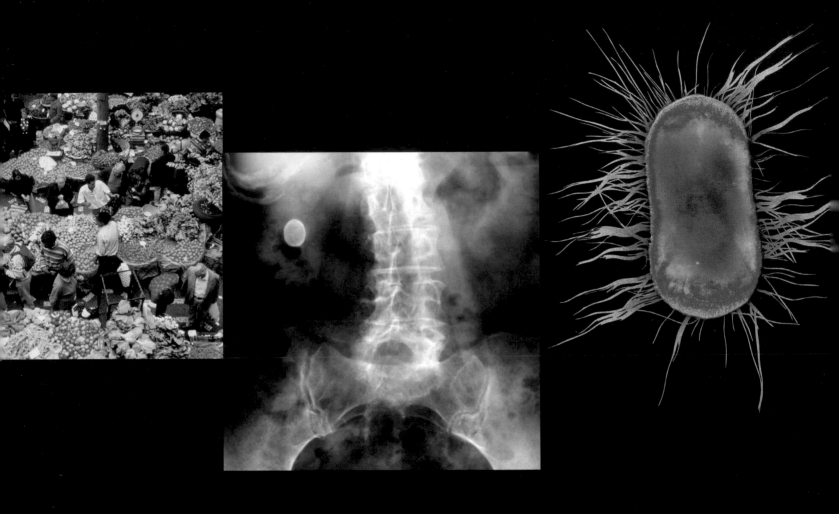

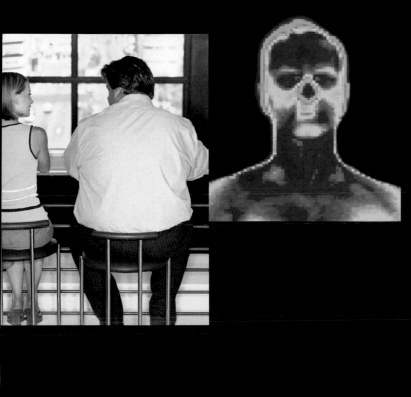

Nourishment and maintenance

Survival demands a means for obtaining nutrients and eliminating wastes. The digestive system processes food, bringing nutrients into the body and disposing of residues. The urinary system adjusts the chemical composition of blood as cell operations add some substances to it and subtract others.

The body in balance	194
From food to nutrients	196
The digestive system	198
The digestive journey begins	200
The multipurpose stomach	202
Enzymes	204
The liver	206
Accessory organs	208
The small intestine	210
The large intestine	212
Hormones and hunger	214
The essential nutrients	216
Body weight: the energy equation	218

The body in balance

Change is the one constant in life. This is as true for us biologically as in other ways. As our trillions of cells carry out the functions that keep them and the whole body alive, they continually take up needed substances, such as nutrients and oxygen, and remove waste products. In a sense each body cell is like an island, obtaining supplies from and offloading unwanted material into its watery surroundings. This extracellular ("outside the cell") fluid includes both the fluid component of blood plasma and the thin film of fluid that bathes each cell. Together these components form what physiologists call the body's "internal environment."

As cells remove some substances and excrete others, the composition of extracellular fluid fluctuates. This flux is potentially dangerous: without any mechanisms to offset the shifts, cells would die. They would either be starved of raw materials or left to wallow in accumulating wastes.

Mechanisms that compensate for changes in temperature and other conditions in the outside world are also important for our survival. In fact, our organ systems continually make adjustments that maintain the necessary chemical balance in the body's internal environment. This balance is a dynamic "steady state" called homeostasis. By way of feedback loops that steadily update the brain and other organs, the body monitors changing conditions, and its different parts adapt their operations to ensure that life will go on.

FEEDBACK LOOPS

Feedback loops help maintain homeostasis in the body. They inform control centers such as the brain about changes and provide a means for the controls to make adjustments as conditions warrant.

Blood sugar levels, the heart rate and a great many other body functions are regulated by negative feedback. In this type of loop, a change in some parameter—perhaps a drop in blood pressure—leads to control adjustments that work to return conditions to a normal level. Negative feedback is often likened to a home thermostat. When the temperature drops below a set point, a sensor detects the change and conveys this information to a control device that switches on the heating system until the temperature rises again to the desired level.

A few body functions are governed by positive feedback, in which a change "feeds on itself," being reinforced until some event finally brings it to a halt. One example of positive feedback is the process of labor, which culminates in the birth of a baby.

→ **Staying alive** is a matter of balance. A wide range of actions and activities, such as eating and digesting food, breathing, the formation and elimination of urine and solid wastes, and changing clothes to accommodate temperature variations help maintain chemical balance in the extracellular fluid.

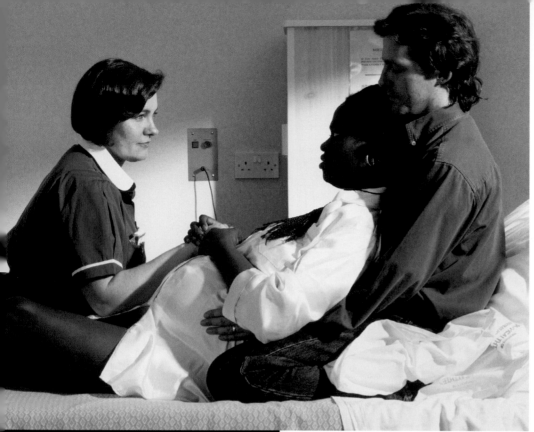

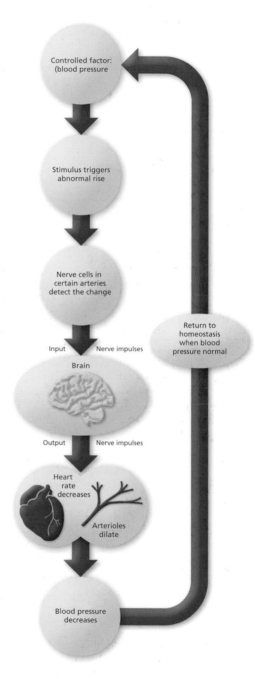

Controlled factor:
(blood pressure

Stimulus triggers
abnormal rise

Nerve cells in
certain arteries
detect the change

Return to
homeostasis
when blood
pressure normal

Input   Nerve impulses

Brain

Output   Nerve impulses

Heart
rate
decreases

Arterioles
dilate

Blood pressure
decreases

↑ **Positive feedback** helps a baby to be born. When labor begins, contractions of muscles in the wall of the mother's uterus are relatively weak and infrequent. They strengthen and become more frequent as labor progresses, then abruptly stop when the uterus expels the newborn infant.

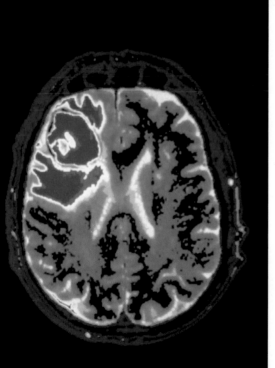

← **Brain disease** can disrupt homeostasis. This microscope image shows a large malignant brain tumor (pink). Because the brain controls so many feedback mechanisms, this type of illness can progressively limit the brain's ability to process and respond to information on how body systems are functioning.

→ **Negative feedback** adjusts blood pressure. When blood pressure rises above a set point, sensors in certain blood vessels relay the information to a control center in the brain. Its response slows the heart rate and dilates arterioles. When sensors register falling blood pressure, brain signals speed up the heart rate and constrict arterioles.

From food to nutrients

Food can mean many things to us, but to the body it is simply a source of the substances that sustain life. Unlike plants, we cannot manufacture most of the nutrients we need but must obtain them from other sources. In addition to vitamins and minerals, the human body requires nutrients from four broad categories of biological molecules—proteins, lipids, carbohydrates and nucleic acids. Most of these chemicals enter the body as relatively complex units that are then disassembled into their constituent parts, which are subsequently reused to meet specific needs. For instance, the proteins we consume in foods such as meat, eggs and beans are broken down into amino acids. In cells, these components are reassembled into new proteins that become either cell parts or "workers," such as enzymes and hormones. Lipids consist of fats and oils. Cells use them as structural building blocks as well as to make signaling chemicals. Cells also store lipids as energy reserves. Carbohydrates in grains and all manner of foods derived from plants are dietary staples. Like proteins and fats, they are used to build cell parts and are also fashioned into energy stores. Nucleic acids are information carriers. They are incorporated into our DNA, the genetic material, among other uses.

→ **Wheat plants** use sunlight to make sugars that serve as their own food. The sugars fuel processes that help sustain each plant's growth, including the formation of seeds. Wheat grain, in the form of flour, is an important source of nutrients for billions of people all over the world. Few do not include wheat, corn, rice or some other type of grain in their diet.

THE BIOLOGICAL MOLECULES IN FOOD

Type	Description	How used
Proteins	Contain 20 kinds of amino acids	Build body proteins
Carbohydrates	Simple and complex sugars and starches	Used by body cells for energy and as building blocks
Lipids	Fats, oils and sterols	Stored for energy; used to build cell parts including the outer cell membrane, and to make molecules such as steroid hormones and cholesterol
Nucleic acids	Substances found in meats and a few other foods	Used to make DNA, RNA and ATP, the basic chemical fuel for cells

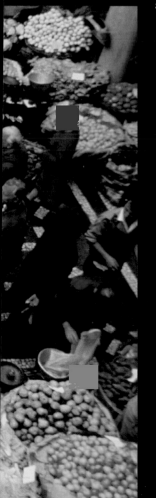

← **This open-air produce market** tempts buyers with its fresh offerings. Good nutrition depends on consuming a variety of fresh fruits and vegetables, which are sources of complex carbohydrates and fiber. Yellow and leafy green types also are sources of important vitamins and antioxidants.

ANTIOXIDANTS

Orange and dark green fruits and vegetables are good sources of antioxidants, including vitamins A, C and E. An antioxidant is a substance that can block the damage caused by oxidation, an everyday chemical reaction that causes iron to rust, wood to burn—and in our bodies produces molecules called free radicals. When free radicals accumulate in a cell, they can disrupt the normal structure of its DNA and other cell components. Research suggests that this sort of damage contributes to aging processes and may be a factor in some types of heart disease, among other ills. Antioxidants turn the tables by changing the chemical structure of free radicals in a way that prevents them from wreaking havoc. In addition to antioxidant vitamins, substances such as the plant pigments alpha carotene and beta cryptoxanthin—both found in yellow-orange foods such as pumpkins and carrots—disable free radicals as well. In general, foods that supply antioxidants fit well into a healthy diet because they are naturally low in fat and packed with vitamins and fiber.

↑ A vegetable omelet, whole-grain toast and juice together supply every major category of biological molecules, including protein, fat, carbohydrates and nucleic acids. These foods are also good sources of vitamins and minerals that are required for many body functions.

The digestive system

Each of our organ systems contributes to homeostasis, but we may be most consciously aware of the operations of our digestive system. Its central feature is the tube-like gastrointestinal (GI) tract, which begins with the mouth and ends with the anus. Along with organs such as the liver, pancreas and gallbladder, this tube's different parts transform the bulk food we eat into nutrient molecules that body cells can use. It also accumulates, briefly stores and eliminates indigestible residues.

The space inside the digestive tube is called a lumen, and ingested food moves from place to place through it in a sequence—from the mouth into the esophagus, on to the stomach and then into the small intestine. At each of these processing stations, incoming food is changed in some way. By the time food reaches the large intestine, most of its usable nutrients have been extracted, and what remains is chiefly unwanted "leftovers" that travel through the large intestine and finally enter the rectum ready to be expelled. Because the digestive tract lumen opens to the outside world at both ends, technically anything that is in it, both usable food and wastes, has not truly entered the body. Our cells gain access to the nutrients we consume only when the nutrients move across the lining of the digestive tract and pass into the bloodstream.

HOW FOOD IS MOVED ALONG

The digestive system is built to break down bulk food into small particles and to move this material in an orderly fashion from one processing region to another. The system's first task is to mechanically break down large chunks of food into smaller ones and then to begin mixing these pieces with digestive enzymes—chemicals that continue the dismantling process.

These tasks fall first to our teeth and muscular tongue. After food is swallowed, however, muscles assume the role of mixing food and moving it onward. This is smooth muscle, not skeletal muscle, and its fibers are aligned in sheets resembling the criss-crossing layers in a tire. One sheet runs lengthwise, and others crosswise or on a diagonal. These strategic alignments allow contracting smooth muscles in the GI tract to knead, squeeze and propel food from the esophagus all the way to the rectum. At three points—where the esophagus joins the stomach, where the stomach opens into the small intestine and again at the terminus of the rectum—sturdy rings of muscle called sphincters control the forward transit of food and residues.

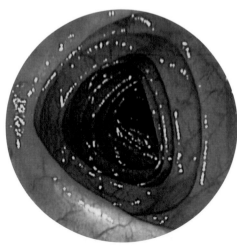

↑ **Some animals lack** a digestive system. These tropical barrel sponges and their kin are the simplest animals, and each of their cells must obtain and digest food on its own. Sponges are porous, so that water can flow through them, bringing food particles close to cells.

← **The wall of the large intestine,** or colon, has accordion-like folds. Due to the arrangement of the muscles in the wall, its lumen appears triangular. The inner wall is lined with a mucous membrane that secretes lubricating mucus.

⇐ **Until children have enough teeth** to chew effectively, they must eat soft food that can pass easily into the stomach without being torn or crushed into smaller bits.

→ **The gastrointestinal tract** is basically a hollow tube through which ingested food moves in one direction. Other digestive system components, such as the liver and pancreas, assist with chemical food processing. Glands such as the salivary glands also secrete digestive enzymes that aid digestion, and produce fluid and mucus that lubricate material moving through the system.

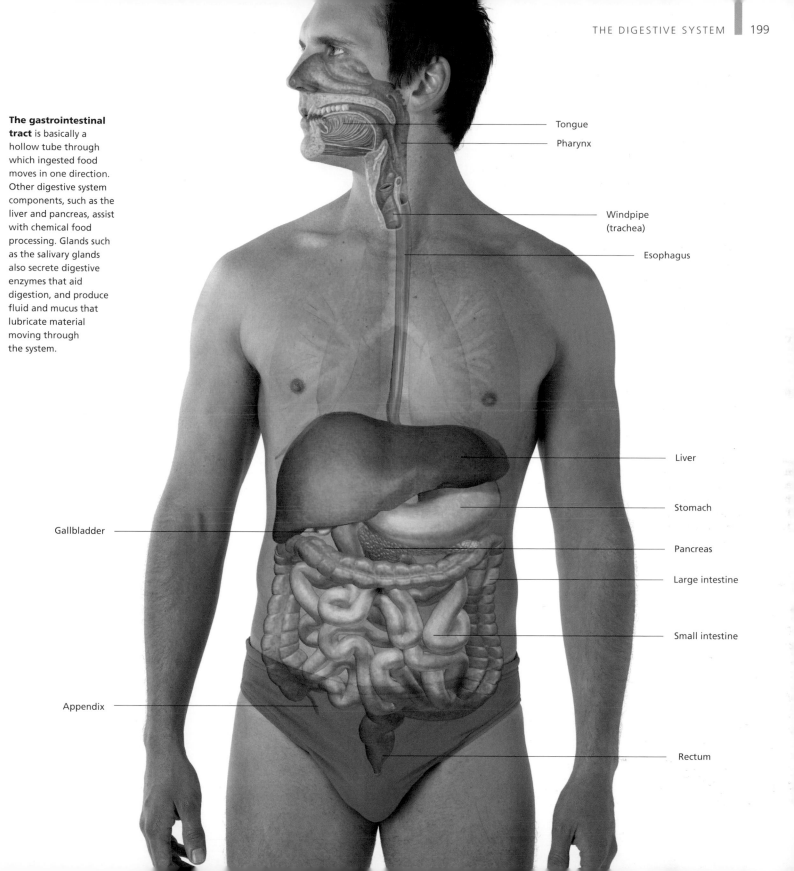

Tongue

Pharynx

Windpipe (trachea)

Esophagus

Liver

Stomach

Gallbladder

Pancreas

Large intestine

Small intestine

Appendix

Rectum

The digestive journey begins

The simple act of eating is the first step toward providing our bodies with the nutrients we need to survive. The mouth is more than just an orifice that receives food, however. It contains teeth and a tongue, both of which immediately go to work to break down solid foods. Biting and chewing are mechanical operations that help break bulk food apart into chunks that can be comfortably swallowed. As we chew, and often when we merely think of food, our salivary glands produce the fluid we call saliva. This complex mixture contains water and gluey mucins that moisten chewed food and bind particles together, as well as enzymes that help break down starches and other simple carbohydrates. Saliva also dissolves chemicals in food so taste buds can detect them. The muscular tongue manipulates this moist mass into a ball called a bolus. Swallowing moves each bolus through the pharynx and into the esophagus for its next digestive stop, the stomach.

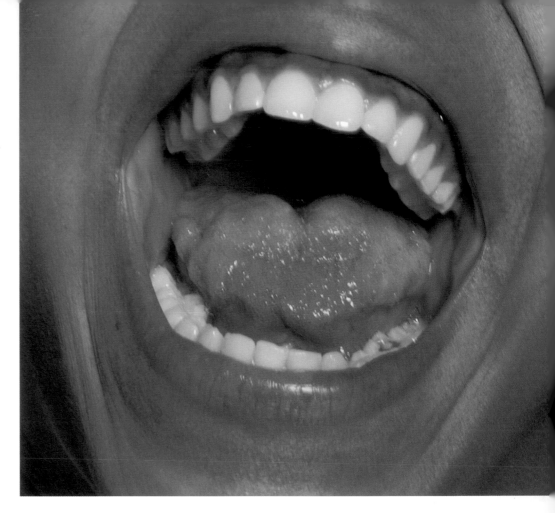

VERSATILE, DURABLE TEETH

Human teeth reflect our heritage as omnivores, animals that are equipped to consume and digest foods from both plant and animal sources. Children have 20 primary or "milk" teeth, while adults normally have 32 teeth, including "wisdom" teeth. In the front of the mouth, our sharp-edged incisors and pointed canines ("dog teeth") are specialized for biting and shearing off chunks of firm foods (such as meat, an apple or corn on the cob), while our flat premolars and molars at the back of the mouth are better suited for crushing and grinding foodstuffs into smaller morsels.

Given proper hygiene and a healthy diet, the teeth are extremely durable. The crown of a tooth—the portion that projects above the gum line—is blanketed with a thick layer of enamel, the hardest substance in the body. Once the enamel has been formed, however, it cannot be replaced naturally if it cracks or chips, because the tissue that produces it degenerates after each tooth erupts.

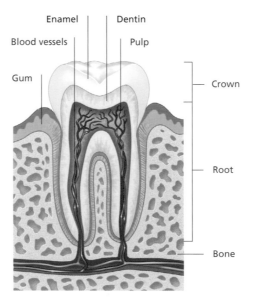

Enamel | Dentin
Blood vessels | Pulp
Gum
Crown
Root
Bone

↑ **Various mouth parts** assist food intake. As the tongue forms a bolus it presses food up against the hard roof of the mouth, or palate. When you swallow, the soft palate and its finger-like uvula rise up, helping shunt food into the esophagus and avoiding the windpipe.

← **This cross-section of a molar** shows the general structure of a tooth. The crown overlies a bony layer called dentin, which in turn surrounds the pulp cavity containing nerves and blood vessels. Each tooth has one or more roots that fill the tooth socket in the jaw and are anchored there by ligaments.

→ **A chameleon** has a different use for its tongue—to capture food. It stalks its insect prey, edging as close as possible, then lunges forward and unfurls its long, sticky weapon. Some species of chameleon can lay claim to a tongue that is as long as 1 foot (30 cm).

← **The digestion of the carbohydrates** in this tart begins in the mouth. Each day our salivary glands make more than a quart (1 liter) of saliva. In addition to keeping the mouth and chewed food moist, it contains enzymes that begin the process of breaking down complex sugars and similar food carbohydrates into smaller chemical units. This chemical digestion will continue in the small intestine.

PHASES OF SWALLOWING

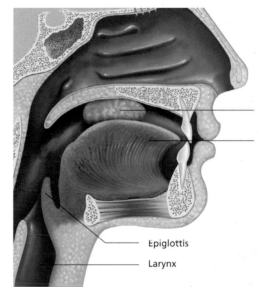

Bolus

Tongue

Epiglottis

Larynx

1. In the mouth phase of swallowing the tip of the tongue comes up against the hard palate, pushing the food bolus back toward the opening of the esophagus.

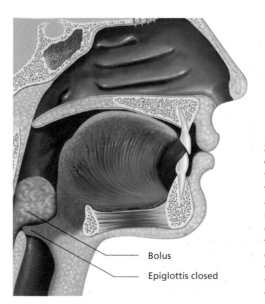

Bolus

Epiglottis closed

2. In the throat phase of swallowing, as the food bolus nears the throat, the uvula and pharynx move up and the bolus moves downward. Muscles in the pharynx contract, closing off the opening to the larynx so the food slides past it into the esophagus.

The multipurpose stomach

Anatomically the stomach is designed to store bulk food and begin the vital process of removing nutrients from it. Positioned in the upper left part of the abdominal cavity, the stomach is a stretchable pouch with thick, muscular walls. Rumpled with folds when empty, an average adult's stomach can expand to hold as much as 2 quarts (about 2 liters) of food. Food digestion, especially of proteins, begins in earnest here. In the stomach's lining are glands that collectively produce a mix of digestive juices, hydrochloric acid and several enzymes. As food enters the stomach, its wall muscles contract and churn the food with this acidic gastric fluid, which kills many bacteria and begins the chemical breakdown of large proteins and complex carbohydrates. The result is a chunky mixture called chyme—a blend of fluid and partially digested food that most people see only when they vomit. Cells in the stomach lining also release hormones, which play a central role in feelings of hunger and satiety. Food that has been processed in the stomach is slowly squeezed onward into the small intestine about a teaspoon at a time. This slow, steady cadence prevents the small intestine from being overloaded with more food than it can handle efficiently. Depending on the size of the meal, it can take from 3 to 6 hours for the stomach to empty.

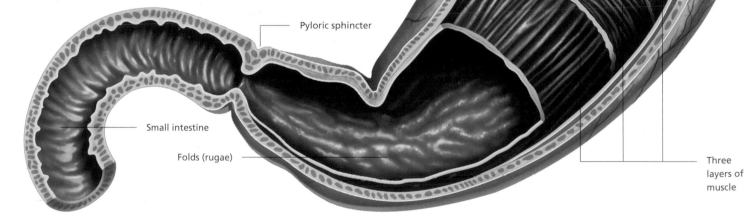

STRUCTURE OF THE STOMACH

Pyloric sphincter

Small intestine

Folds (rugae)

Three layers of muscle

↑ **The stomach** has sheets of muscle in its walls that allow it to mix and propel food. The folds, or rugae, on the inside of an empty stomach's walls are covered by the stomach's lining—a mucous membrane packed with various types of glands. Partially digested food passes from the stomach into the small intestine through the pyloric sphincter.

→ **Peristalsis moves food** into, through and out of the stomach. Peristalsis means "tightening around," and in the esophagus and stomach wall belt-like muscles alternately contract to tighten and then relax. With each sequence, a small amount of food is mixed and squeezed along. Peristaltic waves also move food through the intestines.

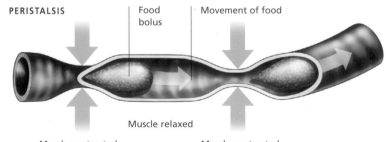

PERISTALSIS

Food bolus

Movement of food

Muscle relaxed

Muscle contracted

Muscle contracted

PROTECTING THE STOMACH'S LINING

Several times each day the stomach produces a caustic brew of acid and enzymes to further the chemical digestion of a meal or snack. Ordinarily, tissues exposed to such a powerful chemical assault would be damaged—a danger that the stomach lining eludes by way of an equally potent chemical barrier. While some cells in the stomach secrete either hydrochloric acid or protein-digesting enzymes, others are pumping out mucus containing a natural antacid called bicarbonate. The mucus coats and usually protects the stomach lining from being digested along with food. The open sores we call ulcers can develop when the bacterium *Helicobacter pylori* releases toxins that trigger inflammation of the stomach lining. Ulcers heal swiftly after an antibiotic cures the infection.

← ***Helicobacter pylori* bacteria** cause most stomach ulcers. Visible here as mottled yellow ovals on cells of the stomach lining, these organisms release toxins that irritate the lining and allow corrosive stomach acid to enter it.

← **Unlike the human stomach, a cow's stomach** has four chambers, which suit its diet and method of digestion. Grass is mostly cellulose, which cattle digest with the aid of bacteria in two of the stomach's compartments. When the bacteria are also digested, they are a source of protein.

↓ **A healthy pink, the stomach's folds** are visible in an endoscopic view of the region where the stomach opens into the small intestine.

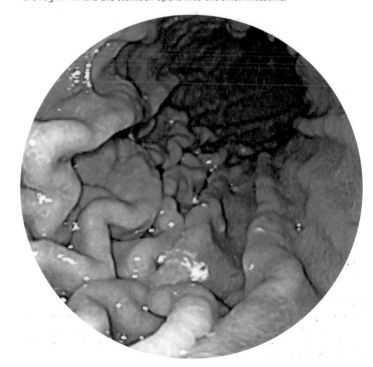

Enzymes

Without the chemicals called enzymes we would starve, no matter how much food we eat. Enzymes are proteins that have the ability to join or break the chemical links that hold complex substances together. Digestive enzymes are charged with breaking food particles into molecules that are small enough to pass into the bloodstream through the lining of the small intestine, where the vast majority of nutrients are absorbed.

Different enzymes break apart proteins, carbohydrates, fats and other lipids, and nucleic acids. Some food molecules are chemically highly complex—an intact protein may consist of hundreds or thousands of amino acids, while complex carbohydrates are made up of hundreds or thousands of sugar units. Our digestive apparatus divides its efforts, with each enzyme operating best under the chemical conditions in different parts of the system. The powerful enzymes needed to begin protein digestion work best in the acid milieu of the stomach, while most enzymes that break down carbohydrates are effective in the more alkaline environment of the mouth and small intestine.

PROTEIN DIGESTION

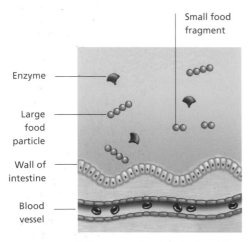

Enzyme

Large food particle

Wall of intestine

Blood vessel

Small food fragment

1. Enzymes break up large food particles.

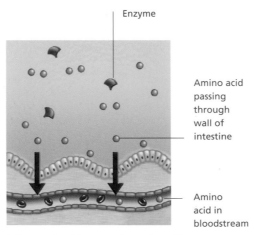

Enzyme

Amino acid passing through wall of intestine

Amino acid in bloodstream

2. Enzymes break fragments into amino acids.

↑ **Eating ice cream** requires the body to break down milk sugar, or lactose. Up to about age 4, children generally have plenty of the required lactase enzyme. Thereafter, in some people production of this chemical declines so much that eating milk products leads to bloating, cramps and diarrhea.

← **Protein digestion** takes place intwo stages. Proteins are chains of amino acids, sometimes thousands of them. The first step in digestion is to break up the chains into smaller units. Several enzymes are involved in this initial step. Later other enzymes break the fragments into amino acids that move across the intestinal lining into the bloodstream.

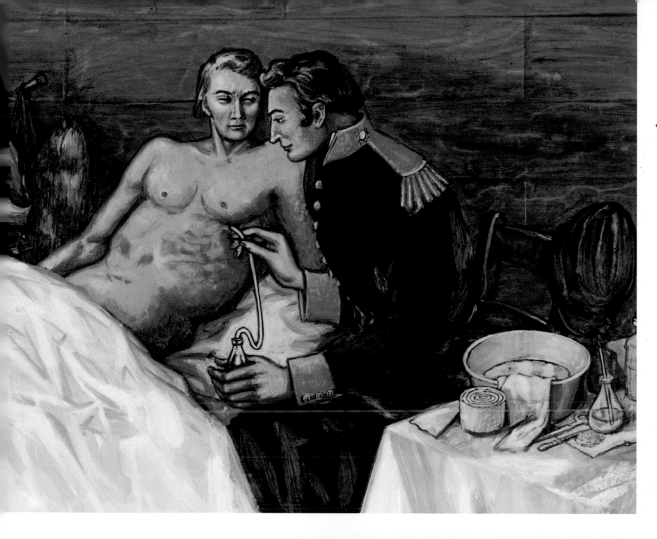

When an incompletely healed gunshot wound left the young Canadian trapper Alexis St Martin with a permanent hole in his stomach in 1822, physician William Beaumont seized the opportunity to study the stomach's functions. For two years he fed St Martin different types of foods, and then analyzed how gastric juice affected each one. Beaumont proved what was then a controversial idea—that the stomach released substances that partially digested certain types of food. We now know those substances as enzymes.

FAT'S DIGESTIVE ODYSSEY

In just a few short hours, digestive enzymes can transform a piece of bread, a bite of banana or a morsel of meat into a largely liquid blend of tiny absorbable molecules. Simple sugars and amino acids are shuttled directly from the intestinal lining into the bloodstream, but fats and other lipids follow a more circuitous route that entails a series of chemical steps.

The fragments created when a fat is first broken apart, called triglycerides, are too large to cross the lining of the small intestine. In an intermediate step involving bile they are repackaged as small droplets, called micelles. These droplets pass into the small intestine lining, where they break apart. The triglycerides re-form and move into vessels of the lymphatic system. Finally, as part of lymph, they are delivered into the bloodstream. Common blood tests that tally a person's triglycerides are measuring fats that have made this complex journey.

Moles eat only insects, such as the large locust being devoured by a Grant's golden mole in this photograph. This diet requires an ample supply of enzymes that break down proteins. Overall, however, it is much more easily digested than the plant food consumed by herbivores such as rabbits and deer or the varied diet of omnivores such as humans. As a result a mole has a simple gut consisting of a stomach and a short, uncoiled intestine.

The liver

At about 3 pounds (1.4 kg) the wedge-shaped, reddish-brown liver is one of our largest organs. It is also one of the body's most multifaceted parts. The liver's direct role in the early part of digestion is the seemingly simple job of making bile, which is stored in the gallbladder and helps package fats destined to be absorbed. Beyond this task, however, the liver plays a major part in managing nutrients after they have been absorbed.

Virtually all the blood that carries digested nutrients from the small intestine is briefly diverted into the hepatic portal system, a network of blood vessels that weave throughout the liver's two lobes. Packed arrays of specialized cells process this blood in various ways. They convert excess blood sugar (glucose) into a storage form called glycogen and they use incoming amino acids to assemble various proteins. Only about 15 percent of the cholesterol in blood is derived from food. The liver makes much of the remainder and uses it as an ingredient in bile. Some proteins assembled in the liver transport triglycerides—the main fats in the blood—to fat (adipose) tissue, where the fat is stored, sometimes to the detriment of our waistline. The liver also inactivates hormones and drugs such as antibiotics, dismantles worn-out blood cells and detoxifies dangerous byproducts of protein digestion, among other tasks.

HEPATITIS AND CIRRHOSIS

At least half a dozen forms of hepatitis can upset and even destroy the liver's ability to cleanse the blood of impurities and help manage blood sugar levels. Hepatitis is an inflammation of the liver. A family of viruses carried in food, blood, water or animal wastes causes most cases, although poisonous mushrooms and some therapeutic drugs are also toxic to the liver's cells. One of the most dangerous forms of the disease is hepatitis C, which is transmitted in infected blood by contaminated needles or transfusions. Chronic hepatitis or alcohol abuse can lead to liver cirrhosis, in which healthy cells are replaced by scar tissue and fat deposits. As a result the liver can lose much of its ability to function.

STRUCTURE OF THE LIVER

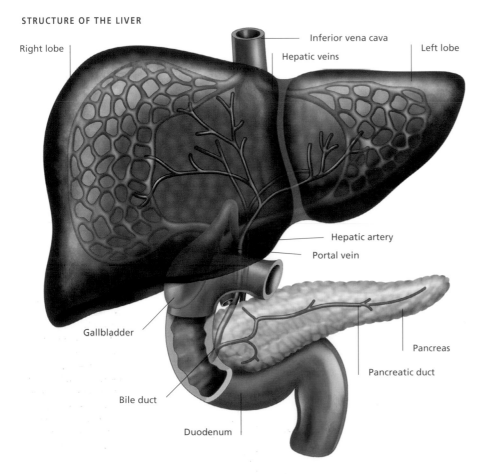

Right lobe

Inferior vena cava

Hepatic veins

Left lobe

Hepatic artery

Portal vein

Gallbladder

Pancreas

Pancreatic duct

Bile duct

Duodenum

← **The liver is located** on the upper right side of the abdominal cavity, just under the diaphragm. It has two main lobes, right and left, and receives blood from the digestive tract through the large portal vein. The bile it manufactures is stored in the gallbladder, which nestles under the right lobe.

KEY FUNCTIONS OF LIVER
Partners with the endocrine system to store and release blood sugar as needed.
Stores fats from ingested food.
Assembles lipoproteins, which carry cholesterol and other needed lipids to body cells.
Manufactures cholesterol and uses some of it to make bile salts (components of bile).
Stores iron, and vitamins A, B_{12} and D.
Forms proteins that make up part of blood plasma.
Converts toxic ammonia (from protein breakdown) into less toxic urea that is excreted in urine.
Breaks down hormones to be excreted.
Recycles substances from dismantled red blood cells.
Processes alcohol and some drugs to forms that can be excreted in urine.

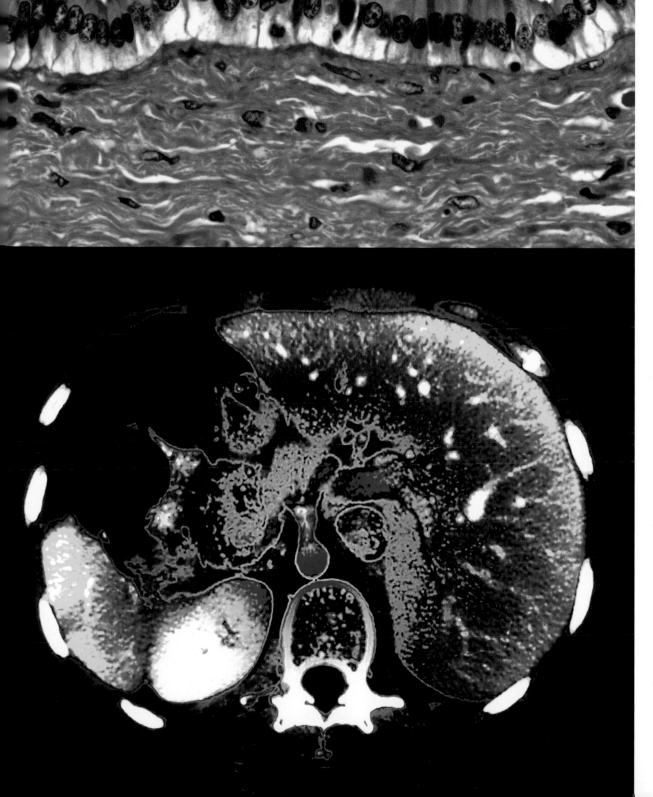

← **Like all ducts,** this bile duct is lined with column-shaped cells. The bile duct channels bile from the liver to the gallbladder. As it manufactures bile, the liver incorporates cholesterol into particles called bile salts, which are used to help process fats in ingested food so they can be absorbed more readily.

← **A healthy liver** appears as a large orange and yellow mass on the right-hand side of this vivid, top-down scan. The patient's spine is at the bottom of the frame, while the nearby spleen appears as a blue and white oval. The yellow oval next to the spleen is the lower part of the stomach, and thick white dashes around the perimeter are sections of ribs.

Accessory organs

The salivary glands, liver, gallbladder and pancreas are all considered to be "accessory organs" of the digestive system. The gallbladder's role is limited to storage, and the other organs also carry out a variety of non-digestive tasks. Of all of them, the pancreas may contribute the most to the actual processing of the food we eat. Although some cells in the pancreas make hormones that adjust blood sugar levels after and between meals, most of this soft, tapering organ is devoted to pumping out vital digestive enzymes when food is in the GI tract.

About 6 inches (15 cm) long, the pancreas is positioned behind the stomach. As partially digested food empties from the stomach, fluid from the pancreas also begins to pour into the small intestine. Each day the pancreas makes as much as 1.5 quarts (1.7 liters) of this "pancreatic juice," which contains enzymes that are capable of breaking down most kinds of sizeable food molecules—including protein fragments, complex carbohydrates, fats and nucleic acids— into a form that can be absorbed. The small, green-colored gallbladder simultaneously adds bile from the liver, which aids in the absorption of any fats. Because the gallbladder is merely a holding tank for bile, people can survive very well without one.

CURIOUS BILE CHEMISTRY

The bile that passes into the small intestine from the gallbladder contains two main ingredients—substances called bile salts and the bright yellow pigment, bilirubin. Bile salts speed the absorption of fats, but bilirubin has nothing to do with digestion. It is a waste produced when the liver dismantles worn-out red blood cells, and bile is a convenient vehicle for getting it out of the body. Normally, bilirubin mixes with the undigested material we excrete as feces. When it is not removed rapidly enough—for example, when a gallstone blocks the channel that carries bile to the small intestine—bilirubin builds up in the body. The result is jaundice, a condition that gives a distinct yellowish tinge to the skin and whites of the eyes.

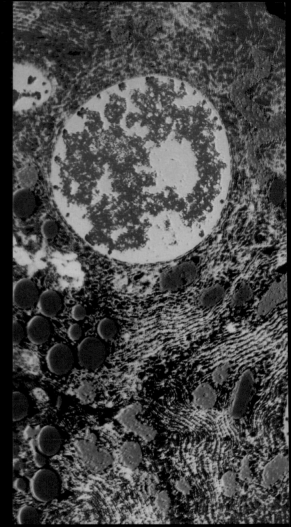

↓ **Most of the enzymes the body uses** to digest food come from the pancreas. Other sources are the starch-digesting enzymes released by the salivary glands and stomach enzymes, such as pepsin, that begin the digestion of proteins.

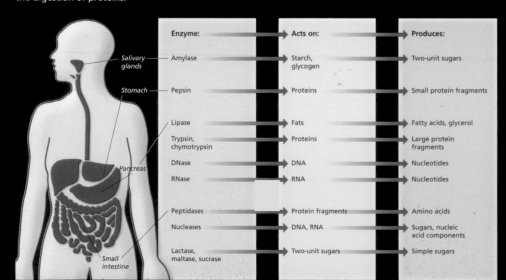

Enzyme:	Acts on:	Produces:
Amylase	Starch, glycogen	Two-unit sugars
Pepsin	Proteins	Small protein fragments
Lipase	Fats	Fatty acids, glycerol
Trypsin, chymotrypsin	Proteins	Large protein fragments
DNase	DNA	Nucleotides
RNase	RNA	Nucleotides
Peptidases	Protein fragments	Amino acids
Nucleases	DNA, RNA	Sugars, nucleic acid components
Lactase, maltase, sucrase	Two-unit sugars	Simple sugars

Salivary glands
Stomach
Pancreas
Small intestine

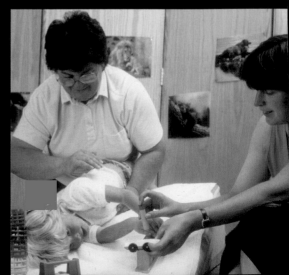

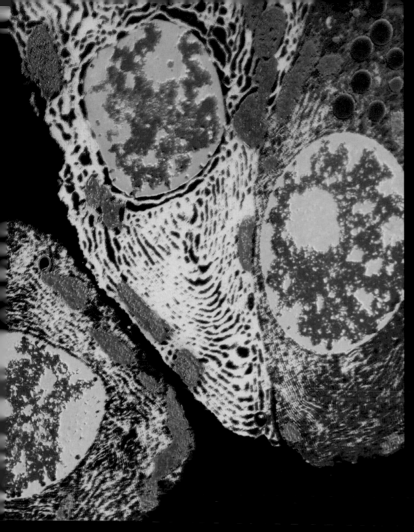

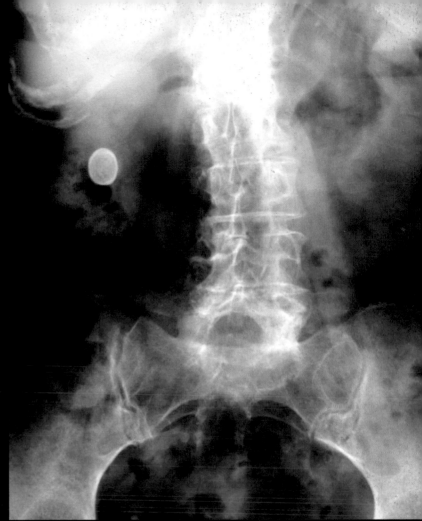

↑ **In these colored pancreas cells** bright red spots are granules filled with stored enzymes. Each of the large yellow and blue structures is a cell nucleus. The pale yellow thread-like masses are membranes where various digestive enzymes are manufactured. The clusters of cells that make the hormones insulin and glucagon are elsewhere in the pancreas.

↗ **A bright spot marks a gallstone** in this patient's gallbladder. Small or large, gallstones can cause excruciating pain when they block or irritate the bile duct. They are usually removed using laparoscopy, in which a viewing tube and instruments are inserted through small slits in the abdomen.

→ **Sliced in half, this large gallstone** reveals the layers of material, mainly cholesterol, that built up as it formed. Most gallstones develop when there is too much cholesterol in the bile, but in some cases the yellow pigment bilirubin is the main ingredient.

← **In a child with cystic fibrosis,** thick mucus builds up in respiratory passages and in the duct that delivers pancreatic enzymes. In addition to receiving back-pounding "percussion therapy," which helps clear mucus from her airways, this girl probably also takes enzyme supplements and follows a special diet.

The small intestine

The small intestine is the workhorse of digestion. In a healthy adult this coiling, labyrinthine tube can measure from 7 to 13 feet long (2 to 4 m) , and in it 90 percent of the nutrients the body will receive from food are broken down and absorbed into the bloodstream. The small intestine's structure is the key to this life-sustaining function. Its inner surface is pleated with thousands of folds, which are covered with millions of smaller finger-like folds called villi. Each villus is covered with even smaller folds called microvilli. Together the villi and microvilli have a surface area similar to that of a small house—about 1500 square feet (140 sq. m). Having such an extensive area available to take in nutrients helps ensure that the body can meet its nutritional needs. People who suffer from disorders that damage the small intestine generally lose weight and may even become malnourished. Once nutrients are inside intestinal villi, most of them move directly into blood vessels and begin their journey to body cells and tissues.

MAKING THE MOST OF FOOD

The vast surface area of the small intestine is certainly a key feature of the digestive system's ability to extract nutrients from food. Another enhancement is a process called segmentation, in which muscle contractions alternately cinch in and release digesting food in a given section of the small intestine. Although we can't feel this squeezing and relaxing, it helps blend food with digestive enzymes and brings it close to villi, where food molecules can move into the bloodstream. On an average day, the small intestine also absorbs more than 9 quarts (10 liters) of water from the material moving through it. Most people do not consume that much water in food and drink. Instead, about two-thirds of it is reclaimed from digestive fluids that were released into the stomach and other parts of the digestive tract.

STRUCTURE OF THE SMALL INTESTINE

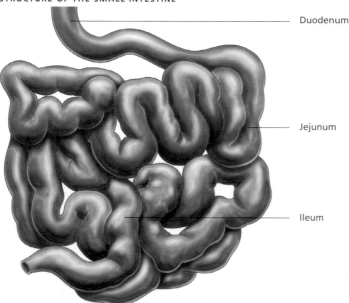

Duodenum

Jejunum

Ileum

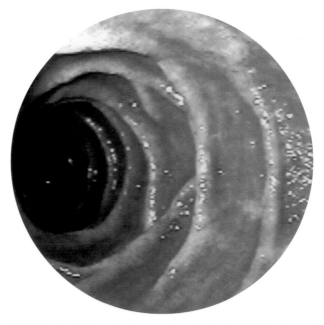

The small intestine is divided into three sections, the duodenum, jejunum and ileum. It is "small" only in that it is about 1 inch (2.5 cm) in diameter. Although the duodenum is only 10 inches (25 cm) long, the jejunum measures about 3 feet (1 m) and the ileum about 6 feet (2 m).

This view inside the duodenum shows its circular folds. Each one extends about half an inch (1.25 cm) into the lumen. The deep folds force chyme to swirl through the small intestine. Enzymes from the pancreas and bile released from the gallbladder enter the small intestine through a duct that opens into the duodenum.

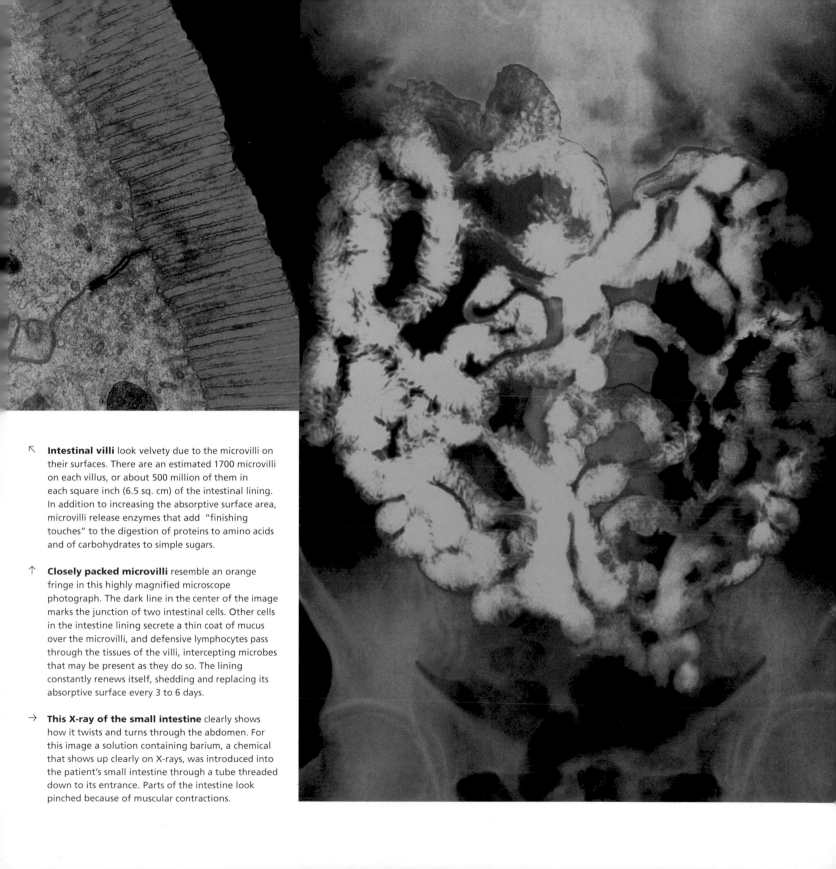

Intestinal villi look velvety due to the microvilli on their surfaces. There are an estimated 1700 microvilli on each villus, or about 500 million of them in each square inch (6.5 sq. cm) of the intestinal lining. In addition to increasing the absorptive surface area, microvilli release enzymes that add "finishing touches" to the digestion of proteins to amino acids and of carbohydrates to simple sugars.

Closely packed microvilli resemble an orange fringe in this highly magnified microscope photograph. The dark line in the center of the image marks the junction of two intestinal cells. Other cells in the intestine lining secrete a thin coat of mucus over the microvilli, and defensive lymphocytes pass through the tissues of the villi, intercepting microbes that may be present as they do so. The lining constantly renews itself, shedding and replacing its absorptive surface every 3 to 6 days.

This X-ray of the small intestine clearly shows how it twists and turns through the abdomen. For this image a solution containing barium, a chemical that shows up clearly on X-rays, was introduced into the patient's small intestine through a tube threaded down to its entrance. Parts of the intestine look pinched because of muscular contractions.

The large intestine

After the small intestine has removed usable nutrients from food, the undigested remains of a meal move on into the large intestine. This part of the digestive tube is only about 5 feet (1.5 m) long, but it is "large" because its interior is much more spacious—the better to accumulate, process and store material that will be eliminated.

The colon makes up most of the large intestine. Contractions of the muscles in its wall soon after a meal can move material through it rapidly , making room for the next round of food residues. In the colon, water and a few more nutrients are absorbed. This processing produces feces, which consist mainly of the solid leftovers from digested food, sloughed cells and a few other components. By dry weight, about one-third of feces is made up of bacteria. Harmless species of bacteria that live in the large intestine do us the service of manufacturing vitamin K and some B vitamins.

COLON HEALTH

A healthy colon can efficiently handle its constant task of storing and moving the remains of our meals and extracting from them the last useful nutrients. Insufficient fiber or too little fluid in the diet, a lack of exercise, and emotional upsets are often factors in constipation. People who habitually eat a low-fiber diet also run a greater risk of developing small pouches, called diverticula, that trap hard feces and then become inflamed. Fecal material can also lodge in the appendix and trigger appendicitis. Many people develop colon polyps as they age. These small masses of tissue may be harmless, but can sometimes lead to colorectal cancer—one of the most common of all cancers in developed countries. A diet rich in fruits, vegetables and whole grains provides the fiber that helps keep the colon in top working order.

→ **This _Escherichia coli_ bacterium** is a normal inhabitant of the human colon. Feces may contain other, disease-causing strains of this and other bacteria, however.

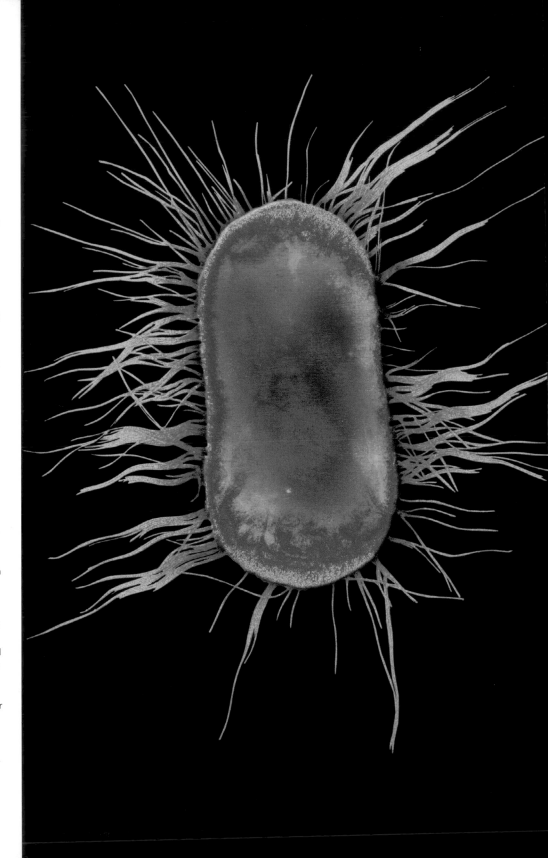

STRUCTURE OF THE LARGE INTESTINE

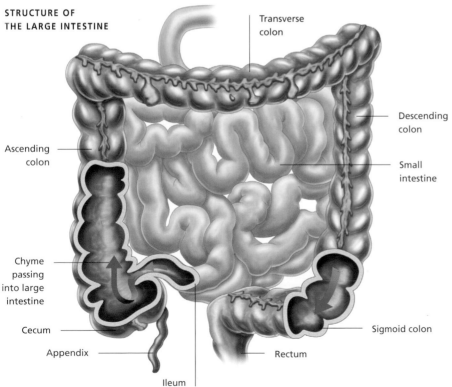

- Transverse colon
- Descending colon
- Small intestine
- Ascending colon
- Chyme passing into large intestine
- Cecum
- Appendix
- Ileum
- Sigmoid colon
- Rectum

↑ **These cells from a colon tumor** have been magnified 15,000 times. Colon cancer can run in families, but more often diet and other factors probably play a role. Some health authorities recommend periodic colon cancer screening for people over the age of 50.

↖ **Fiber-rich foods** are good for the colon. Foods such as whole-grain bread, beans and leafy green vegetables provide roughage that speeds the passage of feces through the large intestine.

← **Like an inverted U, the large intestine** arcs up and around the small intestine. Near its beginning is a small pouch called the cecum, to which the slender, worm-like appendix is attached. Beyond the cecum are the ascending, transverse, descending and sigmoid (S-shaped) regions of the colon. The short, final stretch of the large intestine is the rectum, where feces are briefly stored before being eliminated through the anus.

Hormones and hunger

The urge to eat is a survival instinct. The sensation we call hunger consists of insistent signals from the brain that the body needs food. Although unpleasant, it fulfills the natural drive to keep the body supplied with food. Appetite, by contrast, is a pleasant anticipation of food we expect to enjoy. Given the choice, most people combine the two by satisfying their hunger with something tasty.

At least four hormones influence hunger and appetite. When the stomach is empty, cells in its wall release a hunger signal called ghrelin into the blood. When this hormone reaches the brain, it acts on a center that monitors satiety, triggering a desire to eat. After you eat, insulin from the pancreas suppresses appetite at the same time as it spurs body cells to take up digested sugar from the blood. When ingested food arrives in the small intestine, cells there secrete PYY, another hormone that dampens our desire to eat. Over a longer time frame, our fat-storing cells release the appetite-suppressing hormone leptin as the amount of stored fat rises. When a person loses fat, they tend to feel hungrier as leptin levels fall.

HORMONES THAT CONTROL APPETITE

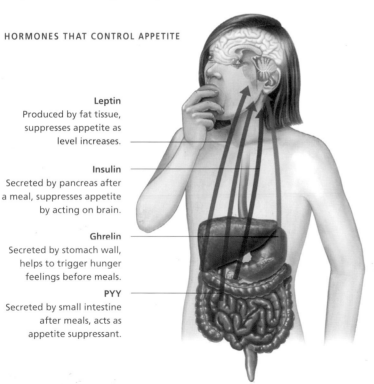

Leptin
Produced by fat tissue, suppresses appetite as level increases.

Insulin
Secreted by pancreas after a meal, suppresses appetite by acting on brain.

Ghrelin
Secreted by stomach wall, helps to trigger hunger feelings before meals.

PYY
Secreted by small intestine after meals, acts as appetite suppressant.

↑ **Hunger and appetite hormones** act in the brain. Carried in the bloodstream to the hypothalamus, they trigger nerve impulses that lead to conscious feelings of hunger or satiety—that is, being "full."

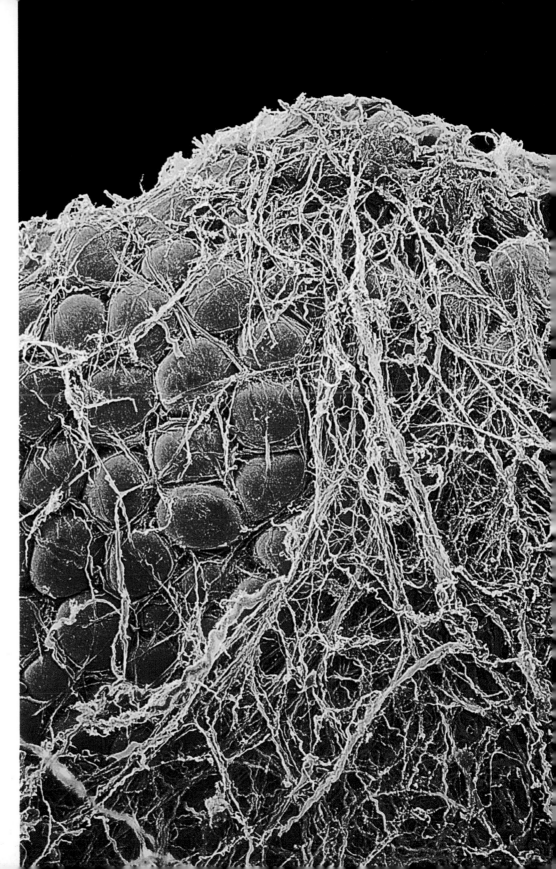

← **People tend to eat more** when in the convivial surroundings of a social gathering, or when large amounts of food are offered "buffet" style. In these situations actual hunger may have little to do with how much food and how many calories a person consumes. Globally the number of severely overweight humans is rising, especially in affluent societies where more food is available and people tend to be more sedentary.

→ **Fat cells store** unused food energy in the form of fat. These human fat cells are wrapped in a mesh of connective tissue. Fat cells fill up or shrivel as fat is added or used. The leptin they produce circulates in the bloodstream and provides the brain with feedback about how much fat the body is storing. This feedback mechanism appears to be faulty in some very overweight people.

← **This woman weighs less** than the amount of weight, mostly fat, that her husband lost after a "stomach-stapling" procedure. Gastric bypass surgery is increasingly common as a treatment for severe obesity. It drastically reduces the size of the stomach and the amount of food that can be consumed at one sitting. It also reduces the secretion of ghrelin, the hormone that influences appetite.

OUR FAT-STORING HERITAGE

Fat cells make leptin, which apparently stimulates appetite when the amount of a person's body fat declines. Some researchers propose that leptin's appetite-stimulating effect is an ancient boon that helped our early ancestors maintain needed fat stores as insurance against leaner times.

In humans and some other animals, a gene governs the ability to make leptin. When faced with situations where our access to a ready supply of calorie-rich foods is not guaranteed, having a gene-based trigger for storing fat can be a life-saving advantage. As a person eats more and the amount of body fat rises again, leptin signals should theoretically taper off, making overeating less likely. Studies suggest that in some people the brain's eating-related controls are not as sensitive to leptin signals, which means that increases in leptin do not help them to moderate food intake.

The essential nutrients

The foods and fluids we consume have a major effect on how well the body functions. The basic nutritional equation is simple: a healthy diet provides all of the raw materials—complex carbohydrates, proteins, lipids and other substances—that our cells use as building blocks and energy supplies. Simple carbohydrates such as refined sugar and corn syrup provide calories, but little else. Complex ones such as whole grains, vegetables, fruits and legumes (peas and beans) generally contain an assortment of vitamins and minerals as well as fiber. Body cells also make their own proteins from about 20 different amino acids. Cells can assemble some of these amino acids themselves, but eight are "essential" because we can get them only from protein foods. Likewise, the liver can use other nutrients to make most of the simple components of fats that cells need, but food is the only source for a few others, such as lecithin. Beyond these general parameters, however, there is no one "best" diet. People the world over have discovered many ways to supply themselves with the nutrients they need.

↑ **A grilled chicken salad** is one dietary solution to obtaining lean protein, complex carbohydrates and a modest amount of fat. Humans need only a spoonful or two of fat each day. Unsaturated fats such as olive oil are the healthiest. Most health authorities recommend limiting saturated fats—found in meat, many dairy products and solid margarine—because they can contribute to cardiovascular disease.

GENERAL GUIDELINES FOR A HEALTHY DIET					
Protein	**Grains**	**Fruits**	**Vegetables**	**Dairy**	**Fats**
Fish, legumes, eggs, nuts, tofu, poultry, lean meat. Limit processed meats such as salami and bologna.	Breads, rice, corn, wheat, barley, pasta, bulgar wheat, couscous. Emphasize minimally processed grains such as brown rice and whole wheat.	A variety of fruits and berries, emphasizing fleshy orange, yellow and "red" fruits such as oranges, peaches, cantaloupe, mango, papaya, strawberries, blueberries. Limit sweetened fruit juices.	A variety of vegetables, emphasizing dark green varieties such as broccoli, dark green leafy varieties such as spinach and chard, and yellow-orange vegetables such as squashes, carrots and sweet potatoes.	Milk, fortified soy milk, yogurt, cottage cheese, other cheeses. Emphasize low- or non-fat varieties.	Emphasize unsaturated types such as olive oil and canola oil. Use nut oils, palm oil, butter, lard and margarine sparingly. Limit intake of added fats.
2 to 3 servings daily, 2 to 3 oz (50 to 80 g) per serving.	6 to 10 servings daily, about 3 oz (80 g) per serving.	2 to 4 servings daily.	3 to 5 servings daily.	2 servings daily for adults, 3 for teenagers and others needing more calcium.	

→ **Eating sugary treats** can produce a "sugar rush." As digested sugar moves rapidly into the blood, insulin levels spike, then plunge as cells take up sugar. After the initial burst of energy the person often feels lethargic.

VITAMINS AND MINERALS

The body needs both vitamins and minerals to survive and operate normally. Vitamins are "organic" chemicals—that is, they are complex substances such as ascorbic acid (vitamin C) and thiamine (vitamin B_1) that contain the element carbon along with other components. Minerals, such as iron, calcium, potassium, sodium and magnesium, are "inorganic" because they do not contain carbon. Many vitamins help assemble or break down substances in cells and tissues, and others are antioxidants. Minerals become components of our teeth, bones, blood, many proteins and substances such as stomach acid. A varied diet that includes protein, complex carbohydrates and lipids may supply adequate amounts of vitamins and minerals. Many physicians now also endorse moderate use of nutritional supplements to combat free radical damage associated with aging and heart disease and to help stave off disorders such as osteoporosis and anemia.

↑ **In the 1700s** Scottish physician James Lind pioneered the use of citrus fruit to combat scurvy among British sailors. Scurvy is caused by a lack of vitamin C. It weakens connective tissue, and in earlier times often led to lost teeth, hemorrhages and even death.

Body weight: the energy equation

In developed parts of the world, tens of millions of people, adults and children alike, carry around an unhealthy amount of body fat. A growing proportion of them are obese—their body is more than 25 percent fat in the case of females, or more than 20 percent fat in the case of males. In a tempting environment of abundant, high-calorie food, evidence is mounting that long-term health correlates closely with maintaining a body weight that does not unduly strain the cardiovascular, respiratory and other body systems. The risks of type 2 diabetes, high blood pressure, heart attack and joint problems all rise as an individual's weight climbs. In a general way a "healthy" body weight relates to a person's height, a fact that is reflected in standardized weight–height charts. Other variables that influence this equation include a person's age, gender, race and genetically determined "build." Even so, many of us find ourselves engaged in the "battle of the bulge," in which the usual culprit is an imbalance between how much we eat and how much energy we expend in physical activity.

DIFFERENT WAYS OF STORING FAT
An "apple" shape—storing fat above the waist, as in a "beer belly"—increases the risk of health problems due to overweight. For unknown reasons, the cardiovascular and other disorders associated with obesity are much more likely to occur in people who carry excess weight in this fashion. The pattern is called android (male-type) obesity because it is more common in men. On the other hand, being "pear-shaped," with excess fat stored around the hips and thighs, is most often seen in overweight women. This gynoid (female-type) obesity is not associated so clearly with major health problems, although it, too, places extra strain on the heart and joints.

↑ **Wild animals never get fat.** This gray wolf spends much of its life hunting the large animals such as elk that are its main food source. A strenuous lifestyle and erratic meals keep this predator lean.

→ **Obesity is increasing dramatically** among young adults and children. Lifestyle factors—being sedentary and eating too much high-calorie food—are a major cause of this shift. For some people genetic differences in "hunger hormones" also contribute to weight increase.

"Pinching an inch" is one way of measuring the relative amounts of fat and lean muscle in the body. About half of stored body fat is just under the skin. A skilled practitioner can use skin-fold measurements, along with formulas set by age and sex, to calculate the person's approximate body composition.

Regular physical activity helps maintain a desirable body weight and a healthy ratio of fat to muscle. Although it weighs more than fat, muscle tissue burns more calories. This cyclist can eat more calories without weight gain than a less muscular sedentary man of the same height can consume.

ENERGY USED IN COMMON ACTIVITIES BY A 154 LB (70 KG) PERSON		
Activity	Kilocalories per hour	Kilojoules per hour
Sleeping	65	272
Sitting, at rest	100	418
Getting dressed	118	494
Word processing on a computer	140	586
Slow walking on level ground	200	837
Cycling	570	2386
Swimming	500	2093
Jogging	570	2386
Walking up stairs	1100	4605

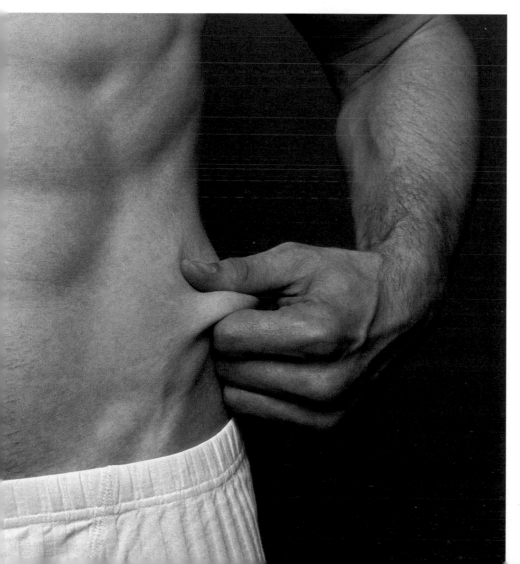

Our fluid interior

Our living cells survive in an internal fluid environment that presents a massive challenge to homeostasis, the vital stability of conditions inside the body. The chemical make-up of our blood and tissue fluids must remain more or less the same, even as life processes are adding some materials to them and removing others. The additions include electrolytes, ammonia—a dangerous waste produced when proteins are broken down—and acids generated during various cell processes. Electrolytes can carry an electric current. Two common ones are potassium and sodium chloride—simple table salt. Both are necessary for nerve cells to generate nerve impulses, among other essential roles. Perhaps the most vital substance in tissue fluid is water. It makes up more than half of the body by weight and fulfills such crucial roles that even a few days without it can amount to a death sentence.

WATER AND LIFE

Water has several features that underlie its importance in body functions. To begin with, some substances can dissolve in water. The majority of chemical reactions in body cells take place after substances have dissolved. Also, water is liquid at body temperature. One result of this property is that our watery blood can flow through our blood vessels. In addition, water can absorb a great deal of heat. Given that our cells constantly produce heat, this property is vital in maintaining proper body temperature. Heat produced in metabolism is transferred to the water in blood and sweat. Our insides— the internal environment—cool off when blood vessels in the skin radiate heat or when water in sweat evaporates.

↓ **A desert environment** challenges the body's water balance. These hikers in the Namib Desert must carry ample water to replenish the 2-plus quarts (more than 2 liters) they will lose during the day.

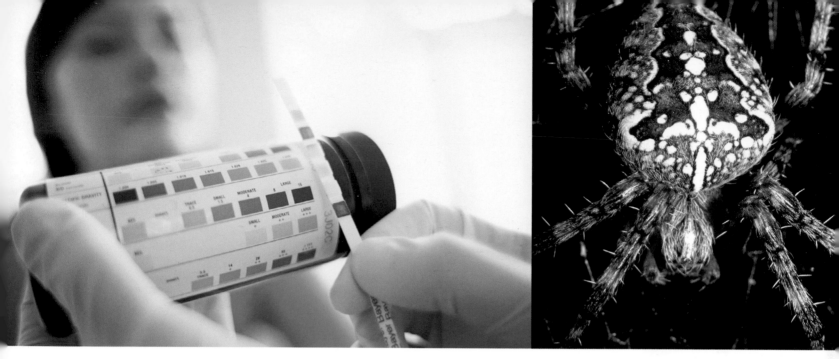

↑ **A urine sample** is a common diagnostic tool. Disease indicators such as excess sugar, proteins, blood and pus all show up in urine, as do hormones associated with pregnancy.

→ **Nephrons in the kidneys** make urine. A Bowman's capsule at the start of a nephron filters water and dissolved substances from blood. This fluid then flows through the nephron's coiled, U-shaped tube. Along the way, wastes and some water are removed to form urine, and useful substances are reclaimed. Urine then drains into the collecting duct and passes into the ureter.

→→ **Each kidney** consists of several triangular lobes that collectively contain millions of filtering units called nephrons.

STRUCTURE OF A NEPHRON

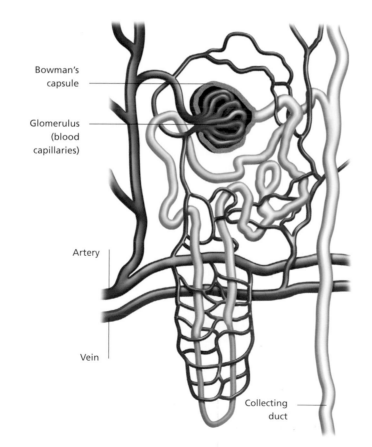

Bowman's capsule

Glomerulus (blood capillaries)

Artery

Vein

Collecting duct

↑ **Spiders have no kidneys.** Specialized tubes in their bodies collect waste-bearing fluid and empty it into the intestine. Special glands retrieve most of the water, so a spider's excreted wastes are nearly dry.

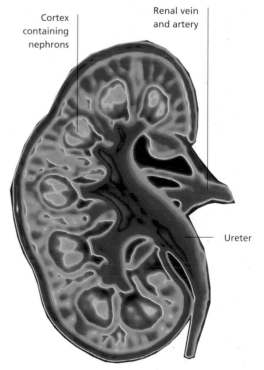

Cortex containing nephrons

Renal vein and artery

Ureter

Maintaining the balance of water and salt

The ability of the urinary system to manage the body's water balance is important because water is a major component of blood and the fluid in our tissues. Pure water tends naturally to move into a solution, such as urine, that contains many dissolved substances. Yet the body must retain most of this water or risk serious harm to its cells. One way this needed water can be kept is by retaining a certain amount of salts, such as sodium, as well. In effect, this is a major part of what the kidneys do. On average, each day you eliminate more than 1 quart (1 liter) of urine, but this is a tiny fraction of the water that has passed through your kidneys. Although an adult's kidneys process between 45 and 50 gallons (about 200–225 liters) of fluid daily, 99 percent of the water and useful materials in it are returned to the blood that flows in the capillaries around a nephron's U-shaped loop. By the time urine reaches the bladder, it contains only unneeded water and wastes.

→ **Eating salty food** increases the body's need for water. As excess salt moves into the forming urine, it is accompanied by water—leaving less water to be returned to the bloodstream.

↓ **A control center** in the brain's hypothalamus monitors the body's water balance. When we need water, it sends signals that we comprehend as feeling thirsty.

↑ **Urine forms in a nephron's U-shaped tube.**
After the glomerulus has filtered blood plasma (minus proteins), water and many needed solutes leave the first part of the tube and move back into capillaries. If conditions warrant, more salt and water can be returned to the blood farther along the tube.

Diagram labels: Glucose · Amino acids · Other substances · Salt · Glomerulus · Water · Water · Water · Salt · Salt · Distal tubule · Salt · Salt · Collecting tubule · Proximal tubule · Salt · Salt · Water · Water · Loop of Henle · Water · Water · Water · Water · Collecting duct · Urine flows to bladder

↖ **Barracuda risk becoming dehydrated** because they take in so much salt. To manage this, ocean fishes drink seawater, urinate little and pump out salt. Freshwater fishes tend to take on water, producing a great deal of urine.

⇑ **Sea birds such as this cormorant** ingest a great deal of salt with their food. Special salt glands in a bird's head remove salt from its blood and secrete it to the outside through the nostrils.

↑ **A kangaroo rat never drinks.** A desert dweller, this small New Mexico rodent relies instead on water in the seeds it eats and as well as water generated by chemical reactions in its cells. Lacking sweat glands, it never sweats and voids only a tiny amount of urine.

URINE—MORE THAN WATER

Although urine is about 95 percent water, it normally contains a wide variety of dissolved substances that are to be eliminated from the body. Much of its load of solutes is urea, the toxic byproduct released when body cells break down the amino acids in proteins. Other common chemicals include salts such as excess sodium, a small amount of glucose, amino acids and hormones. Residues from therapeutic drugs, pesticides and other foreign materials are also present in our urine, as are pigments from the hemoglobin in dismantled red blood cells, which give urine its color. Depending on how much fluid a person takes in, the kidneys may produce urine that is dilute and pale, concentrated and dark yellow, or somewhere in between.

The blood in balance

The things we eat, do and are exposed to all challenge the kidneys' ability to make needed adjustments to the blood. A key gauge of healthy chemical balance in the blood is its acidity, for having too much or too little acid in the blood can lead to death. The pH scale measures acidity. It ranges from zero for extremely acid substances such as the hydrochloric acid in gastric fluid to 14 for extremely alkaline ones such as oven cleaner. Day in and day out, our blood and tissue fluid must remain in the center of this range, at about 7.4, and the kidneys either excrete acid or return it to the bloodstream as conditions demand. Food-poisoning or other infections that cause severe diarrhea can also be life-threatening, especially in children and the elderly. They can flush so much water out of the body that blood pressure plummets, the kidneys fail and the heart stops beating.

Strenuous exercise increases the acidity of blood because it requires muscle cells to work harder. The term "pH" stands for "percentage of hydrogen." Metabolically active cells produce a great deal of hydrogen and release it into blood and tissue fluid.

The pH scale is chemical shorthand. A change of one unit reflects a tenfold change in acidity. The comparative pH values of some common solutions are shown here, ranging from the high acidity of gastric juices to extremely alkaline oven cleaner at the other end of the scale.

In this rack of test tubes, color reflects the different pH values of various fluids. The pale green fluid in the center has a neutral pH with a value of 7. The fluids to the left of it are progressively more acid, those to the right increasingly alkaline.

THE PH SCALE

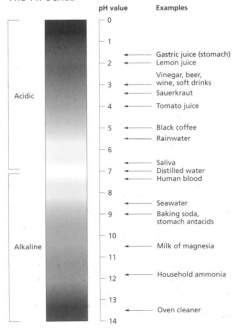

pH value	Examples
0	
1	
2	Gastric juice (stomach) / Lemon juice
3	Vinegar, beer, wine, soft drinks / Sauerkraut
4	Tomato juice
5	Black coffee / Rainwater
6	
7	Saliva / Distilled water / Human blood
8	
9	Seawater / Baking soda, stomach antacids
10	Milk of magnesia
11	
12	Household ammonia
13	Oven cleaner
14	

Acidic

Alkaline

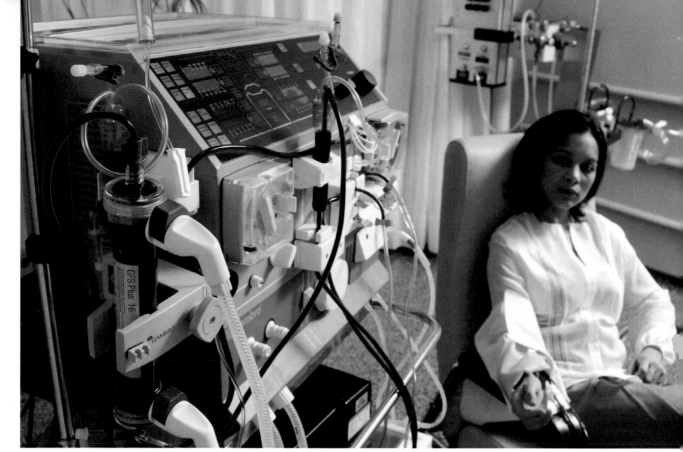

→ **People whose diseased or damaged kidneys** falter or fail cannot live long without regular dialysis—a procedure in which the blood is filtered externally before being returned to the patient's body.

↓ **Water-borne bacteria** can lethally disturb the body's fluid balance. After Hurricane Katrina flooded much of New Orleans in 2005, rescue workers sought to remove stranded citizens before contaminated water supplies could trigger outbreaks of severe diseases such as dysentery and cholera.

ELECTROLYTES, THE ESSENTIAL SALTS

Salts, especially sodium, potassium and calcium, are important factors in the blood's chemical balance. The body needs only small amounts of these substances, but they are crucial because they are electrolytes that function in muscle contraction and many other basic processes. Our cells generate electrolytes, and we also consume them in food. We lose electrolytes in sweat, urine, solid wastes and vomited material, which is why they can become depleted by overheating, food poisoning and other gastrointestinal upsets.

The effects of electrolyte imbalances run the gamut from nausea and feeling tired to convulsions and coma. For example, when sodium is lacking the body retains water. In the short term, the kidneys reduce their urine output. With a long-term imbalance, the brain may swell and the heart may be unable to pump blood effectively. Stressed or diseased kidneys can fail to retain or eliminate enough of other electrolytes as well. Potassium is vital for the operations of both the nervous system and muscles. When the blood carries either too much or too little potassium, one consequence can be sudden heart failure.

A warm-bodied heritage

Like all mammals, human beings are homeotherms—that is, our bodies maintain a relatively constant core temperature. At about 98.6 °F (37°C), this internal temperature is generally much warmer than that of our surroundings. We are also endotherms, which means that our warm body temperature is sustained "from within" by heat generated in our incessantly working cells. These thermal features correlate with our status as highly active animals, and they helped our early ancestors survive in cool weather and cold places.

A steady, warm core temperature also imposes some stringent demands. We must eat fairly often and in sizeable amounts because many nutrients in food—and those stored as fat—fuel our rapid metabolism and so keep our cells producing heat. A fast metabolism also requires us to have some means of dissipating excessive heat, such as sweating, and demands relatively rapid breathing to keep working cells supplied with oxygen. In addition, because water leaves the body in both sweat and exhaled air, to replenish it humans must consume more water than would otherwise be necessary. Finally, having so much of our total energy budget allotted to staying warm and active is one reason why, compared to many other animals, we develop and grow slowly.

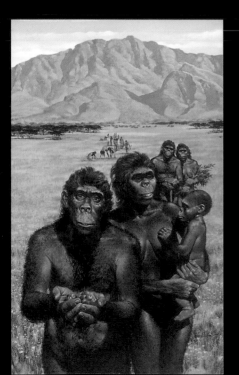

← **Neanderthals,** an early species related to modern humans, lived in Europe and parts of western Asia some 250,000 years ago. Their short, stocky bodies helped them to conserve heat. They became extinct roughly 30,000 years ago, about the time that modern humans emerged.

→ **Body heat is visible** in a thermogram, which shows infrared radiation given off by the skin. The warmest parts of the body appear white, followed by red, yellow, green, light blue, dark blue to purple, and black. Skin temperature generally accords with the amount of blood in various body areas.

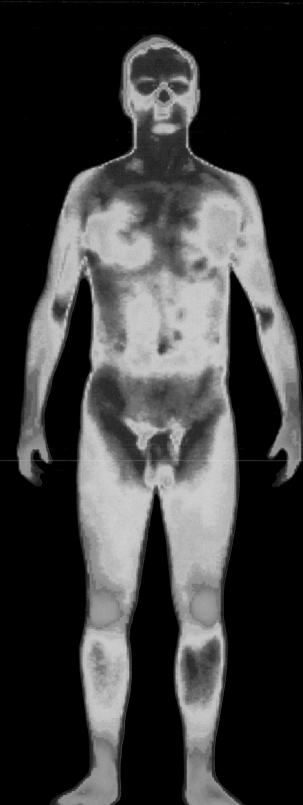

SURVIVING THE EXTREMES

The catalytic proteins called enzymes participate in nearly every chemical reaction that keeps us alive, and they stop operating when core body temperature rises above or falls below the normal range. At around 88°F (31°C) muscles begin to lose their ability to contract. If the temperature drops below 80°F (27°) our metabolism slows to a crawl. At 70° (21°C) the heart stops beating. At the "hot" end of the temperature spectrum, enzymes also stop working or are destroyed. The maximum survivable core temperature is around 108°F (42°C).

These biological facts mean that our early human ancestors had either to live in environments that were not subject to temperature extremes, or to discover ways of coping with cold and heat. Lacking the protection of thick hair, outside of the tropics they would have had to fashion clothing, devise shelters and adapt to their climatic envronment in other ways. One hallmark of our humanness is the ability to make cultural changes such as these. Based on fossils, ancient cave paintings and other evidence of human activity, anthropologists believe that by about 15,000 years ago humans were living on every continent except Antarctica.

↖ **In the coldest months,** this grizzly bear will spend most of its time inactive and dozing in a protected den. Its "winter sleep" is not true hibernation, because the bear's body temperature drops only a few degrees.

↑ **Like humans, birds generate** their own body heat. Unlike humans, the core temperature of many species of birds fluctuates significantly, so that their body temperature—and their need for food—drops markedly when they are inactive. This beautiful kingfisher gains insulation from its colorful feathers.

↗ **Bluefin tunas** have specialized networks of blood vessels around the brain, eyes and swimming muscles, as do some species of sharks and other fast-swimming, predatory fishes. Heat from the vessels keeps those parts of the fish much warmer than the surrounding sea.

→ **Reptiles have** a slow metabolism that does not generate much heat. Alligators bask in the sun to elevate their body temperature, which will allow them to move much faster.

Body heat

We have a variety of options for keeping our body temperature within normal bounds. Some of them are intentional—pulling off clothing when it's hot or bundling up when the air is cool. Other coping mechanisms are involuntary and automatic. For instance, heat stress triggers sweating, which dissipates heat as perspiration evaporates. In another response, as the blood becomes overly warm a center in the brain's hypothalamus detects the rise in temperature and signals blood vessels in the skin to dilate. Heat then radiates out through the skin just as it does from a radiant heater. People who are heat stressed also move less and so generate less body heat.

The skin provides the first notice of downward shifts in temperature. Thermoreceptors in the skin send signals to the hypothalamus, and then blood capillaries in the skin constrict. This reduces the flow of blood to its surface, conserving heat. At the same time, the sympathetic nervous system may trigger the release of the hormone norepinephrine, which speeds up metabolism, so more body heat is generated. The hypothalamus can also order shivering, spurring the skeletal muscles to contract as often as 20 times per minute, which in turn produces heat.

EXPOSURE TO EXCESSIVE HEAT OR COLD

Overall the human body is not well equipped to deal with extremes of heat and cold. When initial strategies for eliminating excess heat have done their utmost, heat exhaustion sets in. Blood pressure drops due to the loss of water in sweat and the dilation of blood vessels throughout the body. Eventually the person may feel weak or faint—a clear signal to seek a cooler environment and rehydrate as soon as possible. The next phase may be heatstroke, which can quickly be lethal as temperature controls in the brain collapse. The body's core temperature soars, and when it reaches 108°F (42°C) the enzymes in cells stop working.

Abnormally low body temperature (hypothermia) leads to the same outcome but by a different path. When core temperature falls below about 90°F (32°C), it is too cold for cell enzymes to function normally. Heart muscle and skeletal muscles cannot contract properly and shivering stops. A person may feel sleepy and not realize the impending danger, but eventually the metabolism will break down and the heart will simply stop beating.

↗ **"Goose flesh" comes** from a response to cold that pulls skin hairs erect. It is not very helpful for humans, but in hairier mammals it creates a layer of still air that helps prevent heat from radiating away.

→ **Sweat cools by evaporation.** Some 2.5 million sweat glands are distributed over the body, but those in the skin of the forehead, in the armpits, on the palms and on the soles of the feet are most active in regulating temperature.

← **An Alaskan trapper** wears a beaver-fur hat to protect his head and ears from freezing weather. His bare skin is vulnerable to frostbite, in which tissues freeze and die unless thawing is carefully controlled. Most land mammals have built-in insulating layers of hair and fat, which protect the skin and help retain body heat.

↙ **Dogs don't sweat much** through their skin, so they pant to help dissipate excess heat. To keep warm, most dogs rely on their thick fur, which includes a down-like layer close to the skin. Canine behaviors also help dogs thermo-regulate. In hot conditions, a dog may sprawl out, exposing its more-or-less hairfree belly so that more heat can radiate away. When a sled dog sleeps outside in the snow, it curls up to help prevent its nose and toes from freezing.

↓ **Mourning doves fluff up their feathers** in the cold. The upper feathers, used for flight, conceal soft "down feathers" underneath that conserve heat. Fluffing also creates air spaces that trap heat and slow its loss. Although birds can withstand a lowered body temperature for short periods, they also use mechanisms like this one to help keep warm.

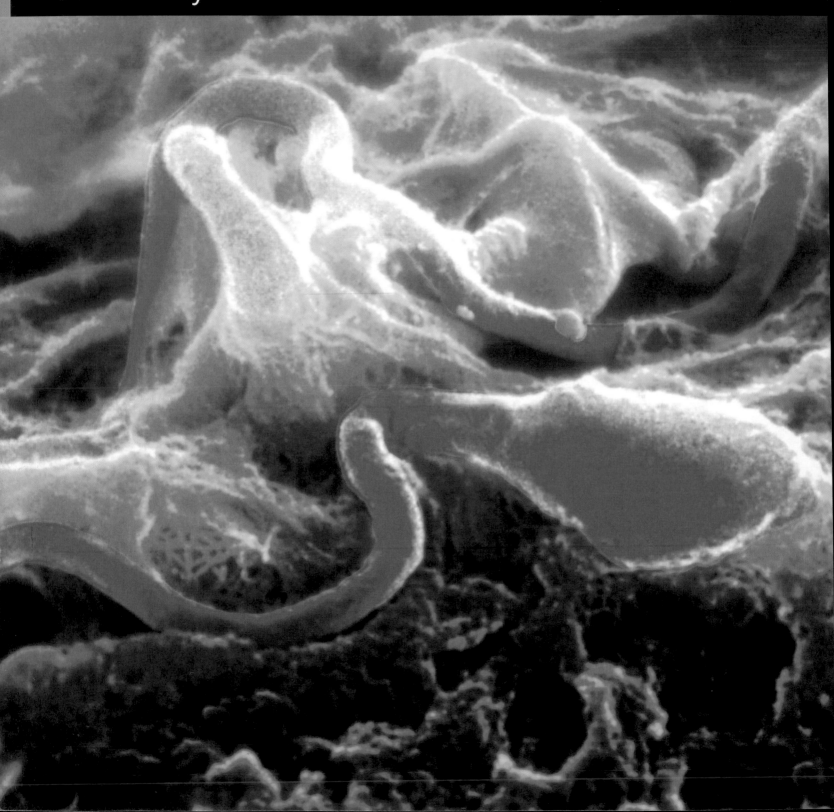

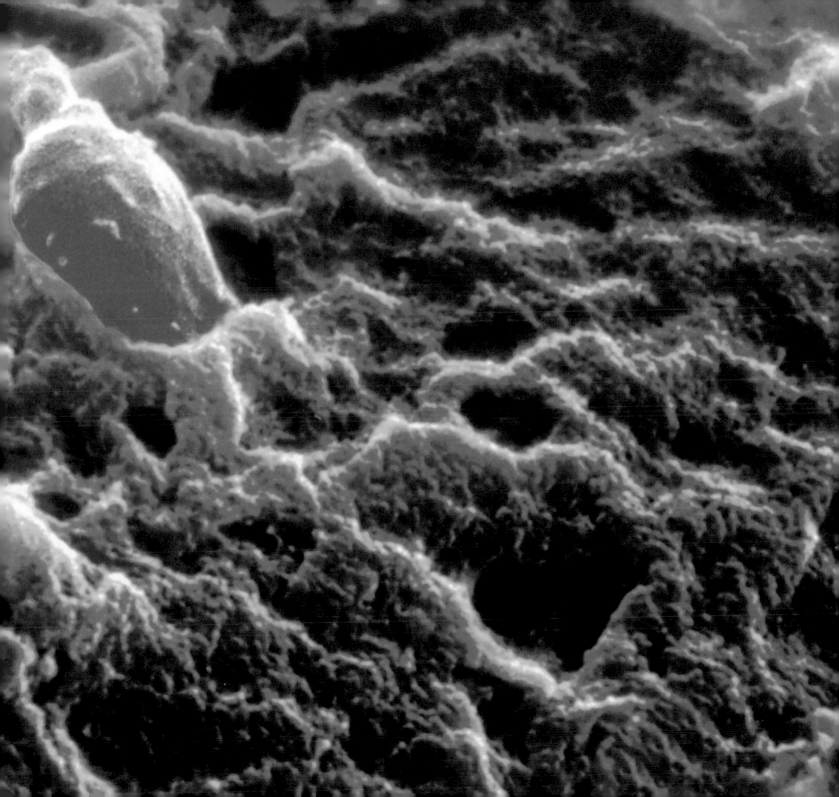

Continuity

Whereas other body systems function in day-to-day survival, the reproductive systems give rise to cells that can unite and launch the development of a new individual. Human genes, bearing information coded in DNA and packaged into sperm and eggs, transmit the traits of parents to children.

The reproductive imperative	236
The female reproductive system	238
The remarkable egg	240
The male reproductive system	242
The remarkable sperm	244
Uniting sperm and egg	246
Pregnancy	248
Development begins	250
The dynamic early embryo	252
Prenatal life	254
Birth	256
Transitions from infancy to adulthood	258

The reproductive imperative

The continuity of human life depends on the process of reproduction—the making of new generations to replace those that inevitably pass away. Reproduction is a basic biological imperative, but not all life forms reproduce as humans do. Bacteria are single cells, and for them the answer is simply splitting in two in a way that provides each new cell with the genetic instructions it needs to operate. For a strawberry plant, reproduction may take the form of sending out a runner to take root nearby. The plant that grows from this new set of roots is a clone—it is genetically identical to the parent plant.

Many other kinds of organisms reproduce through the union of special sex cells produced by parents—eggs and sperm. When these cells unite, they provide the necessary genetic information to build and operate a new individual. This is the strategy employed by humans and other complex animals, as well as by many trees and other plant species. No one mode of reproduction is "best," for the sole and simple objective of reproduction is merely to ensure that life will go on.

WHY HAVE SEX?

In some ways, reproducing without sex, as bacteria and many other life forms do, is an extremely efficient strategy. There is no need to find a mate, and the new generation will be genetically identical to its parent. This means that if the parent is well adapted to its environment, the offspring will be also. From a biological perspective, this simple approach to reproduction raises the question: why do it sexually at all? For an answer, all we have to do is look at our families. None of us looks exactly like our parents, and if we have children, neither are they carbon copies of us. Sexual reproduction mixes genes, producing variations in the traits of new generations. It is likely that such variations became the genetic raw materials for characteristics—such as upright walking and a complex brain—that allowed our ancient ancestors to thrive.

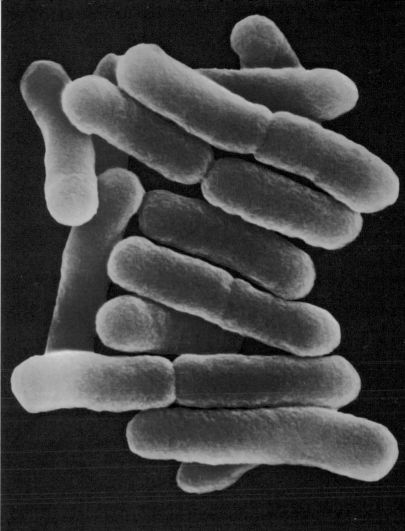

← **Easter lilies** reproduce sexually. Each flower has both male and female sexual parts. The lily's reddish anthers produce pollen, the male sex cells, and its pale, central carpel houses eggs.

↑ **Sexual "chemistry"** between lovers helps fuel natural reproductive urges. Our survival as a species depends on behaviors that result in mating and the development of offspring.

↗ **These *Salmonella typhi* bacteria** will reproduce themselves by binary fission—dividing in half more or less across the middle. Some bacteria can divide as often as every 20 minutes.

→ **Strawberry plants** can reproduce asexually. A new plantlet grows from a runner extended by its parent. and will survive on its own if the runner is severed.

The female reproductive system

For the first 12 or so years of a female's life, her reproductive organs are not active. Then, as she reaches puberty, her ovaries begin to secrete hormones that signal the development of secondary sexual characteristics, such as breasts, and the beginning of her reproductive years. For most women this major life phase will last for at least three decades, during which time their ovaries will expel eggs, usually one during each monthly menstrual cycle. A newborn girl's ovaries each contain an estimated one million eggs. Most of them will be absorbed by her body during childhood, but by the time she reaches sexual maturity there will still be several hundred thousand remaining. An egg released from an ovary must cross a very small gap to enter the nearby oviduct, but once inside the oviduct it can be fertilized by sperm from a male. If a pregnancy ensues, a woman's uterus will nurture the developing fetus and ultimately deliver a baby. Key to this function is the uterine lining, called the endometrium, where an early embryo implants. The lower part of the uterus, called the cervix, produces mucus, which acts as a barrier to bacteria during pregnancy.

THE FEMALE REPRODUCTIVE ORGANS

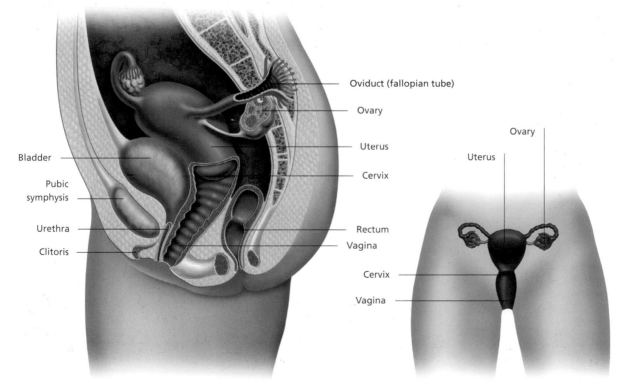

Bladder

Pubic symphysis

Urethra

Clitoris

Oviduct (fallopian tube)

Ovary

Uterus

Cervix

Rectum

Vagina

Ovary

Uterus

Cervix

Vagina

↑ **A female's life** has three reproductive stages. Up to about age 10, a girl's reproductive system is not active. Her reproductive years usually begin around age 12 with menarche, the first menstruation, and continue until her late 40s or early 50s. Menopause, when menstruation ceases, marks her post-reproductive life stage.

← **The female reproductive system** consists of the ovaries, oviducts (or fallopian tubes), the uterus, the cervix and the vagina. A female's external genitals—collectively called the vulva—have their own important role to play in sexual arousal.

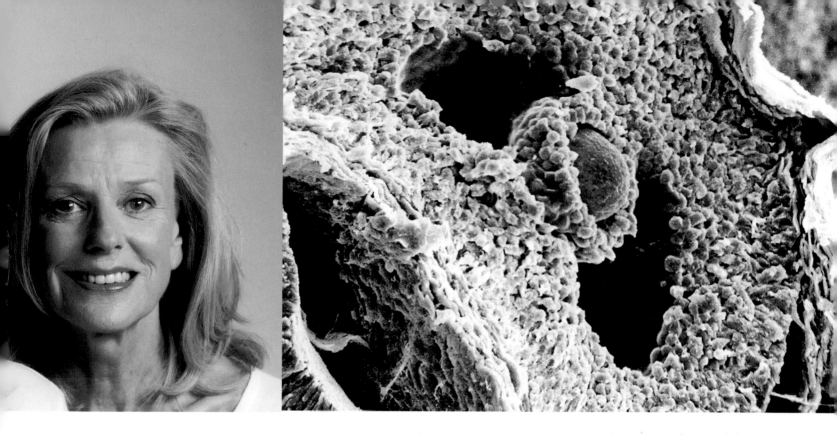

THE MENSTRUAL CYCLE

For nearly all women, the reproductive years include regular menstrual cycles of about 28 days. During each cycle, fluctuating hormones coordinate changes in the uterus with events in the ovaries. The cycle's first few days are marked by menstruation, when all but the deepest layer of the endometrial lining of the uterus is shed in a blood-infused fluid. At the same time, however, an egg is starting to mature in a sac-like follicle in an ovary. In due course it will be released, swept into an oviduct, and possibly fertilized. An endocrine structure called a corpus luteum ("yellow body") is also developing in the ovary. About mid-way through the cycle it releases progesterone and estrogen, which signal the endometrium to prepare for the possible arrival of a fertilized egg. After menstruation the endometrium regrows and thickens, setting the stage for a fertilized egg to implant. If this does not occur, a new cycle begins.

→ **In the first phase of the menstrual cycle** most of the endometrium is shed. In the next phase it begins to rebuild, influenced by estrogen from the ovaries. After about day 14, in the third phase, ovarian hormones—especially progesterone—prime the endometrium to receive a fertilized egg.

↑ **A human follicle and egg** develop together. This microscope image shows an egg (red) surrounded by the follicle wall (pale pink). During each monthly ovarian cycle, the follicle enlarges as fluid builds up inside it. The egg also undergoes key changes that prepare it to be fertilized. Eventually the follicle bursts open and releases the egg in a rush of fluid.

THE MENSTRUAL CYCLE

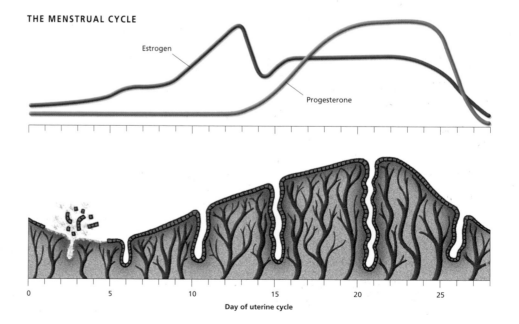

Estrogen

Progesterone

Day of uterine cycle

The remarkable egg

By puberty, a girl's ovaries contain several hundred thousand eggs, or oocytes, all in a state of arrested development. Ultimately about 400 of these will be ovulated. An egg is still immature when it is released into the oviduct and contains only half the DNA needed for a normal embryo to develop from it. A sperm cell will provide the rest of the DNA, while fertilization will stimulate the egg to mature. The egg also contains biochemical instructions that will guide the earliest steps of embryonic development if conception should occur.

A human egg emerges from an ovarian follicle coated with a thin, clear membrane. This delicate wrapping is aptly called the zona pellucida, meaning "transparent boundary." Healthy human sperm have special enzymes at their "heads" that can quickly penetrate this membrane. Once a sperm has broken through, the zona pellucida changes so that no other sperm can do so. Although vulnerable, a fertilized human egg is sheltered inside the mother's body as it develops. Animals that lay eggs, such as birds and reptiles, must invest valuable resources in providing their eggs with a great deal more protection.

THE OVARIAN CYCLE

The cycle begins when an immature oocyte that has been dormant for years shifts into a new development phase. A zona pellucida and a layer of cells develop around it. The cells will form the fluid-filled follicle that protects and nourishes the egg until it is ovulated. Over a period of days, follicle-stimulating hormone (FSH) and luteinizing hormone (LH) from the pituitary gland spur the growth of the follicle and the egg within it. Just before ovulation a surge of LH causes the follicle to balloon with fluid and burst, releasing the egg for its journey through the oviduct and into the uterus. The remains of the follicle form the corpus luteum, and this produces the hormones progesterone and estrogen, which prime the uterine lining for a pregnancy. If none occurs, the corpus luteum disintegrates, the production of progesterone and estrogen declines, the uterus lining breaks down and menstruation begins.

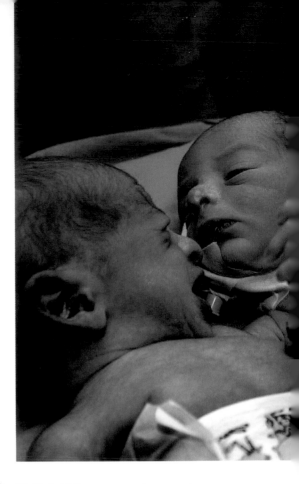

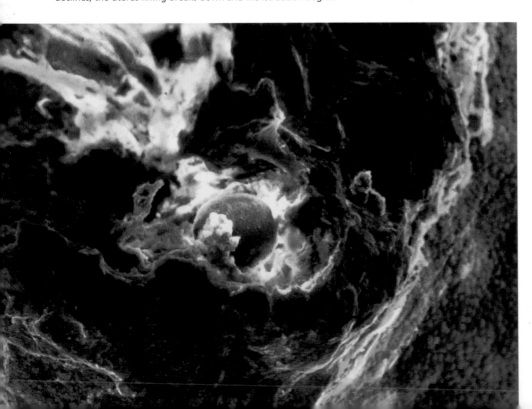

↑ **Multiple births** can occur when several eggs are ovulated at once. Born by Caesarian section, these Egyptian quintuplets, three boys and two girls, were two days old when photographed. Up to 20 follicles begin to develop during each month's ovarian cycle. Natural multiple births are relatively rare because usually only one of the follicles ovulates an egg.

← **An egg erupts** from an ovarian follicle. In this microscope view the egg appears red and is surrounded by fluid and other material from the ruptured follicle. It is shed into the abdominal cavity and must find its way into the oviduct. An ectopic, or "outside-the-uterus," pregnancy occurs when a fertilized egg implants before it enters the uterus.

→ **The ovarian cycle** is controlled by hormones. LH and FSH from the pituitary gland stimulate the follicle to develop, and the growing follicle releases estrogen. The effects of estrogen include thinning cervical mucus—an aid to swimming sperm—and helping to ready the endometrium for pregnancy.

→ **A leatherback sea turtle** lays dozens of shelled eggs, but few, and possibly none, of her offspring will survive. Like other reptiles, sea turtles do not protect their eggs once they have been laid, and predators (including humans) may dig up the eggs or capture the hatchlings.

→ **Deposited on a leaf,** tiny, soft butterfly eggs are doomed. Caterpillars that would otherwise hatch from the eggs would feast on the plant's leaves, so ants living on the plant crush the eggs to preserve their home.

THE OVARIAN CYCLE

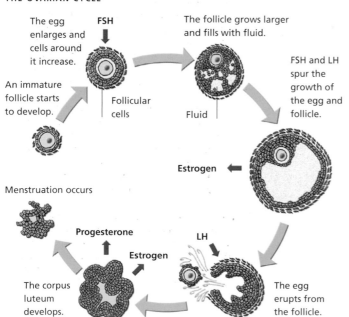

The egg enlarges and cells around it increase.

FSH

The follicle grows larger and fills with fluid.

An immature follicle starts to develop.

Follicular cells

Fluid

FSH and LH spur the growth of the egg and follicle.

Estrogen

Menstruation occurs

Progesterone

Estrogen

LH

The corpus luteum develops.

The egg erupts from the follicle.

→ **Birds' eggs** are protected by a hard, watertight shell and are usually nurtured in a nest by one or both parents. Inside a bird egg, the membranes that surround and support a developing chick are very similar to those that surround and support an early human embryo.

The male reproductive system

A male's reproductive system is designed to generate and deliver the sperm that may fertilize an egg. This system consists not only of the five main reproductive organs, including the penis and testes, but also accessory glands such as the prostate that collectively produce the thick fluid called semen. The penis operates in both the reproductive and urinary systems. It is designed to deposit sperm into the female reproductive tract by copulation, while sperm exit through the urethra, which also has the task of carrying urine.

The testes are dedicated to reproduction. Each one is about the size and shape of a large olive and serves two functions: to make male sex hormones—especially testosterone—and to produce sperm. Suspended in the pouch-like scrotum, the testes jointly may make millions of sperm a day inside coiled thread-like tubes called seminiferous tubules. Their vulnerable location outside the abdominal cavity helps keep the testes several degrees cooler than normal body temperature, which is too hot for sperm to develop properly in most mammals. Although modern reproductive technologies allow sperm to fertilize eggs in a laboratory setting, without his testes to make sperm a male could not reproduce. Similarly, because sex hormones guide the development of sexual features, without testosterone a male embryo could not develop male reproductive organs in the first place.

↓ **Of the five major organs** in the male reproductive system, four—the testes, vas deferens, epididymis and seminal vesicles—come in pairs. Sperm mature in the epididymis. The two vas deferens are long, looping ducts that carry sperm to the urethra. Vasectomy severs them so that no sperm can enter ejaculated semen.

THE MALE REPRODUCTIVE ORGANS

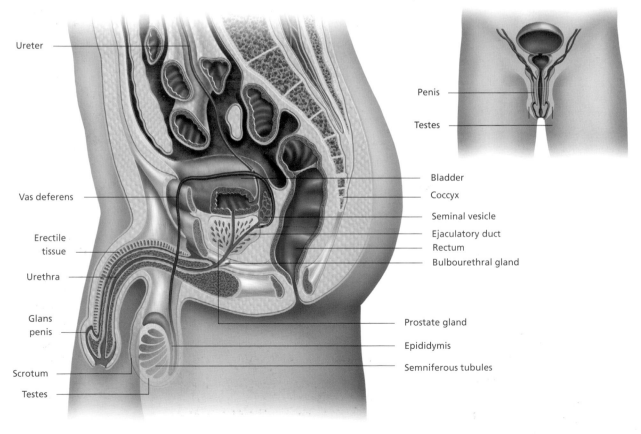

Ureter

Penis

Testes

Vas deferens

Bladder
Coccyx

Seminal vesicle
Ejaculatory duct
Rectum
Bulbourethral gland

Erectile tissue

Urethra

Glans penis

Prostate gland

Epididymis

Semniferous tubules

Scrotum

Testes

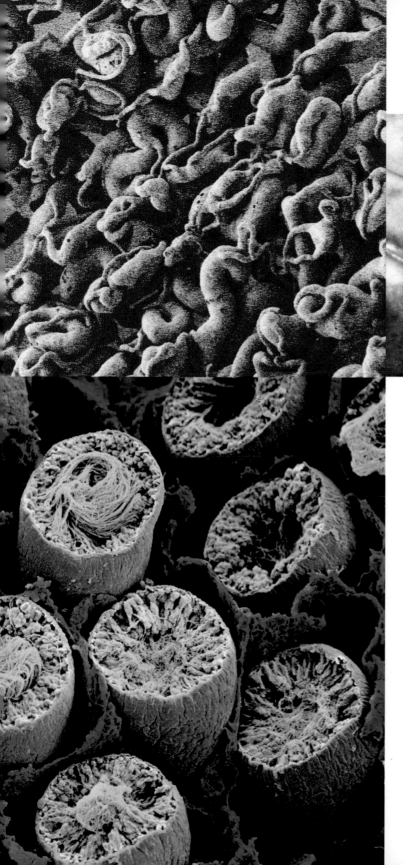

↑ **A male shark's claspers** are muscular appendages that are somewhat analogous to a penis. During mating the male uses one of his claspers to insert packets of sperm into a female's reproductive tract.

↖ **Blood vessels that lace** through erectile tissue in the penis are vividly shown in this resin cast. Three cylinders of spongy erectile tissue run through the penis shaft. When a male is sexually aroused, the penis becomes erect as blood from the vessels fills spaces in the tissue.

← **In this cross-section of semniferous tubules** in the testis, the sperm-forming cells appear blue. Cells lining the tubules make substances that nourish developing sperm and produce chemical signals to guide sperm formation.

DUCTS AND ACCESSORY GLANDS

Delivering a viable sperm to where it needs to be to fertilize an egg requires several glands and ducts to work in tandem with a male's reproductive organs. Much of the fluid semen that transports sperm to the outside when a male ejaculates is manufactured in seminal vesicles, glands that sit behind a male's bladder. This seminal fluid also contains prostaglandins, which are thought to strengthen sperm-propelling muscle contractions in a female's reproductive tract. Below the bladder, the walnut-sized prostate gland produces a pale white fluid that is also a major ingredient in semen. In addition to substances that help activate sperm, this fluid contains prostate-specific antigen—a chemical marker that is measured by clinical tests for prostate cancer. Mucus secreted by a pair of small bulbourethral glands below the prostate makes semen less acid, which in turn allows sperm to swim more effectively. When sperm enter the urethra they mix with all these substances, and thus are prepared to move efficiently toward fulfilling their biological purpose—fertilizing an egg.

The remarkable sperm

A typical ejaculation may contain as many as 400 million sperm, each one capable of fertilizing an egg. When an adolescent boy is about 14 years old, sperm begin to form in the approximately 400 feet (120 m) of seminiferous tubules in his testes, a process that will continue for the rest of his life. Each sperm is a masterpiece of bioengineering—a basically spherical cell streamlined to move rapidly in its liquid surrounds. The oval sperm head contains DNA and has a cap of enzymes that can break through the egg's outer covering to allow the sperm to enter. Behind the head is a mid-piece packed with mitochondria, the cell's powerhouses, followed by a long tail. Fueled by chemical energy provided by the mitochondria, the tail can help propel a sperm cell a remarkable distance—all the way from a female's vagina, through her uterus and up into the oviducts. Once they are ejaculated, sperm generally do not live long, but nature has ensured a generous supply. Whereas women stop ovulating at menopause in late mid-life, a man can produce viable sperm into old age.

HOW SPERM ARE MADE

It takes more than two months for sperm to develop. The process begins inside the walls of seminiferous tubules in testes. Signals from testosterone and other hormones cause forerunner cells called spermatogonia to start dividing. Over the ensuing weeks immature sperm are formed and released into each tubule's hollow interior channel. These testicular sperm cells have all the essential sperm parts—head, mid-piece, and tail—but they are not yet able to move independently and fertilize an egg. For the finishing touches, the sperm are moved along into the epididymis, which curves over and behind each testis. There they mature, and become fully mobile and ready to assume their reproductive task. This process goes on more or less continually, waxing and waning as hormone signals boost sperm production or slow it down. At any given time, a sexually mature male's testes contain sperm in every stage of development.

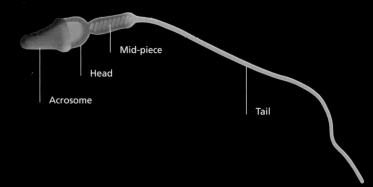

Mid-piece

Head

Acrosome

Tail

↑ **A human sperm's tapered head** is mostly sheathed by the cap of enzymes (acrosome) that will digest a path through the egg's covering, the zona pellucida.

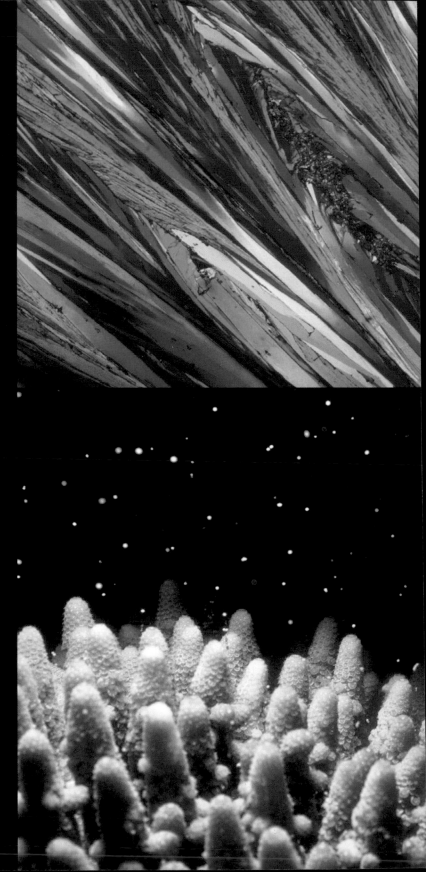

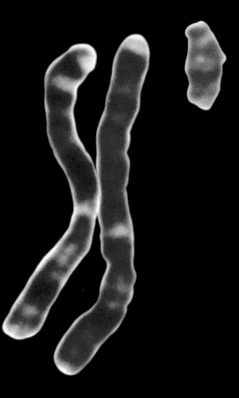

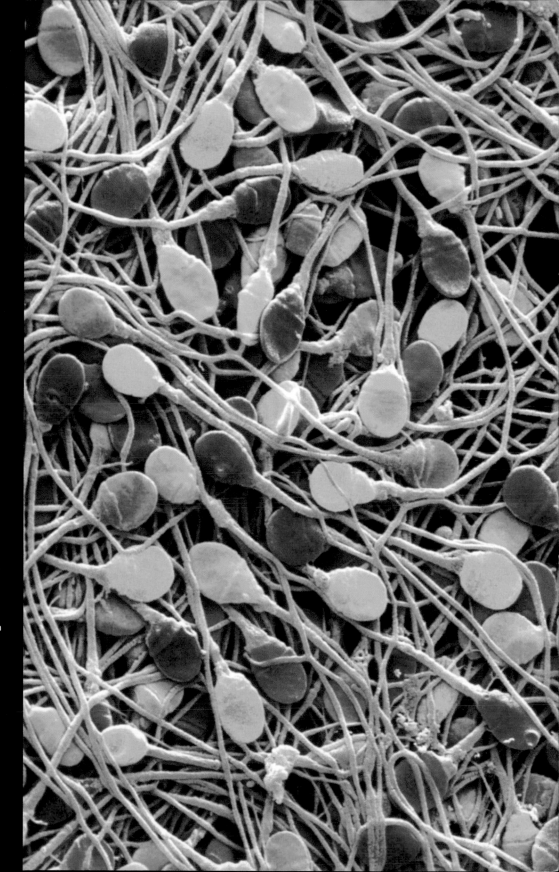

↖ **Crystals of testosterone** gleam in polarized light under a microscope. Among other roles in males, testosterone is the direct trigger for sperm formation. When a male's sperm count is high, a feedback mechanism reduces the blood level of testosterone and fewer sperm form.

↑ **Males have** one X (left) and one Y (right) sex chromosome, shown here during cell division. The instructions for maleness are carried by the Y chromosome. An unfertilized egg has an X, but sperm may carry either an X or a Y. When sperm and egg unite, the sperm determines whether offspring will be XY or XX—male or female.

→ **Human sperm** are mobile reproductive units. Sperm must be in fluid to move, as the tail whips back and forth to propel the cell. Research shows that sperm follow a chemical trail to locate an ovulated egg. Of the millions of sperm in each ejaculation, many may reach the egg but only one will fertilize it.

← **Colonies of reef-building corals** like this *Acropora* species live inside the limestone structures they secrete. Although they are among the Earth's simplest animals, most corals reproduce sexually. Periodically they release clouds of sperm and eggs that may unite and become the foundation for a new generation.

Uniting sperm and egg

The human life cycle begins with fertilization, when a sperm and an egg unite. The egg and sperm are sex cells, and unlike other body cells they each contain only half of the DNA needed to build a new individual. When they later combine, the two halves will make a cell with one full set of genetic instructions, half from each parent.

Fertilization is a wondrous series of physical and chemical events. When a sperm first encounters an egg, the sperm's head—which contains the father's share of DNA arranged on chromosomes— becomes attached to the egg's outer surface. Before reaching the egg, the sperm has undergone capacitation, a process that allows its head to penetrate the egg's protective covering and move inward. Usually, when one sperm makes it past the barrier, chemical changes in the egg block other sperm from entering.

Once inside the egg the sperm sheds its tail, which degenerates, leaving only the head—the sperm nucleus—to fuse with the nucleus of the egg. When this is accomplished, the once-separate cells have merged into one cell, which contains chromosomes from both mother and father. This cell, called a zygote, has all the genetic instructions required to build a new human being.

A SPERM'S JOURNEY

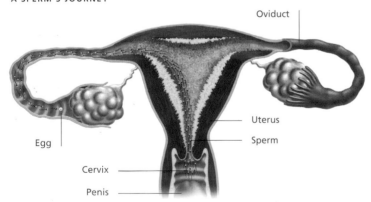

↑ **The reproductive journey of a sperm** may take several days. After making its way through the cervix, a sperm swims onward through the uterus. Relatively few sperm will survive to reach the upper part of an oviduct, where fertilization usually takes place.

→ **Sperm attempt to penetrate** a human egg, which appears brown in this microscope image. If for some reason enzymes are not released from the acrosome blanketing the sperm's head, the sperm will be unable to penetrate the egg and fertilize it.

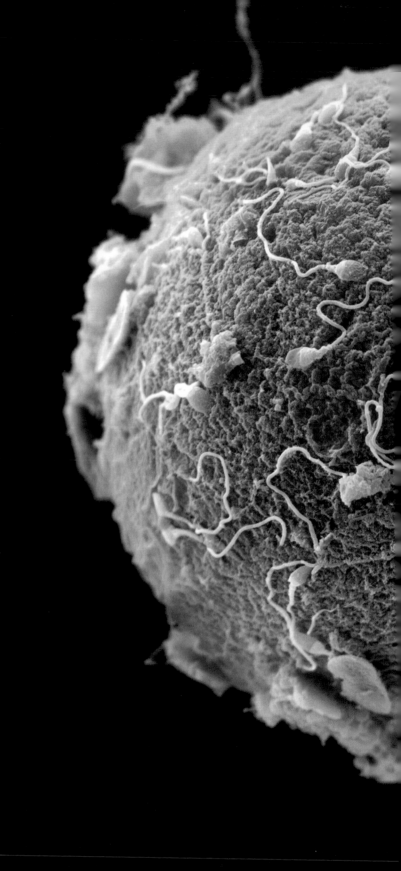

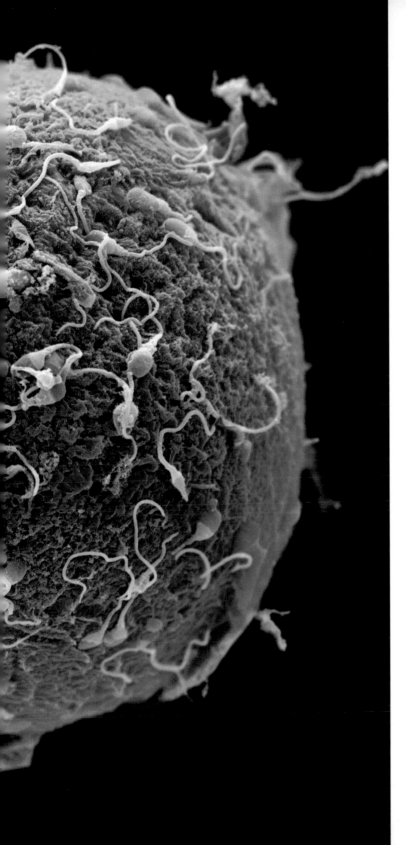

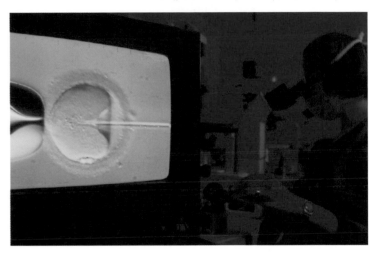

↑ **A male ladybug** uses an organ called an ovipositor to place sperm into the female's reproductive tract. Insect eggs have shells, and a sperm must pass through a tiny pore in an egg's shell in order to fertilize it.

↑ **Reproductive technologies** include in vitro fertilization (IVF), in which a sperm cell is injected into an egg. A monitor provides an enlarged view of what the physician sees in the microscope. The egg (in the center of the monitor) is held in place by suction applied through an instrument called a pipette (left) and is pierced by a hollow microneedle (right) containing sperm.

THE IMPORTANCE OF NUMBERS

The body is geared to function with genetic instructions provided by exactly 46 chromosomes, 23 from each parent. When an embryo receives an abnormal number of chromosomes, it is virtually impossible for it to develop normally. An example is Down syndrome, in which either an egg or sperm carries three copies of chromosome 21 instead of the usual two. After the egg and sperm fuse, the extra chromosome is passed on to each of the cells of the new individual. Several disorders result from having too many sex chromosomes (X or Y). In other cases, complete extra sets of chromosomes may be present in the embryo. Nearly all these pregnancies end in miscarriage, while affected newborn babies survive only briefly. When a fertilized egg chances to lack one or more chromosomes, it too will die, often before it reaches the mother's uterus.

Pregnancy

From a biological perspective pregnancy begins about seven days after fertilization, when a ball of cells, or blastocyst—an early embryo—begins to embed in the uterine lining, the endometrium. The implanted embryo begins to release a hormone called human chorionic gonadotropin (HCG). This stimulates the corpus luteum to survive and continue production of estrogen and progesterone, which prevent the endometrium from being sloughed away. HCG in a woman's blood or urine is a sure sign of impending motherhood.

Over the next few weeks some of the implanted cells form a series of membranes around the embryo. One of them, the yolk sac, gives rise to the embryo's first blood cells, its rudimentary digestive tract and cells that will one day be the forerunners of eggs or sperm. A membrane called the amnion forms the fluid-filled amniotic sac around the embryo and a third membrane, the allantois, generates blood vessels that will be enclosed in the umbilical cord. The fourth membrane, called the chorion, surrounds and protects the embryo and these other structures, and helps form the placenta.

→ **During pregnancy** a woman's uterus expands to fill most of the abdominal cavity and her breasts also enlarge in preparation for suckling an infant.

↓ **Implantation** has three main stages. First, cells on the blastocyst's surface begin to burrow into the endometrium. In phase 2, the endometrium closes over the blastocyst, and the chorion, the embryo's part of the placenta, begins to form. In phase 3, the embryonic disk divides into three basic tissue layers—the ectoderm, endoderm and mesoderm—which will give rise to all body tissues and organs.

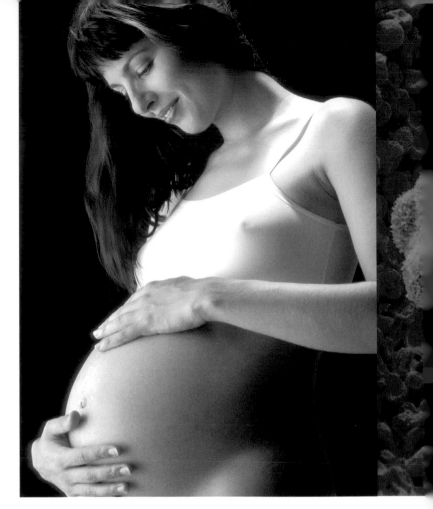

IMPLANTATION OF THE EMBRYO

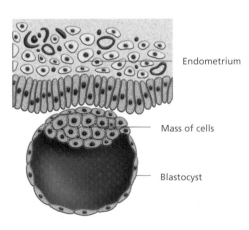

Endometrium

Mass of cells

Blastocyst

Phase 1

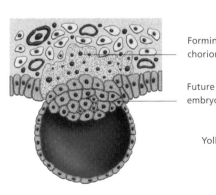

Forming chorion

Future embryo

Phase 2

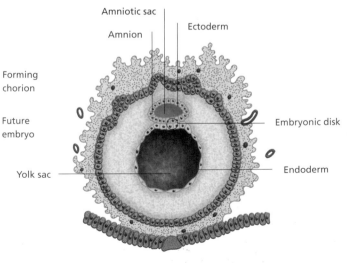

Amniotic sac

Amnion

Ectoderm

Yolk sac

Embryonic disk

Endoderm

Phase 3

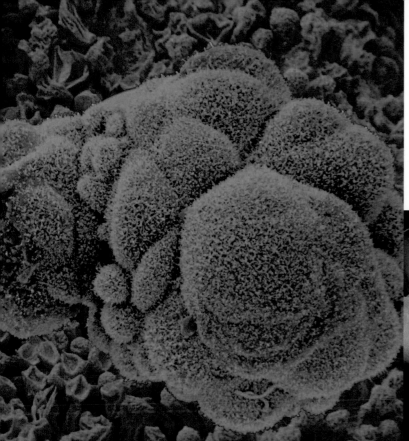

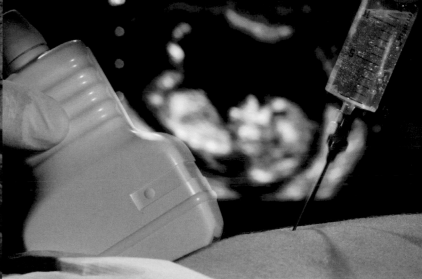

← **An embryo** (orange) grows protected in the womb. After the first nine weeks of gestation, when organ systems begin to form, the embryo is called a fetus.

↓ **In amniocentesis,** fluid containing cells shed by the embryo is withdrawn from the amniotic sac, guided by ultrasound. Later analysis reveals the sex of the embryo, as well as any disorders.

THE NURTURING PLACENTA

The placenta consists of extensions of the chorion called chorionic villi, which are surrounded by blood from the mother's bloodstream. The villi take up oxygen and nutrients for the embryo and empty out wastes for disposal by the mother's body. This provides for the embryo's needs while preventing its blood from mixing with that of its mother. As the tissues of the chorion and embryo are identical, a test called chorionic villus sampling (CVS) can identify signs of genetic disorders. With time the placenta also takes over producing estrogen and progesterone, which maintain the pregnancy.

→ **In a fully developed placenta,** blood vessels from the embryo extend into the chorionic villi, which are exposed to the mother's blood. Although this arrangement bars many substances in the mother's blood from reaching her unborn child, a number of harmful substances, including caffeine, drug residues, alcohol, toxins in tobacco smoke—as well as viruses such as HIV—can all cross the placenta.

FULL-TERM PLACENTA

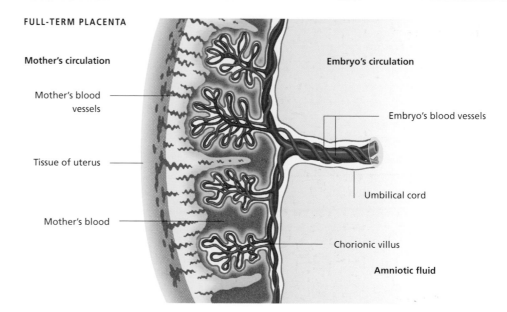

Mother's circulation

Embryo's circulation

Mother's blood vessels

Embryo's blood vessels

Tissue of uterus

Umbilical cord

Mother's blood

Chorionic villus

Amniotic fluid

Development begins

The development of a new individual is a gene-guided sequence in which a fertilized egg divides, early tissues form the embryo, and then tissues becoming increasingly specialized and arranged into the body's organs and organ systems. A month after fertilization, even though the embryo is only about the size of a match head, the basic plan is in place. Although it does not yet look human, it has a head, eyes, buds where limbs will form and the beginnings of a heart and nervous system. To reach this stage, the embryo has grown rapidly and its cells have been rearranged to produce the body's head-to-toe axis and bilateral symmetry—having right and left sides that are rough mirror images of each other. Three basic tissue layers—ectoderm, mesoderm and endoderm—have also formed. As organs begin to form, endoderm will give rise to much of the digestive and respiratory tracts and the cardiovascular system. Ectoderm will produce parts such as the nervous system, skin and many glands, while mesoderm is the source of tissues in the heart, liver, muscles, bone and blood vessels.

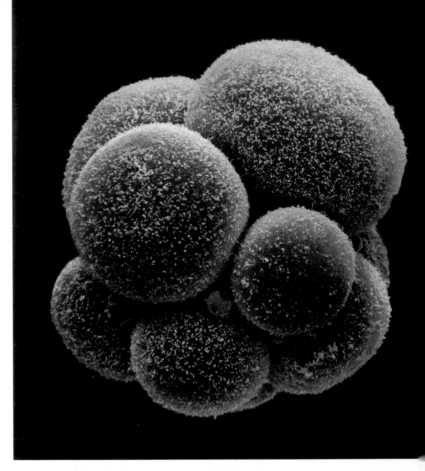

→ **Not yet an embryo,** this ball of 10 cells is only about 4 days old. By the time the ball grows to 32 cells, it will look like a tiny berry.

↓ **Cloning produces identical copies** of a parent. Researchers have cloned animals such as sheep, mice and cows , but most cloned embryos die or have a variety of disorders if they survive. Human cloning is highly controversial, and some governments have taken steps to outlaw it. Even so, several enterprises have announced human cloning programs.

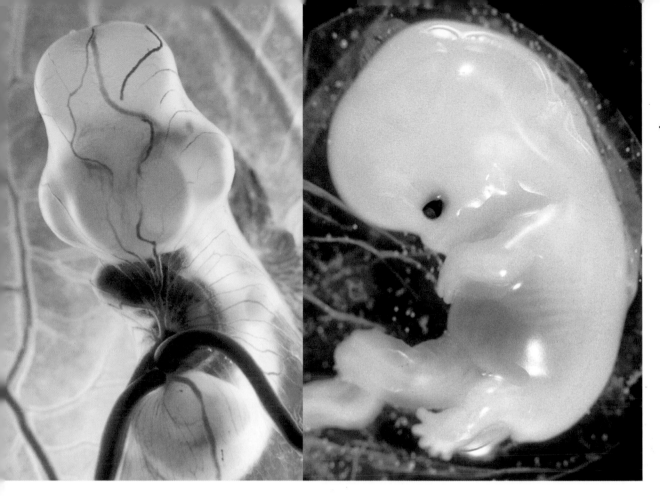

A **4-day-old chick embryo** (left) and 4-week-old human embryo (right) look remarkably similar. For instance, both have a tail and a series of arches and pouches in the neck region. The chick will hatch with a small tail, but the human newborn will have only a small, internal tailbone or coccyx. In humans different parts of the embryonic neck pouches develop into the middle ear, the Eustachian tubes, the thymus, thyroid gland and parathyroids.

FINGERS, TOES AND TAILS

At the start of embryonic development, the overall outlines of a human body are established first. Specific morphogenesis—taking shape—occurs much later, with the development of smaller structures such as ears, fingers, toes and eyelids. Apoptosis (pronounced a-poe-TOE-sis) is a key tool in this body shaping, involving the genetically programmed death of cells in strategic places. For example, the hands of a 4-week-old embryo look initially like paddles, with connective tissue "webs" between the digits. Separate fingers emerge only weeks later after cells between the embryo's fingers die. Our toes, upper and lower eyelids, and some other parts are completed in the same fashion.

In its early development a human embryo closely resembles the embryos of other vertebrates such as mice and chickens. For the first several weeks of life, the embryos look very much the same, complete with a bulbous head, a curving line of segments that will become the backbone, and a tail. Only later, when different genes begin to act, does the body acquire the distinctive features of a human being.

Identical twins result when a fertilized egg splits in two and develops into two genetically alike babies. Fraternal twins, by contrast, develop from different fertilized eggs that implant in the uterus more or less at the same time. They are no more alike than any other two children their parents may produce.

The dynamic early embryo

Pregnancy is a 37-week developmental sprint. After the embryo's first few weeks in the womb, its parts rapidly begin to be defined. Stubby little limbs gradually lengthen, genitals take on the features of a male or female, dark spots of pigment begin to color the eyes and the face features a broad nose and the hint of a brow ridge. The prominent tail that was present at 4 weeks begins to shrink and soon disappears. Embryonic blood vessels invade the forming umbilical cord, and that lifeline to the placenta begins functioning. The most dramatic growth is reserved for the embryo's head, which around week 5 begins to enlarge rapidly as the first brain tissues are arranged. At the end of week 8, the embryo is slightly longer than 1 inch (2.5 cm) and preliminary versions of all its organ systems are in place. With these changes, it now has the biological status of fetus and looks distinctly human.

INFECTIONS AND OTHER THREATS

An embryo is extremely vulnerable to certain infections and to various types of drugs that can cross the placenta. Although the mother's IgG antibodies help protect the embryo from bacteria, pathogens such as HIV and the rubella (German measles) virus can reach the embryo's bloodstream. Some sedatives, and anti-acne drugs such as retinoic acid, can cause misshapen skull, facial and limb bones. Antibiotics that cross the placenta can do harm. For example, streptomycin, which is widely prescribed in many parts of the world, can cause nervous system defects. Spina bifida, the "divided spine" defect that leaves part of an infant's spinal cord dangerously exposed, is caused by a lack of folic acid during embryonic development. It is easily prevented if the mother's diet includes foods that are fortified with folic acid.

→ **This 8-week embryo** is beginning to look like a human baby, but is still too small for the mother to feel its movements. It now receives all its nutrition by way of the umbilical cord and placenta.

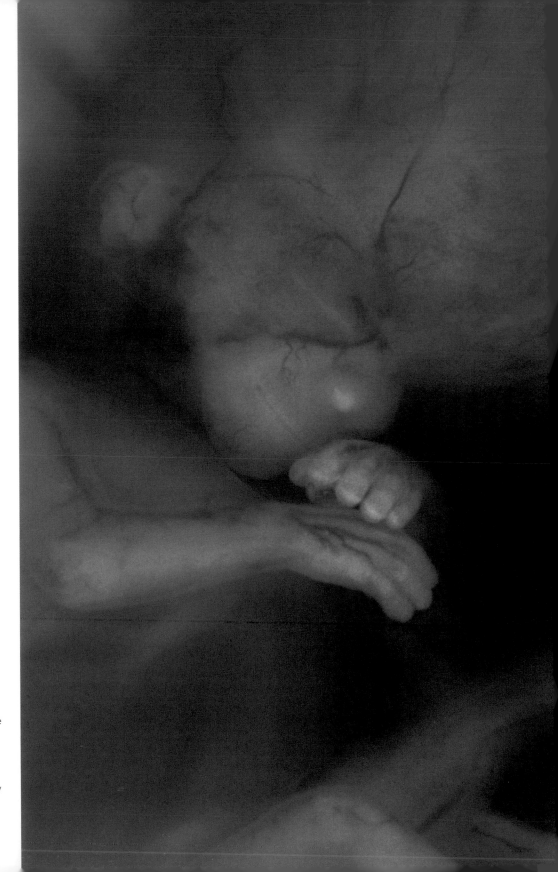

DEVELOPMENT OF THE EMBRYO

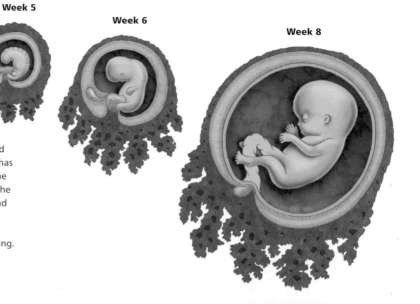

Week 2

Week 3

Week 4

Week 5

Week 6

Week 8

↑ **During its first 8 weeks,** an embryo changes dramatically. At 14 days it consists of a two-layered disk attached to a large yolk sac. Ten days later it has elongated and a neural tube, the forerunner of the brain and spinal cord, is taking shape. By week 4 the embryo is about one-quarter inch long (0.6 cm) and its heart and limbs are forming. By week 6 the embryo has doubled in length, rudimentary eyes and ears are visible and the umbilical cord is forming. At 8 weeks the embryo looks human, with clearly defined limbs, fingers and toes, and all its major organs are developing.

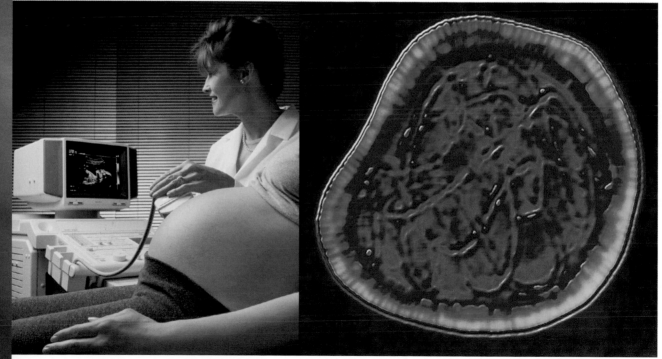

↑ **A doctor uses ultrasound** to check the health of a developing baby. The resulting images can give a real-time snapshot of how well the fetal heart and other organs are functioning.

↑ **A rubella virus** can cause birth defects if it infects an early embryo. In developed countries women who plan to become pregnant may be vaccinated against rubella as a precaution. In this image the outer portion of the viral particle appears yellow and its inner core of genetic material appears red.

Prenatal life

From the third to the ninth month of pregnancy the fetus grows to its birth size and its organ systems mature. At 12 weeks it is nearly 5 inches (12 cm) long and is covered by soft, fine hair. A white, cheesy substance, the vernix caseosa, protects its skin, and inside its body the maturation of its organ systems is proceeding full-bore. At 16 weeks connections between its nervous system and muscles are established, and its mother often can begin to feel movements of its arms and legs. By the end of the seventh month the fetus can suck its thumb and swallow amniotic fluid and, before long, its eyes will open. It measures nearly 12 inches (about 30 cm) in length. A fetus that is delivered at this time may survive, but usually not without intensive care. Even though its organ systems are fairly well developed, it still may be unable to breathe normally, in part because its lungs may not yet produce enough surfactant, a chemical that helps the lungs expand. On the other hand, a full-term infant, which may be almost twice as long and weigh nearly twice as much, is well prepared for independent life.

A (USUALLY) TEMPORARY HOLE IN THE HEART

A pregnant woman not only nourishes her unborn child but breathes for it as well. By way of the placenta and umbilical cord, oxygen from her blood crosses into the fetal bloodstream and waste carbon dioxide from the fetal metabolism is removed. As a result, while a fetus is in the womb, little of its blood circulates to its tiny lungs. Instead, much of the blood that enters the right side of a fetus's heart moves directly to the left side, passing through a hole in the inner heart wall. When a newborn starts breathing on its own at birth, a valve-like flap of tissue normally closes the hole, which then gradually seals permanently. When this process goes awry, an infant's tissues do not receive enough oxygen and the opening must be closed surgically.

↓ **A premature baby's breathing** is aided by a ventilator, and electrodes attached to the infant's chest connect to a device that monitors the heartbeat. Five to 10 percent of babies are born before 37 weeks of gestation.

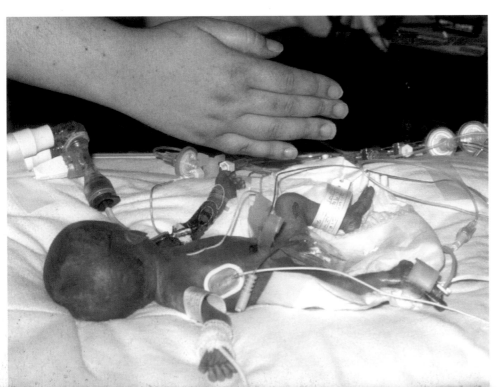

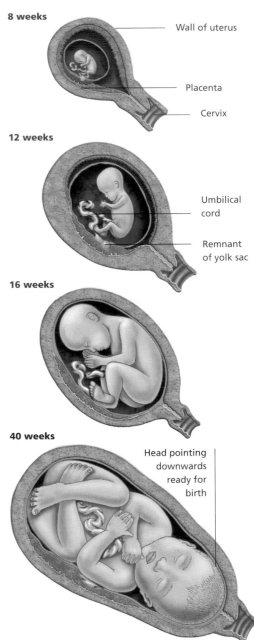

8 weeks

Wall of uterus

Placenta

Cervix

12 weeks

Umbilical cord

Remnant of yolk sac

16 weeks

40 weeks

Head pointing downwards ready for birth

↑ **The size and proportions of a fetus** change dramatically during the second and third trimesters of pregnancy. At 16 weeks it weighs 10 times what it did 8 weeks before, and that pace of growth continues until birth.

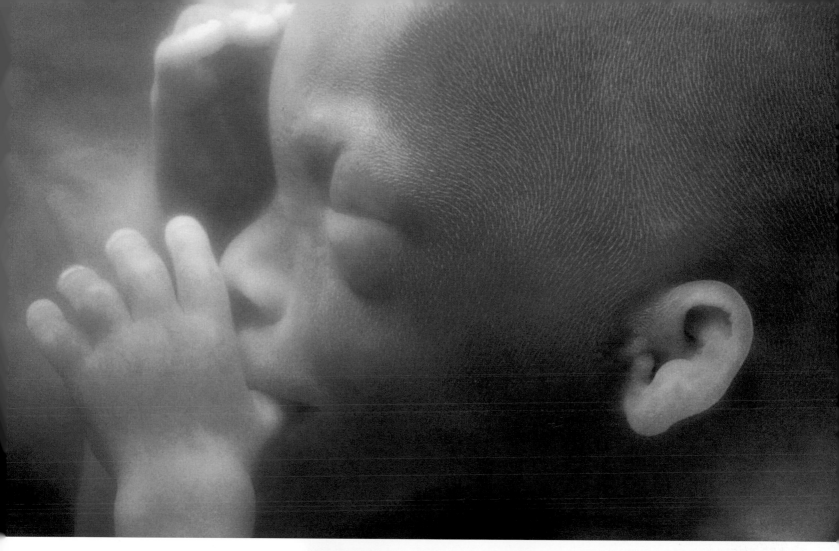

↑ **This 19-week fetus** is a boy. The fine hair (lanugo) on his skin is readily visible, although it will be another month before he has eyelashes and brows. He can already suck, a skill that will be essential to his ability to feed after birth.

→ **Much of the skeleton** has formed by the end of the fourth month of gestation, including limb bones, the rib cage and vertebrae in the spine. As a fetus develops, bone forms over a cartilage model; the hard bone tissue appears yellow in this image. The gaps at the end of the long bones are joint cavities which are not ossfied.

→→ **Smoking endangers** the health of a fetus. Babies born to women who smoke daily are usually underweight. Studies show that women who smoke while pregnant also have a greater risk of miscarriage or delivering their baby prematurely.

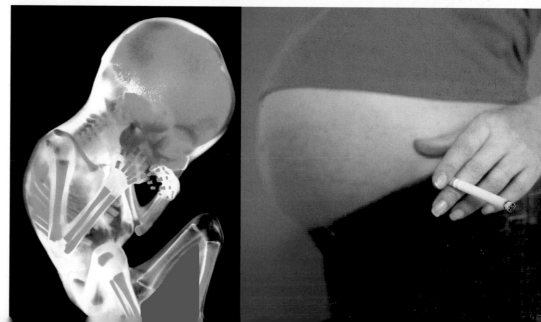

Birth

Approximately 39 weeks after an egg is fertilized, birth marks an infant's entry into the world beyond its mother's womb. Labor, or the birth process, begins when muscles in the uterus start to contract—slightly at first, then with increasing strength. We still do not fully understand what triggers the birth process, but it seems to be instigated by the fetus itself. At almost any time toward the end of the ninth month of pregnancy, the fetal endocrine system abruptly starts to release hormones that act on the placenta. The placenta responds by boosting its production of estrogen, setting in motion a rising chemical barrage that includes oxytocin and prostaglandins. These substances cause the muscles of the uterus to begin contracting. Over the ensuing hours the amniotic sac will rupture, the woman's cervix will dilate, and eventually the infant will be expelled from its mother's body.

CUTTING THE CORD

When the umbilical cord tethering a newborn to the placenta is cut, the baby's body must finally begin to function entirely on its own. Perhaps most important, its small lungs, which have been collapsed ever since they developed, must suddenly inflate so that breathing can begin. Stressed by being squeezed by labor contractions and then pushed out into a cold, open and often bright and noisy new environment, for the next few hours the baby will breathe fitfully, its heart will beat more rapidly than normal, and its body temperature will fall. Over the next few days these physiological functions stabilize, and by the time it is 2 weeks old it will be well adapted to the new world it has entered.

→ **Birth is an adjustment** for a baby. Initially alert, for its first few hours in the outside world a newborn will alternate between alertness and sleep until the usual 3 to 4 hour pattern of waking (usually to feed) and sleeping is established.

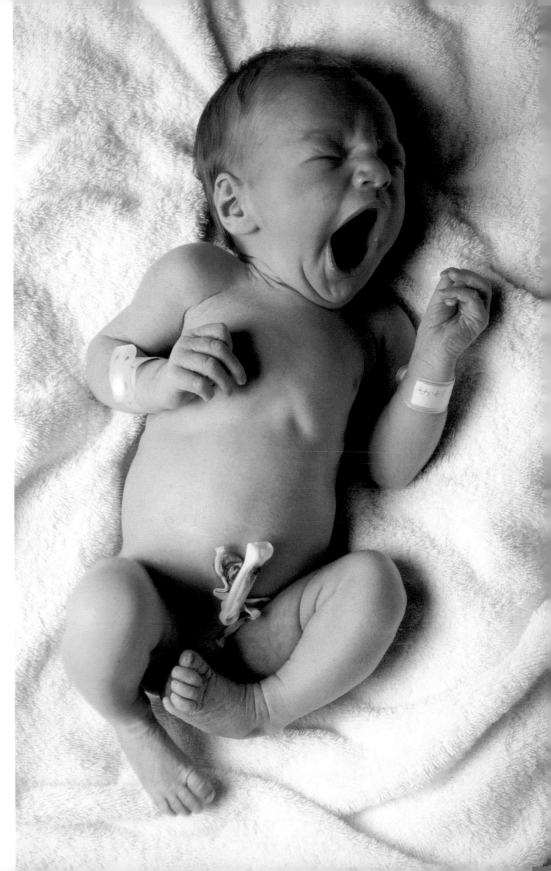

Labor is **physiologically stressful** for both mother and baby. A monitoring device attached to the mother's abdomen tracks the infant's heart rate and other vital signs, and can alert clinical staff to developing problems.

A woman's breasts begin to produce milk about 2 to 3 days after she gives birth. Before that, the breasts make the high-protein fluid, colostrum. This also contains maternal antibodies that may help the newborn fight bacteria that enter its digestive tract.

→ **At the onset of labor** contractions begin and force the infant's head (usually) against the mother's cervix, which gradually dilates. At first 15 to 30 minutes apart, the contractions may continue for 6 to 12 hours. As stage 2 begins, contractions are stronger and arrive every 2 to 3 minutes, until the baby is forced out. In the final phase, contractions expel the placenta, or afterbirth.

STAGES OF LABOR

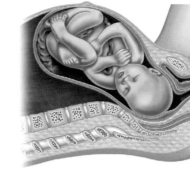

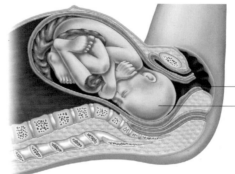

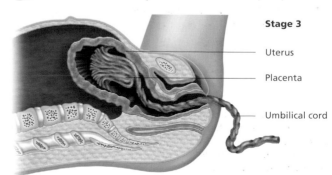

Early stage 1

Cervix slightly dilated

Late stage 1

Cervix almost fully dilated

Baby's head rotated toward mother's back

Stage 2

Stage 3

Uterus

Placenta

Umbilical cord

Transitions from infancy to adulthood

From the moment that an embryo forms, the body will grow for at least two decades. During this long period of marked physical change, the body's proportions will shift dramatically and various parts will be subtly or obviously remodeled. The most profound physical changes take place in the embryo and fetus as the body's framework and internal organs are formed.

In infancy—the period from birth to about 15 months—a child's shape changes as the body's trunk grows faster than the arms and legs. During this time also, organs and organ systems, especially the central nervous system, are growing and maturing at a rapid pace.

Childhood spans the period from the end of infancy to adolescence. When adolescence arrives, it brings puberty and the most striking physical changes since birth. Puberty takes place between the ages of about 10 and 13 in girls, and between about 12 and 15 in boys. While sex hormones trigger the maturation of the reproductive organs and secondary sexual characteristics develop, a growth spurt also occurs and long bones rapidly elongate to their final, adult size. Females develop "curves" as fat is deposited in their hips and breasts, and males may develop a more robust chest and shoulders as they become more muscular. By the time we reach our early twenties, the person we see in the mirror has overall body proportions that will be with us until old age and death.

← **This baby's eyes** are already fully grown. The eyes are among the few organs that reach their maximum size very early in life.

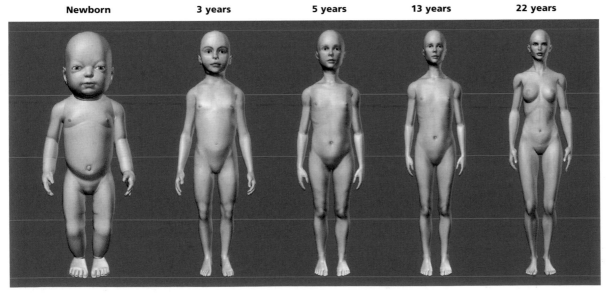

| Newborn | 3 years | 5 years | 13 years | 22 years |

← **Throughout early life** the relative proportions of body parts change. Initially the skull and torso, which contain vital internal organs, are large compared to the limbs. In fact, the skull and brain have grown to near-adult size by age 10. From infancy to adolescence the limbs steadily lengthen while growth of other body parts slows, until eventually the body attains its adult form.

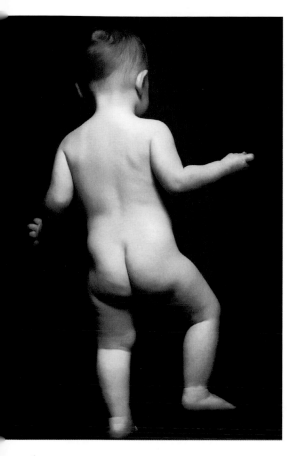

↑ **Walking** is a developmental milestone that most babies achieve by about 15 months.

→ **From late adolescence** until our early thirties the body is at its physical peak, when our muscles, heart and other organs function most efficiently.

GROWING BABIES
After about age 3, children gradually grow larger until adolescence, but for infants it's a different story. Studies show that parents are not hallucinating when they say a baby seems to have "grown" overnight. When babies were monitored over a period of months, researchers found that their height may be stable for several weeks, then suddenly increase as much as three-quarters of an inch (1.9 cm) in the space of 24 hours. Overall, an infant's brain is its fastest growing organ. A newborn's brain weighs about 12 ounces (340 g) —already fully 25 percent of its adult size. In six months the brain doubles in weight, and keeps on growing, although more slowly, until the teen years.

Aging

A gradual, body-wide decline in physical functioning is the natural, inevitable process called aging. In our early thirties, body cells, tissues and organs begin to undergo structural changes that are evidence of accelerating aging—skin sags and wrinkles as the connective tissue beneath it is altered, muscles and joints ache after an active day, and more fat accumulates in places that once were mostly muscle. In our forties and fifties we may notice a decline in our endurance, suffer wear-and-tear injuries such as arthritic joints, and be aware of changes in the functioning of our digestive system, as well as changes in reproductive and sexual functions as the levels of sex hormones fall. It also takes noticeably longer for injuries to heal.

By age 60, a woman has passed through menopause and it takes longer for a man to achieve an erection. Blood pressure rises, the immune system may become less efficient at combating threats, and short-term memory may not be as sharp as it once was. But while all of us will eventually experience many or most of these signs of advancing age to some degree, the body is amazingly responsive to a healthy diet, ample exercise and positive mental stimulation. Studies show that the more active we are and the more connections we keep with family and friends, the more successfully we can maintain our health and vitality well into old age.

→ **This tattooed Thai man** was more than 100 years old when he was photographed in 1994. The oldest documented age for a human is 122 years, although some centenarians have claimed to be much older. People who live to advanced age often have been physically active for much of their lives and typically have a slender build. Researchers have not identified any infallible "recipe" for longevity, and genetic factors almost certainly play a role.

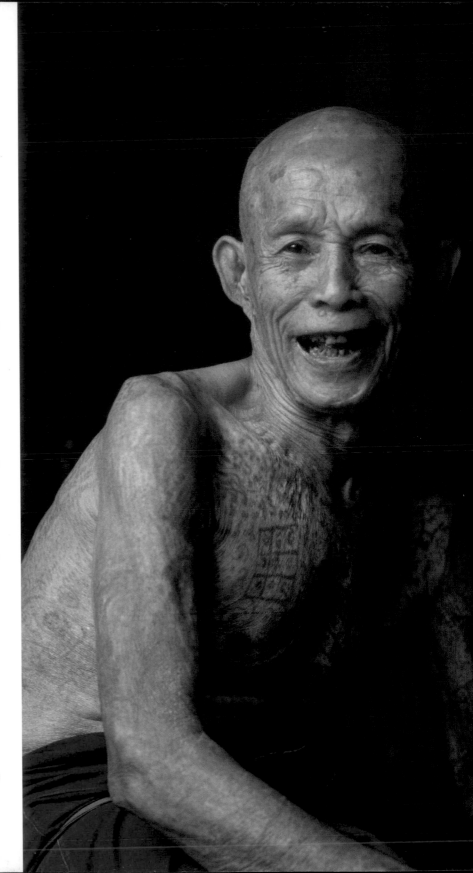

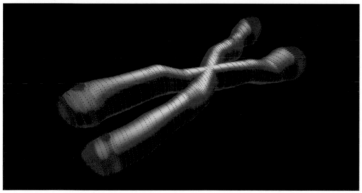

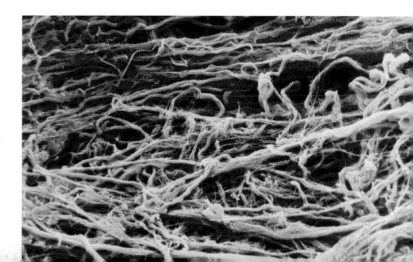

↑ **The practice of tai chi** is one option for gentle exercise that helps keeps muscles and other organ systems in shape. Tai chi involves slow, deliberate movements that help tone and strengthen muscles and also require considerable mental concentration—a "workout" for the brain.

← **Telomeres are sections of DNA** at the ends of a cell's chromosomes, colored red in this illustration. One hypothesis proposes that every time a cell divides, bits of telomeres are chipped away, causing genetic damage that eventually kills the cell.

↓ **Collagen is a basic component** of connective tissue in the skin and elsewhere. As we age, its fibers, shown here in golden yellow, cross-link and become less flexible. This is one reason why aging skin wrinkles and blood vessels become stiffer over time.

WHY DO WE AGE?

There are several theories about the causes of aging, and all of them involve our genes. One idea is that the body has a gene-programmed maximum life span of about 120 years. Research in support of this view has shown that most kinds of human cells can divide only about 90 times. It's an established fact that as we age, the cell divisions that replace lost skin slow considerably. On the other hand, many of our cells never divide after the body has stopped growing, and so as time goes by they may become less efficient at performing their specialized tasks. Some scientists believe that age-related changes are due to unrepaired damage to our DNA. In this "cumulative assaults" hypothesis, factors such as free radicals—rogue chemicals that are produced in normal cell operations—break down our cells' DNA over time. Built-in mechanisms sometimes can repair damaged DNA, but they are not 100 percent effective. As damage mounts through the years, our cells and tissues may gradually lose the ability to carry out their assigned functions.

The genetic basis of life

From the time we are embryos until we die, our cells, tissues and organs carry out a vast number of intricately organized and coordinated activities, all orchestrated by the chemical called DNA, or deoxyribonucleic acid. As the molecule of heredity, DNA carries the genetic instructions necessary to build and operate the body through each phase of life. These instructions are written in a chemical language that tells a cell how to assemble all the different proteins it uses, whether as raw materials in building cell parts, as enzymes and signals such as hormones, or in other ways. Each molecule of DNA consists of two strands of chemical units called nucleotides, which are derived from the nucleic acids in food. Four kinds of nucleotides are used to build DNA, which is a long, slender, twisted molecule. The nucleotides form two parallel strands that link up in a ladder-like structure. The nucleotides on one side of the ladder pair up with those on the other side, following strict chemical rules about which ones can interact. The properly matched pairs form the ladder's "rungs." When all these parts are in place, the ladder coils like a bedspring, forming a double helix. Humans have about 30,000 genes, and each one is simply a segment of a DNA molecule. Different genes carry different hereditary instructions because the nucleotides in each segment are lined up in a distinctive order, known as the "human gene sequence."

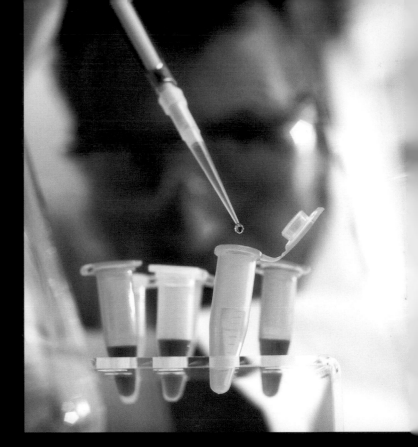

↑ **Scientists use DNA samples** such as those shown here in medical studies, to establish identity, and for basic research. Many modern insights about inherited diseases come from DNA analysis. DNA is also used to shed light on a person's parentage, and to positively identify or exonerate crime suspects.

← **Genes can be isolated** by gel electrophoresis. In this technique a solution containing fragments of DNA is added to a gel made from agar. When an electric current is passed through the gel, the fragments separate by size. After the gel is stained and exposed to ultraviolet light, the DNA fragments are seen as violet bands. Each fragment can then be analyzed to determine whether it contains a gene of interest.

THE GENETIC CODE

The four nucleotides in DNA—A, T, G and C—make up the genetic code. Remarkably, these nucleotides can provide every bit of information needed to construct and operate the body. Different combinations of them provide instructions for making the 20 different amino acids the body draws upon to build various proteins. A cell's protein-making machinery reads the letters in groups of three, and each "triplet" corresponds to an amino acid. When a protein is forming, the amino acids in it are lined up like beads on a string. Genetic instructions dictate the number, type, and order of these beads, and billions of combinations are possible. In practice, the body uses its genetic code to make several hundred thousand different proteins. Studies have shown that the genetic code is almost 100 percent uniform across the whole spectrum of life—humans and other animals, plants, algae, fungi and even bacteria use the same basic genetic directions to build proteins.

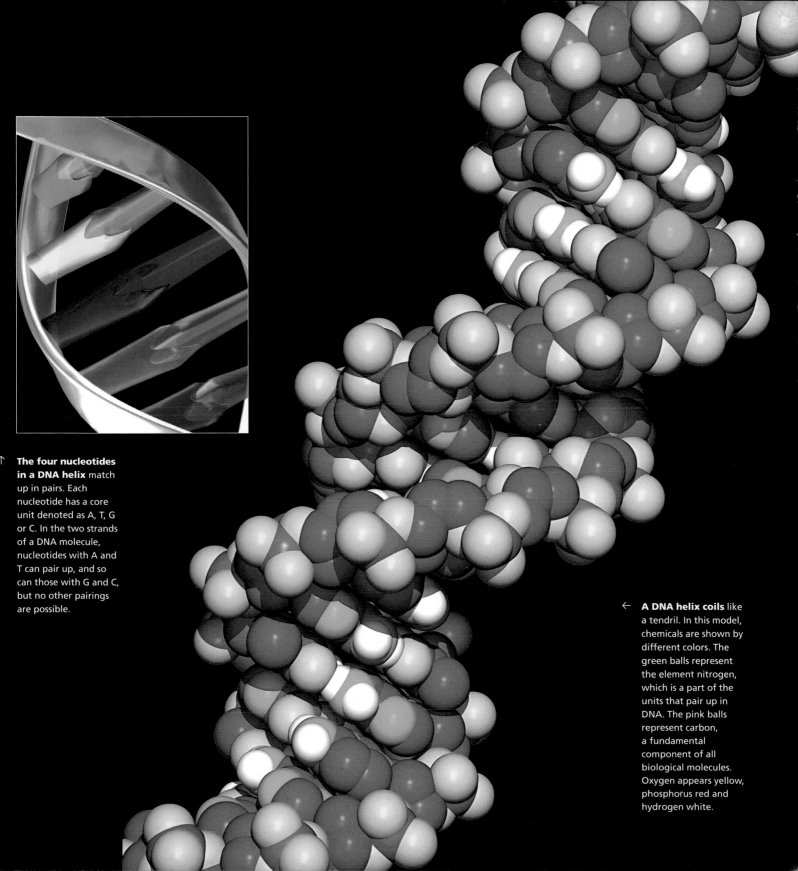

The four nucleotides in a DNA helix match up in pairs. Each nucleotide has a core unit denoted as A, T, G or C. In the two strands of a DNA molecule, nucleotides with A and T can pair up, and so can those with G and C, but no other pairings are possible.

A DNA helix coils like a tendril. In this model, chemicals are shown by different colors. The green balls represent the element nitrogen, which is a part of the units that pair up in DNA. The pink balls represent carbon, a fundamental component of all biological molecules. Oxygen appears yellow, phosphorus red and hydrogen white.

Dividing cells, multiplying chromosomes

The 30,000+ known human genes are shared among our chromosomes. A chromosome is a single, long molecule of DNA, with genes arranged on it like cars on a freight train. In all, about 6.5 feet (2 m) of DNA is packed into the nucleus of each body cell.

We inherit 23 chromosomes from each parent; these 23 pairs make a total of 46 chromosomes in all cells except sperm and eggs. The two sex chromosomes, XX or XY, jointly determine a person's sex. The rest, called autosomes, carry genes for traits unrelated to sex. For a body part to grow or for a wound to be repaired, the cells involved must divide so as to deliver a full set of chromosomes into newly formed cells. Likewise, the cell division process that produces sperm and eggs must parcel out chromosomes so that each sex cell receives only half a full set. This "reduction division" sets the stage for fertilization, when the two halves come together in a single cell. When genes are being "read out," the chromosomes of a cell look like long, loose threads. When a cell divides, however, each chromosome is duplicated and then condenses into a tightly coiled mass, somewhat like a miniature pipe-cleaner. Photographs of human chromosomes nearly always show them in this condition and ready to be parceled out into a new generation of cells.

→ **These chromosomes** are from a human female. There are 23 pairs, with one chromosome of each pair being from each parent. Some chromosomes are naturally larger than others, like the paired X chromosomes at the lower right. This type of display is called a karyotype, and a geneticist can use it to identify certain kinds of genetic abnormalities.

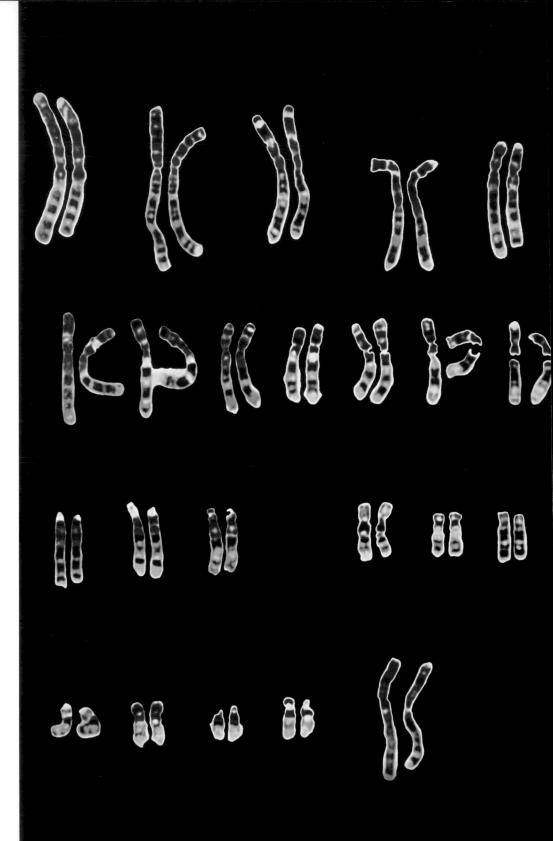

HOW CELLS DIVIDE

Depending on their role in the body, our cells divide in two ways. In the ovaries and testes, germ cells—those that produce eggs or sperm—divide by a process called *meiosis*. It yields four new cells, each of which has 23 chromosomes—half the number in other types of cells. All other body cells divide by *mitosis*, which yields only two cells, each with a full set of 46 chromosomes. Gene-based abnormalities that are transmitted from one generation to the next are rooted in DNA that is distributed into sperm or eggs during meiosis. Anything that alters the DNA in a germ cell can lead to a harmful mutation and inheritable genetic defects. Since our germ cells are vulnerable to random changes in genes, as well as to environmental threats such as radiation and chemical toxins, genetic disorders sometimes occur in individuals with no family history of the condition.

→ **These cells are dividing** by mitosis. Two sets of chromosomes, visible as dark clusters, are lined up in the center of each cell. They will be separated into two equal groups that are moved to opposite ends of the cell. When the cell pinches in two, each new cell will have a full chromosome set.

↘ **The children in this family** were conceived when sex cells created by meiosis in the parents' reproductive organs united into a fertilized egg. The body grows as its cells divide by mitosis. New tissues that heal wounds come about in the same way.

↓ **Mitosis and meiosis** both start by duplicating the parent cell's chromosomes. The duplicate sets are moved to different parts of the cell before it pinches in two. In cells that will produce sperm or eggs there is a second round of division. This time chromosomes are not duplicated, and each new cell receives only a half set of chromosomes.

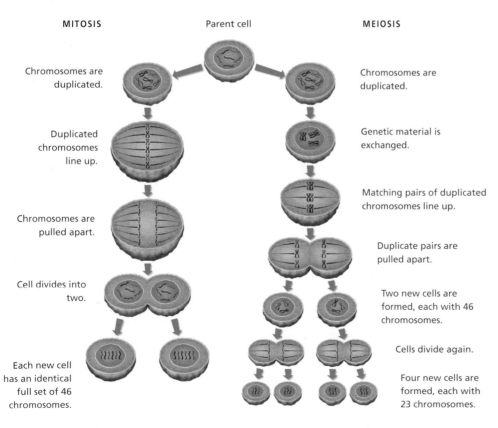

MITOSIS	Parent cell	MEIOSIS
Chromosomes are duplicated.		Chromosomes are duplicated.
Duplicated chromosomes line up.		Genetic material is exchanged.
		Matching pairs of duplicated chromosomes line up.
Chromosomes are pulled apart.		Duplicate pairs are pulled apart.
Cell divides into two.		Two new cells are formed, each with 46 chromosomes.
		Cells divide again.
Each new cell has an identical full set of 46 chromosomes.		Four new cells are formed, each with 23 chromosomes.

Genes and inheritance

Our bodies are organized and operated according to chemical information encoded in the genes we inherit from our parents. Each of our traits comes about as a result of this transformation of DNA instructions into parts and processes, and to understand it we must look deep inside a cell. The transformation unfolds in two steps, beginning in the cell nucleus when a segment of DNA in a chromosome is "read out" by enzymes. Like medieval scribes, the enzymes rewrite the information in that stretch of DNA into a slightly different chemical "language"—RNA (ribonucleic acid). This step is known as transcription. The RNA carries exactly the same instructions that were coded in DNA but, unlike DNA, RNA can leave the nucleus and move into the cell's cytoplasm where the cell's machinery for making proteins is located. There, in a second step called translation, the RNA is "read" and a protein is assembled according to its instructions, which were first extracted from DNA. The new protein may be a structural building block needed by the cell or a component of some biological process there. While every cell has a full set of chromosomes, only a small part of the total genetic code will be translated in it—the part of the code that relates to that cell's function in the body.

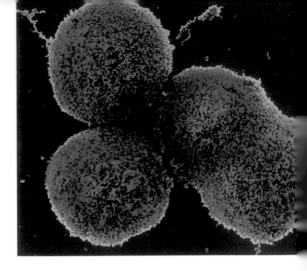

Liver cells decode genes that are linked to processing nutrients and breaking down certain wastes. They form bile used in digestion and help synthesize certain kinds of proteins.

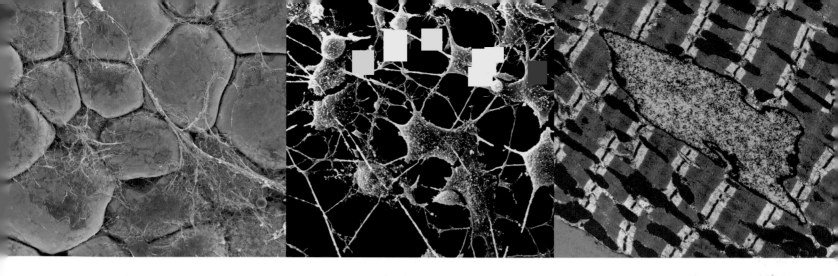

Fat cells use genes that allow them to take up excess sugar from the blood and store it as fat. When the fat stores are needed, other genes provide instructions for releasing fat into the blood.

In nerve cells, active genes include those that provide instructions for making neurotransmitters. A neuron's nucleus, where DNA is located, is in the plump body of the cell (pink).

A heart muscle cell taps genes that guide the development of contracting units called sarcomeres, whose ends are shown here as red lines along the yellow bands. The large nucleus appears gray.

VARIATIONS ON A THEME

The "half set" of chromosomes in sperm and eggs provides one copy of each human gene. Thus, when we inherit one chromosome from each parent, we possess two copies of each gene. Both copies code for the same trait, but they may or may not be exactly alike. Most genes come in slightly different chemical versions, just as a pudding may be chocolate, vanilla or some other flavor, but in the end is still pudding. For example, the gene that carries information about the shape of the hairline has a "straight" version and a "widow's peak" version. With some traits, geneticists have discovered three, four or more possible versions of the gene involved, although a given person can carry only two versions, one on each chromosome in a pair. All these factors contribute to the variety of features we see in ourselves and those around us.

← **Everyone's genetic profile** is unique, except for identical twins. This "DNA fingerprint" can be compared to that of another person and reveal differences between them. Among other uses for genetic profiling, some physicians now try to match a patient's genetic make-up to the properties of therapeutic drugs.

← **Family resemblance** is not inevitable. We usually see a mix of traits in families, in part because parents may pass on different versions of the genes responsible for many features.

Single-gene and polygenic traits

Each physical or operational body trait—the length of your eyelashes, your ability to curl your tongue —is governed by at least one gene and quite often by more. In fact, most human characteristics, including intelligence, the color of a person's skin, hair and eyes, and the genetic components of disorders such as diabetes, are polygenic: they reflect the activity of multiple genes that interact in complex ways.

Some polygenic traits are also multifactorial, which means they are influenced not only by gene activity but by environmental factors as well. Skin color is a case in point. Like eye color, it is determined by the interactions of several genes that affect the amount of the pigment melanin present in skin. We see the effects of external factors when a person's skin is exposed to sunlight or wind, or develops certain disorders, such as the reddening associated with rosacea. Tallness is another polygenic, multifactorial trait. A person's adult height may be influenced not only by genes but by hormonal and nutritional factors.

GENE INTERACTIONS
The different versions of a gene may be dominant or recessive. A dominant one overrides the effect of a recessive one—someone who inherits the dominant freckling gene, for example, will have freckles. Genetic analysis shows, however, that some disease genes, including dominant ones, do not always have the predicted effect. For instance, a dominant gene causes muscles in the "pinkie" finger to attach incorrectly to the finger bones. In some people the result is a permanently bent little finger, but other affected individuals have completely normal fingers. Unfortunately, many disease genes exert their harmful effects in a much more predictable fashion.

← **Tallness** is a polygenic trait that shows continuous variation in a group. Although there are some very tall individuals and some very short ones, taken as a whole height differences in the human population vary in small increments, rather than forming starkly defined groups.

→ **Eye color** is determined by genes for melanin. The iris is the colored part of the eyeball. Blue and green eyes have very little melanin, so the iris readily reflects those wavelengths of light. Brown and hazel shades have somewhat more melanin, and black-colored eyes have the most.

↑ **Both polygenic and single-gene** traits are displayed in this girl's complexion. A person's overall skin color appears to be determined by at least four genes, which collectively influence how much melanin epidermal cells make and how it is distributed. Freckles, by contrast, are produced by the action of just one gene. This girl inherited a version of the gene that causes clumps of melanin to develop in the skin.

↑ **Pattern baldness** is among the traits that are influenced by an individual's sex. Pattern baldness is much more common in males, because expression of the gene responsible for it is influenced by the level of testosterone in a person's blood. Women also can develop pattern baldness, but usually later in life when the female body makes less estrogen and the relative level of testosterone rises.

↑ **A fingerprint** shows the pattern of ridges created by rows of skin cells. Several genes determine this pattern, which develops during early embryonic life. External factors in the womb also influence the pattern. The effects of these factors are so individualized that not even identical twins, who are genetically identical and developed side by side in their mother's uterus, have identical fingerprints.

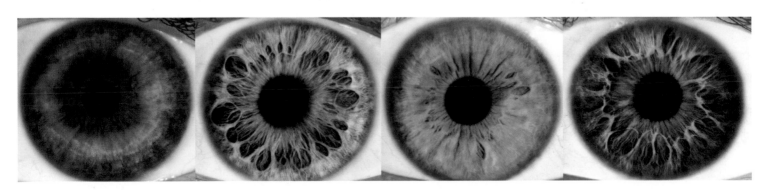

Dominant and recessive traits

We have all inherited two copies of each of our genes. The copies may represent only some of the possible versions of a gene, and they may be alike or different. If they are alike, both may be dominant or both may be recessive. If they are different, then one is dominant and generally overwhelms its recessive counterpart. Where genes are concerned, many people associate dominant with "good" and recessive with "harmful," possibly because recessive genes are the culprits in most human genetic disorders. Yet some diseases are caused by dominant genes, and many ordinary traits, such as having attached earlobes, are coded by recessive genes that do no harm.

A curious feature of human genetics is that some traits are the result of an interaction between two "co-dominant" versions of the same gene. Blood types are a prime example. They are determined by a single gene that comes in three contrasting versions—A, B and O. When a person inherits one A and one B form of the gene, both are "dominant" and so the person's blood type is designated AB. People with the other blood types inherit two copies of either the A, B or O form of the gene.

GENES AND BEHAVIOR
Some aspects of human behavior may be controlled or heavily influenced by genes. Researchers studying mood disorders such as schizophrenia, depression and bipolar disorder have discovered links between these conditions and patterns of inheritance within families. Tourette syndrome, which affects both speech and behavior, is also likely to affect members of the same extended family, and scientists are actively seeking a genetic component to autism and related developmental disorders.

→ **Cheek dimples** are encoded by a dominant gene. Depending on whether this woman carries one or two of the dominant gene forms, any child she bears will have a 50 to 100 percent chance of having dimples.

Father and son actors Kirk and Michael Douglas both have a chin dimple, a dominant trait. Relatively few people have the trait, however, because the dominant gene is rare in the general population. Instead, most humans inherit two copies of the recessive form of the gene, which calls for a smooth chin.

Actor Verne Troyer was born with achondroplasia, in which the arm and leg bones are abnormally short. It develops when one dominant and one recessive version of the gene involved is inherited.

Musician Woody Guthrie developed Huntington's disease (HD), which is caused by a single dominant gene that becomes active in adulthood. Parts of the brain degenerate, leading to jerky muscle movements, dementia and personality changes. Affected people generally die within 10 to 15 years.

If your earlobes are detached (top), you inherited a dominant version of the gene from at least one of your parents. With attached earlobes (below) a recessive gene is inherited from each parent.

Gene mutations

When the DNA sequence in a gene or chromosome is changed in some way, the result is a mutation—a genetic change that can be passed to future generations. A mutation can lead to a positive change, such as resistance to a particular disease, or to innocuous inherited differences, such as hair on the outer ear or a cowlick that grows counterclockwise. More often, though, a mutation upsets the biochemical workings of cells in a way that is harmful. A mutation can come about when DNA is damaged by radiation, a toxin, a viral infection or simply when a mistake occurs while a germ cell is replicating its DNA before the cell divides. Some genes seem to be particularly prone to mutations. For example, the recessive disease cystic fibrosis is caused by mutation in a gene called CFTR, which is on chromosome 7. More than 500 different mutations have been discovered in this gene. A child who inherits any two of them, one from each parent, will develop the disease.

→ **This person developed seven toes** as a result of inheriting the dominant gene for polydactyly, having more than five digits on a hand or foot.

↘ **Britain's Queen Victoria** carried a mutant, recessive gene for hemophilia A, in which blood does not clot. Of her nine children, three inherited the gene, including two daughters who passed it on to Russian and Spanish royalty.

↓ **Some of these red blood cells** have the characteristic sickle shape associated with sickle cell anemia and do not normally carry oxygen.

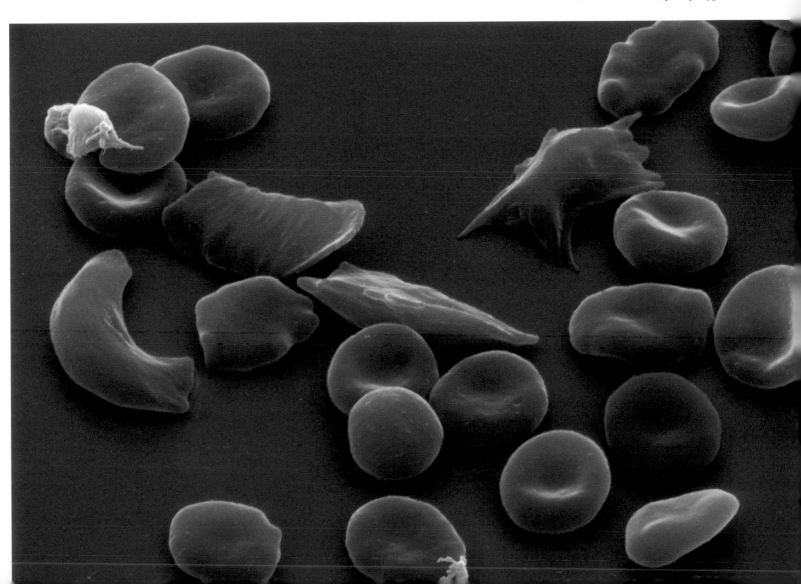

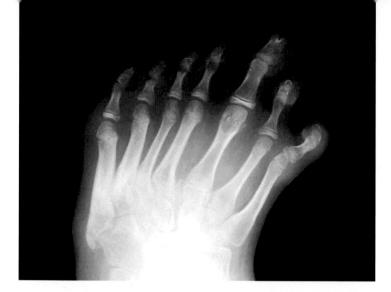

MUTATED GENES WITH MULTIPLE EFFECTS

Occasionally a mutated gene will have an impact on wide range of body parts. Marfan's syndrome is caused by a mutated dominant gene for a protein in connective tissue, and seriously weakens connective tissue in blood vessels, the skeleton and the eyes. People with the condition are usually very tall and lanky, with long, thin arms and fingers. About one-quarter of cases come from families not previously affected, an indication that the normal gene mutates easily. Some people with Marfan's die without warning when their weak aorta suddenly bursts. Another mutation with widespread effects is the one that causes sickle cell anemia. Affected people have two copies of a mutant gene for the oxygen carrier, hemoglobin. Their red blood cells tend to collapse into a sickle shape when oxygen levels in the blood fall. In addition to being extremely painful, sickle cell anemia affects the heart, kidneys, brain and many other body parts and functions.

Mapping the human genome

Scientists have determined the overall sequence of chemical units—the nucleotides A, T, G and C—that make up the human genome, the complete set of genes each person carries. The sequence consists of a string of roughly three billion nucleotides. Of these, however, it appears that only a small percentage actually carries useful genetic information, and we also know that different genes are built of different numbers of pairs. Thus a massive international effort has been launched to sort out the actual genes from "filler" DNA and then to discover each gene's role in the body. This effort has vast implications for our understanding of the variations in individuals' traits, of the genetic basis of inherited disorders, and how new mutations may cause disease. Ultimately human genome research may reveal the biological underpinnings of human life itself. Already we know the general functions of about two-thirds of the human genome, including the finding that at least 3 percent of our genes are proto-oncogenes, which can trigger cancer if they mutate. Discovering the functions of thousands of other genes will take many more years.

→ **Genes underlie** the variation we see in the traits of human populations. To be complete, a DNA sequence that represents the human genome must include all the varying forms of genes for each trait.

↓ **Human chromosome 21** is small and so was one of the first to have its gene sequence analyzed. The resulting chromosome "map" reveals the locations of genes responsible for several diseases and disorders.

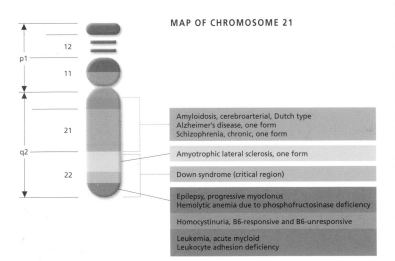

MAP OF CHROMOSOME 21

p1	12	
	11	
q2	21	Amyloidosis, cerebroarterial, Dutch type Alzheimer's disease, one form Schizophrenia, chronic, one form
		Amyotrophic lateral sclerosis, one form
	22	Down syndrome (critical region)
		Epilepsy, progressive myoclonus Hemolytic anemia due to phosphofructosinase deficiency
		Homocystinuria, B6-responsive and B6-unresponsive
		Leukemia, acute myeloid Leukocyte adhesion deficiency

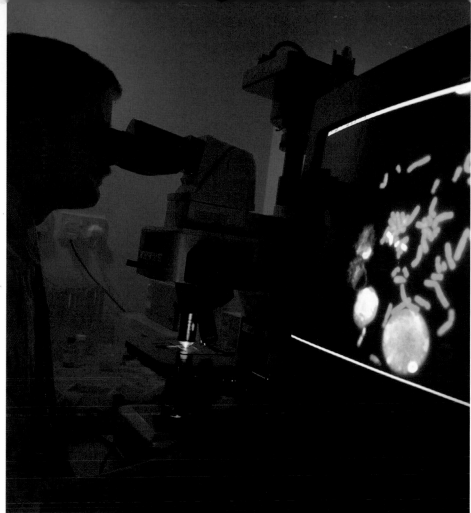

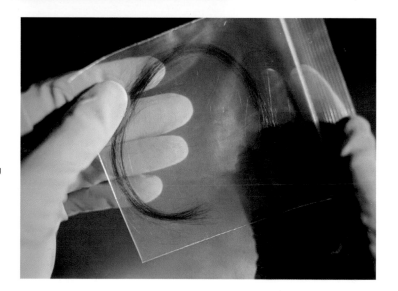

← **Researchers can now pinpoint** the locations of genes on chromosomes. Here the chromosomes from human cells appear as pale blue rods. They have been stained with a dye that makes them fluoresce under light from the microscope. The round objects are cell nuclei, which contain a living cell's chromosomes.

↓ **A lock of hair** contains genetic information. Hair, blood, semen, saliva and cells from the inside of the cheek are some common biological materials that are analyzed for their DNA content to gain genetic insights about the person they came from.

GENETIC TESTING

As we learn more about human genes, we can identify people who have, may develop or may transmit a genetic disorder. A carrier is usually defined as someone who is healthy but has one copy of a disease-causing gene along with one normal copy. If the disease gene is recessive, as with cystic fibrosis, then a child will develop the disease only if both parents pass on a copy of the gene. If the gene is dominant but has an adult-onset pattern, as with Huntington's disease and polycystic kidney disease, an affected parent may transmit it before developing the telltale symptoms. Genetic testing determines whether someone has the particular gene of concern, and people who opt for testing also receive counseling. For couples who are both carriers, any child they produce will have a 50 percent chance of developing the illness. Members of a carrier's extended family may also have the gene, a situation that is often challenging for families.

Cancer: a genetic disease

Each type of body cell is genetically programmed to divide on a certain schedule, after a period of days, weeks, months or even years. If for some reason this programming falters, the affected cell can be the beginning of cancer, producing renegade descendants that also divide outside normal controls. Such cells and the tumors they form can spread, or metastasize, to other parts of the body in the process. Cancer risk rises as we age in part because with advancing years the natural controls over cell division may begin to falter. Usually, a series of genes—two, three or more—must mutate. The most dangerous mutations are those altering genes that normally suppress the development of cancerous tumors or that regulate the repair of damaged DNA. For example, a tumor suppressor gene called p53 is mutated in roughly 60 percent of all cancers. In some families an inherited mutation may mean that cancer takes hold more easily.

↓ **A breast cancer tumor** is colored green in this mammogram. Monthly self-examination and regular screening by mammography are vital to detecting breast cancer before it spreads.

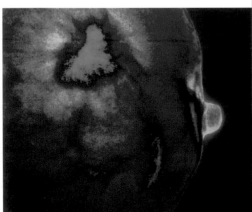

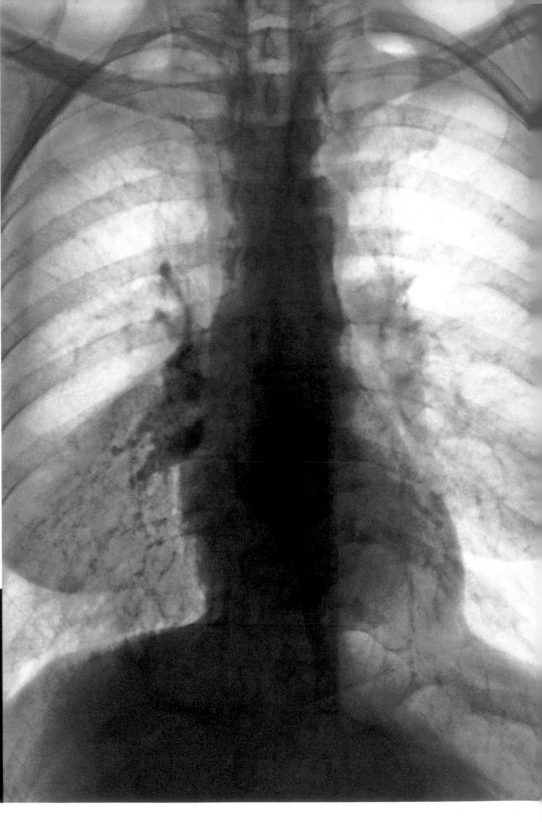

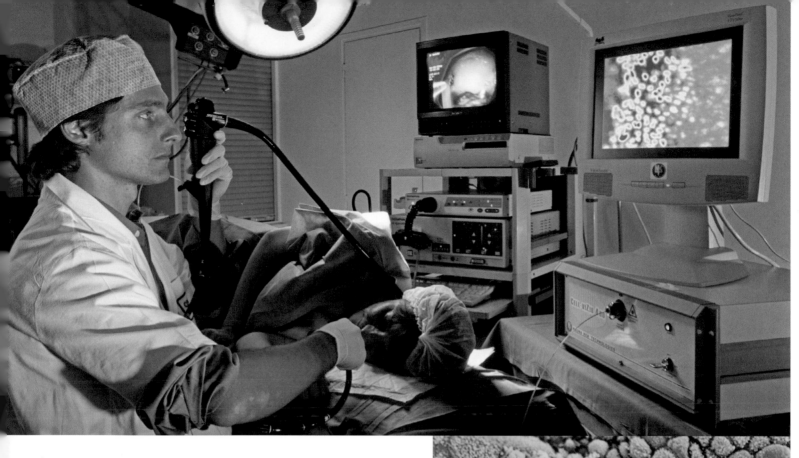

← **Lung cancer** is one of the top cancer killers. This patient's lung tumor appears as a bright orange mass. Carcinogens in tobacco smoke and industrial emissions are implicated in a large number of lung cancers.

↑ **Advanced technology** simplifies cancer diagnosis. This physician is using a sophisticated endoscope, a long tube containing thousands of optical fibers, to obtain a detailed view of one of the patient's internal organs. The microscopic image is relayed to a computer and displayed on a monitor.

→ **Like the cancerous cells** shown here, malignant tumors in the intestines may grow undetected until they cause a blockage or other symptoms, such as bleeding. By then the cancer may have spread to other organs, such as the liver.

CANCER AND THE ENVIRONMENT

Some scientists ascribe at least 50 percent of human cancers to carcinogenic (cancer-causing) factors in the environment. These include chemical carcinogens, such as toxins in tobacco smoke that can trigger mutations in the DNA of smokers or people who inhale the smoke secondhand. Certain agricultural herbicides and pesticides, asbestos, polyvinyl chloride (PVC) used in a range of plastics, and industrial chemicals such as benzene are other potent and widespread substances that can cause genetic mutations. In about 5 percent of cancers, a viral infection is the trigger. Radiation can also alter the normal structure of a cell's DNA, whether it comes in sunlight, from radon in soil, from medical X-rays or from some other source. Long-term, excessive exposure can outstrip the natural repair capabilities our cells have to fix this kind of damage. Sometimes an impaired cell simply dies, but in other cases it becomes the seed from which cancer develops.

Gene therapy

There are more than 15,000 human disorders known to result from one or more missing or malfunctioning genes. As scientists have learned how to use biotechnology to manipulate genes, the potential of gene therapy—treating or curing genetic disease by supplying normal copies of a missing or faulty gene—has become important. The most direct method is to chemically or mechanically insert a corrective gene into target cells, such as T cells of the immune system. Another avenue for placing a desired gene into the body is to use a transfer agent, or vector, typically a modified virus. The gene of interest is inserted into the virus and then the virus is allowed to "infect" body cells and carry the new DNA with it. To date these approaches have met with only modest success. Often, the new gene becomes inserted into too few cells to make a difference. It can also be extraordinarily difficult to ensure that a new gene will be incorporated into the receiving cell's DNA in a way that allows it to function normally, without triggering serious side effects or disrupting the operations of other genes.

↓   **A cervical cancer cell** (pink) undergoes cell division. As its cells evade cell division controls, a growing tumor requires an expanding blood supply. More than 70 percent of US gene therapy trials target these processes.

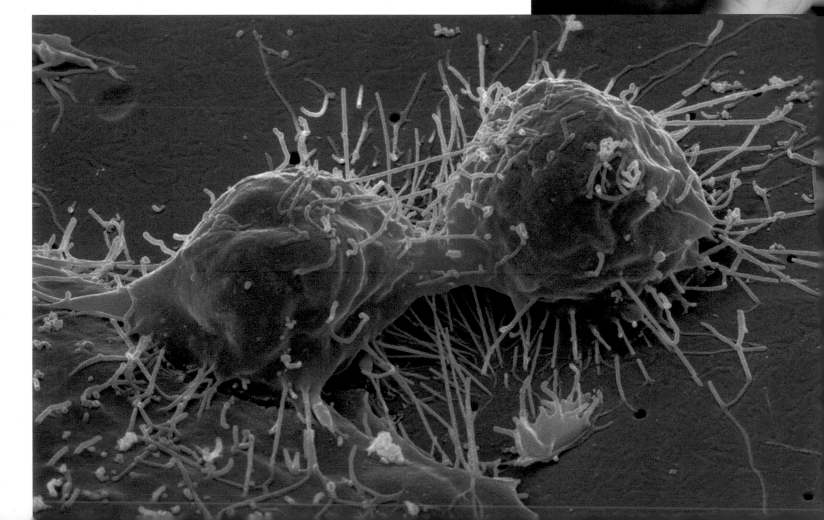

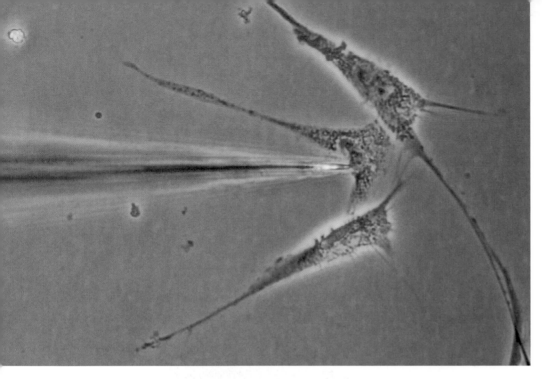

CHALLENGES TO RESEARCH

The promise of gene therapy comes with tremendous challenges. One is financial, for the technology required to isolate genes and genetically modify cells is extremely costly. In addition, many experimental treatments have not led to positive improvements and others have produced serious side effects. In some cases the patient's body mounts an immune response against the virus used as a vector. As a result, early efforts to develop genetic therapies for some disorders have been abandoned. In the campaign against other genetic flaws, researchers are working to develop safer vectors that are also more efficient in delivering normal genes to target tissues. At the same time, gene therapy may become a powerful tool for fighting some kinds of cancers. Research is focusing on boosting the activity of genes that disrupt the division of cancer cells or suppress the growth of blood vessels that nourish tumors.

A rare genetic disease, Crigler-Najjar syndrome, has caused the whites of this girl's eyes to turn yellow. As with other uncommon genetic disorders, efforts to correct it with gene therapy have been largely abandoned.

A needlelike pipette transfers corrective DNA into a patient's defective T cells. When the genetically modified T cells are transfused back into the patient, they help restore some immune functions.

An automated system reproduces genetically modified DNA. A robotic device (left) selects specific DNA sequences and uses them to inoculate colonies of yeast cells, which multiply rapidly and replicate the DNA.

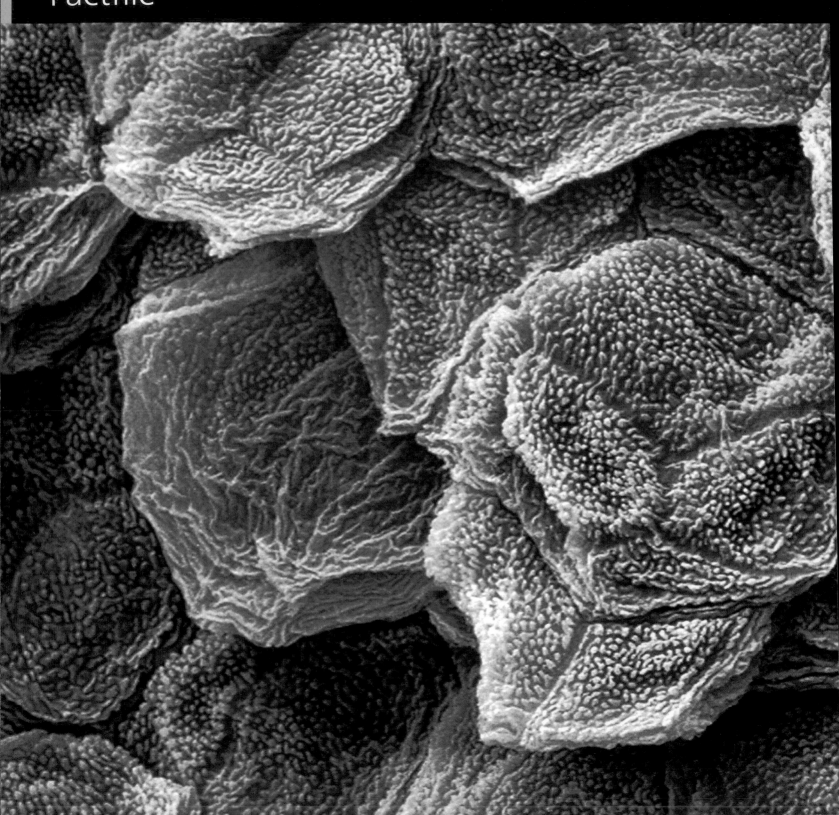

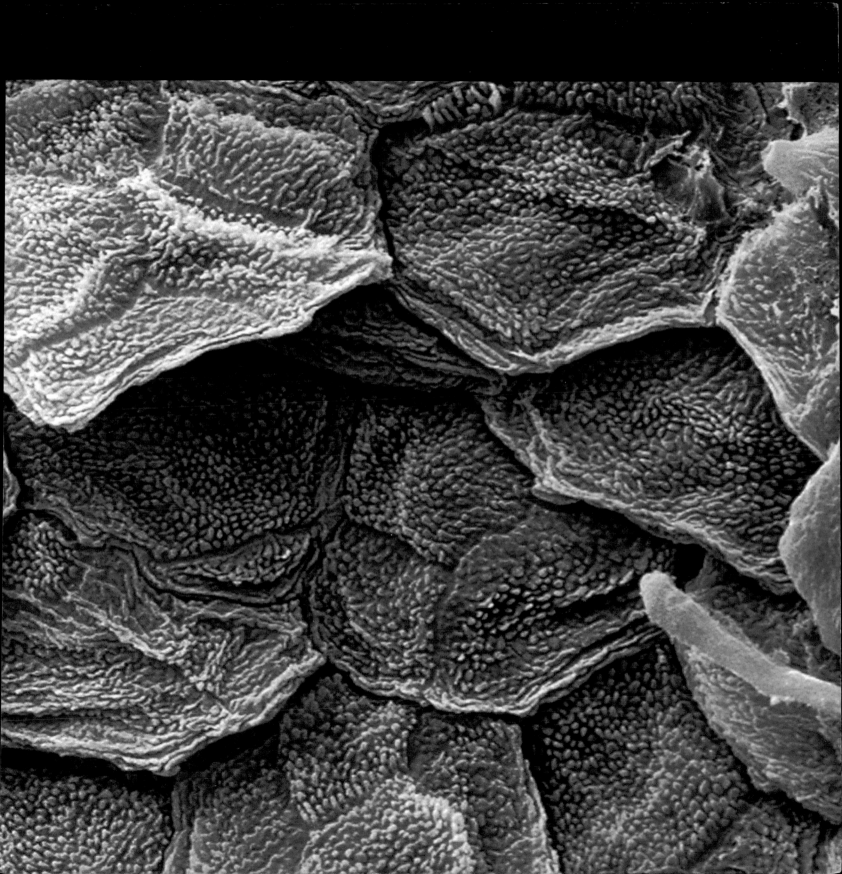

Factfile

Major skeletal muscles 282

Major body systems 282

Major glands and hormones 283

Vitamins and minerals 284

The digestive process 285

From birth to old age: how we change 286

Diseases and disorders 288

Major infectious diseases 288

Genetic disorders 288

Common medical problems 289

Milestones in understanding

the human body 290

MAJOR SKELETAL MUSCLES

Muscle	Location	Movement controlled
Triceps brachii	Upper arm (back)	Straightens arm and extends elbow
Pectoralis major	Upper chest	Moves arm toward body. Assists in lifting arm and pushing
Serratus anterior	Side of trunk	Moves shoulder blade forward. Assists in lifting arm and pushing
External oblique	Side of trunk	Pulls in abdomen. Helps turn trunk from side to side
Rectus abdominis	Abdomen (front)	Pulls in abdomen. Bends trunk forward
Adductor longus	Upper thigh (front)	Pulls thigh toward body and moves it from side to side
Sartorius	Lower thigh (front)	Bends leg at hip and at knee. Turns thigh outward
Quadriceps femoris	Lower thigh (front)	Bends leg at hip and extends it at knee
Tibialis anterior	Lower leg (front)	Bends foot toward shin
Biceps brachii	Upper arm	Bends arm at elbow
Deltoid	Shoulder (back)	Moves shoulder and lifts arm
Trapezius	Upper back	Supports head. Lifts shoulder blade; pulls head back
Latissimus dorsi	Back	Rotates arm; pulls arm back and toward body
Gluteus maximus	Buttocks	Extends and turns thigh when walking, running and climbing
Biceps femoris (hamstring)	Upper thigh (back)	Pulls thigh back; bends knee
Gastrocnemius	Calf	Bends knee when walking; extends foot when jumping

MAJOR BODY SYSTEMS

System	Tissues and organs	Main function
Nervous	Brain, spinal cord, sensory organs, peripheral nerves	Detects, controls and coordinates external and internal stimuli; integrates activities of organ systems
Endocrine	Pituitary, thyroid, parathyroid, pineal and adrenal glands, pancreas, testes, ovaries	Releases hormones that control and regulate body functions
Muscle	Skeletal muscle, smooth muscle, cardiac muscle	Moves and supports the body and its parts; generates heat
Skeletal	Bones, cartilage	Supports and protects the body; marrow produces red blood cells; stores calcium and phosphorus
Reproductive	Female: ovaries, oviducts, uterus, vagina, breasts. Male: testes, sperm ducts, accessory glands, penis	Female: produces eggs and hormones, protects and nurtures embryo, fetus and newborn Male: produces sperm and hormones; introduces sperm into the female
Digestive	Mouth, esophagus, stomach, intestines, liver, rectum, anus, pancreas (mostly digestive), gall bladder	Takes in and digests food and breaks it down; absorbs and stores nutrients, then makes them available to body cells; eliminates wastes
Respiratory	Airways, lungs, diaphragm	Takes in oxygen and delivers it to body cells; removes carbon dioxide wastes
Circulatory	Heart and blood vessels	Transports oxygen, carbon dioxide, nutrients and hormones around the body; helps regulate internal pH and body heat
Lymphatic	Lymph and lymph vessels, lymph nodes, spleen	Collects and returns tissue fluids to the blood; helps defend the body against infection and tissue damage
Skin (integumentary)	Skin, sweat glands, sebaceous glands, hair	Protects the body from injury, dehydration and some invading organisms; regulates body temperature; excretes some wastes; has receptors for external stimuli
Urinary	Kidneys, bladder, ureter, urethra	Controls and maintains the composition of tissue fluids; excretes waste products

MAJOR GLANDS AND HORMONES

Gland	Hormones produced	Targets of hormones	Main functions of hormones
Pituitary: posterior lobe	Antidiuretic (ADH)	Kidneys	Influences reabsorption of water; conserves water
	Oxytocin	Uterus	Promotes contraction
		Breasts	Stimulates milk production
Pituitary: anterior lobe	Growth (GH)	Many organs	Affects growth and metabolism
	Adrenocorticotropic (ACTH)	Adrenal glands	Assists in body's response to stress; influences secretion of adrenal cortical hormones such as cortisol
	Thyroid-stimulating (TSH)	Thyroid gland	Influences the secretion of thyroxine; involved in metabolism, circulation and growth
	Luteinizing (LH)	Gonads	Triggers ovulation and the formation of corpus luteum in ovarian cycle in females; promotes secretion of testosterone in males
	Follicle-stimulating (FSH)	Gonads	Stimulates production of sperm in males; encourages development of ovarian follicles in females
	Prolactin (PRL)	Breasts	Stimulates production of milk
	Melanocyte-stimulating (MSH)	Skin	Influences color change in reptiles and amphibians; function in mammals not known
Thyroid	Thyroxine	Most cells	Regulates metabolic rate; controls growth and development
	Calcitonin	Bone	Prevents calcium loss from bone, so lowering level of calcium in blood
Parathyroid	Parathyroid	Bone, kidneys, digestive tract	Raises level of blood calcium by spurring bone breakdown; promotes calcium reabsorption in kidneys; activates vitamin D
Adrenal medulla	Epinephrine (adrenaline) and norepinephrine (noradrenaline)	Smooth muscle, cardiac muscle, blood vessels	Trigger responses to stress; increase heart rate, blood pressure, metabolic rate and blood flow to vessels; raise level of blood glucose; mobilize fat
Adrenal cortex	Aldosterone	Kidney tubules	Regulates balance of sodium and potassium
	Cortisol	Many organs	Helps in toleratation of long-term stress; increases level of blood glucose; fat motabolizm
Pancreas	Insulin	Liver, skeletal muscles, fatty tissue	Decreases level of blood glucose; influences storage of glycogen in liver
	Glucagon	Liver, adipose tissue	Raises blood glucose level; stimulates breakdown of glycogen in liver
Ovary	Estradiol	General	Spurs development of secondary sex characteristics in females
		Female reproductive organs	Triggers growth of sex organs at puberty and assists in preparing uterus for pregnancy
	Progesterone	Uterus	Completes preparation of uterus for pregnancy
		Breasts	Influences development of breasts
Testis	Testosterone	Many organs	Triggers development of secondary sex characteristics in males and growth spurt at puberty
		Male reproductive organs	Influences development of sex organs; stimulates the production of sperm
Pineal gland	Melatonin	Gonads, pigment cells	Function not well understood; influences pigmentation in some vertebrates; may control biorhythms in some animals; may influence onset of puberty in humans

Vitamins and minerals

Vitamin/mineral	Food sources	Main functions	Indications of severe long-term deficiency	Indications of extreme excess
FAT-SOLUBLE VITAMINS				
A	As vitamin A found in beef, liver, fish liver oils, egg yolks, butter. Precursors (carotenoids) found in dark green leafy vegetables, yellow and orange fruit and vegetables	Maintenance of normal vision, defense against infections, maintenance of skin and other outer body surfaces	Night blindness, skin thickening, dry eyes, infection susceptibility, permanent blindness	Skin changes, hair loss, bone thickening and joint pains, nerve and liver damage; if pregnant, fetal abnormalities
D	As D_3 found in fish liver oils, egg yolks, fortified milk—formed in the skin when exposed to sunlight	Needed to absorb calcium and phosphorus from the gut, bone mineralization, growth and repair	Abnormal bone growth and repair —rickets in children, osteomalacia in adults	Poor appetite, nausea, vomiting, excess urination and thirst, calcium deposits in soft tissues throughout the body
E	Whole grains, green leafy vegetables, vegetable oils, egg yolks	An antioxidant that helps stop cell damage	Rupture of red blood cells, nerve damage	Nausea, flatulence, diarrhea
K	Made by micro-organisms in the gut; also found in green leafy vegetables	Necessary for normal blood clotting	Bleeding	Can interfere with anti-clotting medication
WATER-SOLUBLE VITAMINS				
B_1 (thiamine)	Lean meat, enriched cereals, nuts, yeast products	Carbohydrate metabolism, involved in nerve and heart functioning	Heart failure, abnormal nerve and brain function	None reported
B_2 (riboflavin)	Milk and milk products, enriched cereals, meat, eggs, yeast products	Carbohydrate metabolism, involved in tissue repair processes, maintenance of mucous membranes	Scaling of the lips and cracks at the corners of the mouth, dermatitis	None reported
Niacin	Yeast products, milk, meat, fish, poultry, legumes, eggs, wholegrain cereals	Carbohydrate metabolism and energy production	Skin rash, inflamed tongue, abnormal intestinal and brain function (pellagra)	None reported although flushing is a common side effect of niacin taken as a supplement
B_6	Most foods—fish, legumes, wholegrain cereals, meat, potatoes, eggs	Amino acid and fatty acid metabolism, nervous system function, blood synthesis	Anemia, nerve and skin disorders	Nerve damage, particularly numbness of the extremities
Pantothenic acid	Most foods—liver, yeast products, egg yolks	Synthesis of amino acids, red blood cells and steroid hormones	Abnormal nerve function, 'burning feet syndrome'	None reported
Folate (folic acid)	Fresh leafy green vegetables, liver and other organ meats, whole grains, legumes, some fruits and nuts	DNA and RNA synthesis, amino acid metabolism, linked to vitamin B_{12}	Anemia; in pregnancy folate deficiency is associated with an increased risk of neural tube defects in the fetus	May mask underlying vitamin B_{12} deficiency
B_{12}	Most animal products—meats, milk and milk products, eggs	Normal function of cells, red blood cell production, DNA and RNA synthesis, nerve functioning	Anemia, abnormal nerve and psychiatric function	None reported
Biotin	Made by micro-organisms in the gut; also found in liver, egg yolks, legumes	Carbohydrate and fatty acid metabolism	Inflamed skin and lips	None reported
C (ascorbic acid)	Citrus fruits, berries, peppers, cantaloupe, cabbage, cauliflower, broccoli	Connective tissue growth and wound repair, an antioxidant (maintains cell health), role in brain and nerve function	Bleeding, loose teeth, gum inflammation (scurvy), poor wound healing, impaired immunity	Diarrhea
MINERALS				
Calcium	Milk and milk products, canned fish with bones, tofu, dark green vegetables	Bone and tooth formation and growth, blood clotting, nerve and muscle function	Poor bone growth, possible osteoporosis, muscle spasms	Constipation; excess intake can interfere with absorption of other nutrients and medications
Chloride	Table salt	Electrolyte balance	Disturbance in acid–base balance	Disruption of acid–base balance
Copper	Shellfish, nuts, dried legumes, organ meats	Formation of red blood cells, bone formation	Anemia, growth retardation, hair abnormalities	Nausea, vomiting, diarrhea, liver damage

Vitamin/mineral	Food sources	Main functions	Indications of severe long-term deficiency	Indications of extreme excess
Fluorine	Fluoridated water	Bone and tooth formation	Increased risk of dental cavities	Mottling, pitting of teeth, bony outgrowths in the spine
Iodine	Seafood, iodized salt	Formation of thyroid hormones	Enlargement of thyroid gland (goiter); if pregnant, risk of abnormal fetal development (cretinism)	Enlarged thyroid (goiter)
Iron	Meat (especially red meat), liver, whole grains, leafy green vegetables, tofu, egg yolks	Main component of red blood cells and muscle cells, formation of enzymes	Anemia, impaired growth and learning ability, nail changes, abnormal intestinal function	Gastrointestinal disorder, liver damage (cirrhosis), heart failure
Magnesium	Legumes, dark green vegetables, nuts, wholegrain cereals	Bone and tooth formation, nerve and muscle function, enzyme activation	Abnormal nerve function	Diarrhea, heart rhythm disturbances
Phosphorus	Meat, poultry, wholegrain cereals	Bone and tooth formation, RNA and DNA synthesis, acid–base balance	Weakness	Calcium deposits in soft tissues such as the kidneys
Potassium	Milk, bananas	Nerve and muscle function, acid–base balance	Muscular weakness, heart rhythm disturbances	Heart rhythm disturbances
Selenium	Meat, wheatgerm, seafood	Antioxidant enzyme synthesis	Muscle pain and weakness	Fatigue, depression, arthritis, hair and nail changes
Sodium	Salt, most processed foods	Acid–base balance, nerve and muscle function	Confusion	Confusion
Zinc	Meat, oysters, liver, eggs and dairy products	Wound repair, growth, taste and smell, component of enzymes and insulin	Impaired growth, impaired taste and smell, impaired wound healing	Nausea, vomiting, diarrhea; impaired iron and calcium absorption

THE DIGESTIVE PROCESS

	Teeth, tongue and salivary glands	Esophagus	Stomach	Small intestine	Liver, gall bladder, pancreas	Gall bladder	Pancreas	Large intestine and colon	Rectum
Role in digestion	Teeth break up large chunks of food into smaller pieces. Salivary glands produce saliva, which softens food and makes it easier to move. Enzymes in saliva begin to break down carbohydrates in food. Tongue forms moistened food into a bolus and pushes it to the back of the mouth to be swallowed.	Wave-like contractions of the muscles in the walls of the esophagus move food from throat to stomach.	Glands in stomach's walls produce hydrochloric acid and enzymes, which break down large proteins, complex carbohydrates and fats. Wall muscles contract to blend food and gastric juices into a watery mixture called chyme. Food is slowly moved on to the small intestine a little at a time, taking 3–6 hours to empty the stomach.	Produces enzymes that together with bile and pancreatic juice break down 90 percent of the nutrients the body receives from food so they can be absorbed into the blood-stream (glucose and amino acids). Fatty acids and glycerol are transported to the liver by the lymph. The undigested remains move on to the large intestine.	Liver produces bile, which empties into the small intestine and assists in the breakdown of fats. Liver also processes nutrients carried by the blood diverted from the small intestine into the hepatic portal system. Makes cholesterol and uses it as a component of bile.	Stores bile and adds it to small intestine when food is eaten.	Produces fluid containing enzymes that break down proteins, carbohydrates and fats into a readily absorbable form.	Converts undigested remains of food into feces through contractions of muscles in the colon's walls. Water and some nutrients are absorbed.	Reflex contractions stimulated by feces push them along the anal canal and out of the anus.

From birth to old age: how we change

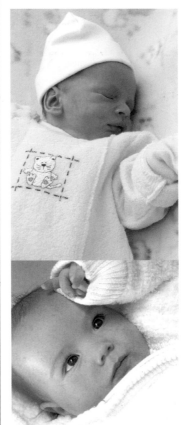

BIRTH TO 2 MONTHS

- At birth babies can see about 8–12 inches (20–30 cm) in front of them but are not able to focus clearly. Their hearing is fully developed and they have primitive reflexes, such as sucking and grasping a finger. They can recognize the shape of a human face and, 3 days after birth, they can distinguish their mother's voice and smell.

- During their first 2 months they learn how to smile and to focus their eyes, and they develop a sense of taste.

2–6 MONTHS

- Babies begin to clearly express different emotions, such as fear, anger and pleasure.

- From 4 months they can recognize objects.

- By 6 months they can coordinate head and eye movements, hold objects in the hand, lift head and chest off the floor when they are lying on the stomach, and roll over.

6–9 MONTHS

- During this time infants learn to sit up with a little support. They begin to crawl and may be able to pull themselves into a standing position.

- They understand that an object still exists even though it is hidden from view and they begin to recognize people's facial expressions, rather than just their features.

- First teeth begin to appear.

- Cooing noises develop to babbling and then to increasingly successful imitation of sounds.

9–12 MONTHS

- From around 9 months infants can stand holding onto furniture and begin to side-step along it.

- By 12 months they may be able to stand alone for a few seconds and can walk with help. They can also say one or two words and can imitate actions.

1–2 YEARS

- Around 12 months toddlers begin to walk unaided and by 2 years can run, kick, walk up and down stairs, walk backwards, and sit in a chair.

- During this time, fine motor skills are rapidly developed, such as stacking blocks, painting with a brush, drawing with crayons and eating with a spoon.

- A 2-year-old has an expanding vocabulary and can put a few words together.

- Toddlers begin to feel emotions such as jealousy, guilt and fear of the dark. Two-year-olds are prone to temper tantrums.

3–5 YEARS

- Three-year-olds develop fine motor skills further, such as drawing squares and circles. They can feed themselves with few spills, and button and unbutton clothes. They begin to play with other children and to understand how to share.

- By age 4 childen are taller and slimmer, muscle having replaced fat. They can copy complete letters and speak in longer, more grammatical sentences. They have a good sense of balance and can twirl, stand on their toes and hop on one foot.

- Between ages 4 and 5 children make closer friends, usually with the same sex.

- By age 5 the brain is three-quarters of its full adult size. Coordination has improved, and children can distinguish letters and write them without copying.

6–10 YEARS

- Around age 6 baby teeth start to fall out and permanent teeth begin to appear.

- Rapid physical, intellectual and psychological growth occurs. Reading and writing skills are refined. Coordination and balance become fully developed.

- From age 7 children begin to understand the concepts of past and future.

10–15 YEARS

- For girls puberty usually begins after age 10 with the development of breasts and pubic hair and a sudden growth spurt. Menstruation starts at about 11–14 years.

- For boys puberty begins at about 11–12 years, when the penis and testes increase in size and pubic hair appears. Around 12–13 years boys grow rapidly in height and weight, and become broader across the chest and shoulders. From around age 13 their voices deepen and underarm and facial hair begins to grow.

15–20 YEARS

- Both sexes continue to grow taller until about age 17 or 18.

- By age 20 the brain has reached full size.

20–30 YEARS

- Both men and women continue to develop muscle and fat. Women's fertility is at its peak in their mid-twenties and child-bearing often begins.

30–40 YEARS

- Body fat tends to concentrate more around the abdomen.

- In women fertility begins to decline and there is an increasing risk of fetal abnormality if they become pregnant.

- In their thirties men may start losing hair, and testosterone levels begin to fall.

40–60 YEARS

- The body's physical functions begin to decline, metabolism slows and there is a tendency to put on weight. Hair gradually turns gray and thins in both sexes. Levels of sex hormones fall.

- In men the prostate gland enlarges and may cause problems.

- In women menstruation ceases and menopause begins around 45–50 years, signaling the end of fertility. Decreasing estrogen levels may cause hot flashes and other symptoms, while there is a decrease in the calcium and mineral content of bones.

60–80 YEARS

- The skin begins to sag and develop wrinkles. Muscles start to waste away as collagen decreases.

- Bone density drops in both sexes, especially in women, who at age 70 have about half the bone density they had at age 30.

- As artery walls become stiffer and thicker, blood pressure increases, and there is a greater risk of heart disease.

- There may be a deterioration in sight and hearing.

- The function of some organs, including the brain, may begin to decline.

- Short-term memory may deteriorate and the immune system becomes less efficient. While many men and women still lead active lives, they may be limited in what they are physically capable of doing.

80 YEARS ONWARDS

- Physical functions decline further.

- Osteoarthritis and dementia are more likely to occur with age.

Diseases and disorders

MAJOR INFECTIOUS DISEASES

Disease	Cause and how spread	Time to develop	Symptoms	Treatment
Chickenpox*	Virus; via direct contact with skin lesions or inhalation of droplets from infected person or person with shingles	11–21 days	Skin rash consisting or widespread, itchy blisters, sore throat, headaches, fever, lethargy	Symptom relief
Cholera*	Bacterium; via contaminated food or water	1–5 days	Severe diarrhea and dehydration	Fluids, antibiotics
Common cold	Various viruses; via hand-to-hand contact or inhalation	1–3 days	Sneezing, blocked or runny nose, sore throat, aching muscles	Symptom relief
Diphtheria*	Bacterium; via contact with infected person or inhalation	2–4 days	Fever, gray membrane in throat, may sometimes damage heart	Antibiotics and antitoxin
Gastroenteritis, various types	Various bacteria, viruses and toxins; via direct contact with infected individual or contaminated food	1–2 days	Vomiting, diarrhea and dehydration	Plentiful fluids, antibiotics if needed
HIV/AIDS	Virus; via sexual intercourse, blood transfusion or from mother to baby	Variable, about 7 years	Weight loss, diarrhea, lethargy, fever, shortness of breath, infections	Infections are treated, but disease incurable
Influenza (flu)*	Virus; via inhalation	1–3 days	Alternating sweats and chills, headaches, sore throat, fever	Symptom relief
Malaria*	Parasitic infection transmitted by mosquitoes	1 week to several months	Sweating, fatigue, bouts of severe fever and chills	Antimalarial drugs
Measles*	Virus; via inhalation	7–14 days	Skin rash of red papules, fever, severe cold symptoms	Symptom relief
Pneumonia*	Various viruses and bacteria	Varies	Fever, cough, shortness of breath, chest pain	Symptom relief, antibiotics if bacterial
Tuberculosis (TB)*	Mycobacterium; via inhalation	Several weeks to several years	Cough producing bloodstained mucus, chest pain; can also affect kidneys, spine and brain	Antibiotics
Typhoid*	Bacterium; via contaminated food or water	7–21 days	Discomfort in abdomen, diarrhea, headaches, high fever	Antibiotics
Whooping cough*	Bacterium; via inhalation of infected particles spread by coughing	12–20 days	Sneezing, sore throat, fever, prolonged fits of coughing followed by vomiting, whooping sound on breathing in	Antibiotics

* Vaccine available

GENETIC DISORDERS

Syndrome	Characteristics
AUTOSOMAL RECESSIVE	
α-antitrypsin deficiency	May cause lung or liver disease
Albinism	Absence of pigment in skin, eyes, hair
Ataxia telangiectasia	Progressive degeneration of nervous system
Cystic fibrosis	Abnormal mucus production blocks lungs, ducts and glands
Galactosemia	Inability to metabolize galactose causes liver and kidney damage, and mental retardation
Phenylketonuria	Inability to metabolize phenylalanine, causing mental retardation and seizures
Sickle cell anemia	Abnormal hemoglobin, causing anemia, spleen enlargement
Thalassemia	Abnormal production of hemoglobin may cause severe anemia
Tay-Sachs disease	Enzyme deficiency causes accumulation of fatty substances that destroy brain and nerve cells
Wilson's disease	Copper accumulation secondary to metabolism disorder causes liver damage and neurological symptoms
AUTOSOMAL DOMINANT	
Achondroplasia	Abnormal bone growth that results in short stature (dwarfism)
Ehlers-Danlos syndrome	Easy bruising, hypermobile joints, hyperelastic skin, weakness of tissues
Familial hypercholesterolemia	High blood levels of low-density lipoprotein cholesterol from childhood, leading to early development of atherosclerotic heart disease
Hereditary non-polyposis colorectal cancer	Tendency to develop bowel cancer
Huntington's disease	From about age 40, progressive degeneration of nervous system
Hypercalcemia	Elevated levels of calcium in blood serum

Syndrome	Characteristics
Marfan's syndrome	Include being tall and thin, with long fingers, arms and legs. Connective tissue disorder causes heart and blood vessel defects
Osteogenesis imperfecta	Connective tissue disorder causing osteoporosis and a susceptibility to fractures
Polycystic kidney disease	Growth of numerous cysts in the kidney leading to kidney failure
Porphyria	Inability to metabolize porphyrins, causing skin problems such as blisters and swelling in sunlight or nerve symptoms such as numbness and mental change
Von Willebrand's disease	One of the body's clotting factors is reduced or absent, causing a tendency to bleed
X-LINKED RECESSIVE	
Adrenoleukodystrophy	Enzyme deficiency that leads to degeneration of the adrenal gland, causing progressive neurological disability and death
Color blindness (green)	Insensitivity to green light: 60–75% of color blindness
Color blindness (red)	Insensitivity to red light: 25–40% of color blindness
Duchenne muscular dystrophy	Progressive muscle weakness that starts in the legs and pelvis, then extends to the whole body
Fabry disease	Fat storage disorder causing skin rashes, impaired heart, kidney and gastrointestinal function
Fragile-X syndrome	High-arched palate, lazy eye, long face, intellectual impairment
Glucose-6-phosphate dehydrogenase deficiency	Benign enzyme deficiency that can produce severe anemia in the presence of certain foods, drugs or illnesses
Hemophilia A	Inability to form blood clots, caused by lack of clotting factor VIII
Hemophilia B	"Christmas disease:" clotting defect caused by lack of factor IX
Ichthyosis	Excessive amounts of dry, fish-like scaly surface skin

COMMON MEDICAL PROBLEMS

Condition	Cause	Symptoms
SKIN		
Eczema (dermatitis)	Inherited, allergy or contact with irritants	Inflammation of skin, with itching and sometimes blisters
Psoriasis	Chronic skin condition that stems from an inherited immune system disorder	Raised red patches covered with silvery scales
Urticaria (hives)	Allergic reaction to a trigger such as a food, insect bite, drug, plant or animal	Itchy, inflamed wheals
Vitiligo	Body attacks own pigment-producing cells—an autoimmune disorder	Loss of color from patches of depigmented skin
SKELETON AND MUSCLES		
Osteoarthritis	Degeneration of the cartilage and bone in joints, especially in knees, hips and hands	Pain, swelling and stiffness of joints
Rheumatoid arthritis	Body attacks own joint cartilage—an autoimmune disorder. Commonly affects wrists, and proximal finger joints	Pain, swelling and stiffness of joints with deformity developing
VISION		
Cataract	Lens of eye becomes cloudy	Increasingly blurred vision
Glaucoma	High pressure of fluid in eye damages optic nerve	Loss of peripheral vision
Macular degeneration	The macula in the center of the retina deteriorates. Smoking accelerates the process	Loss of central vision
HEART AND CIRCULATION		
Anemia	Can be secondary to chronic blood loss, chronic illness, iron deficiency, underlying abnormalities such as thalassemia	Fatigue, lethargy, pale skin, shortness of breath on exertion
Angina	Insufficient blood supply through coronary arteries to heart muscles. Usually related to atherosclerosis—fatty deposits in coronary arteries	Chest pain; may also have pain in jaw or left arm
Hypertension (high blood pressure)	Cause unknown in most cases. May occur with kidney disease	Often none, but chronic hypertension can increase risk of stroke, heart attack and heart failure
Sickle-cell anemia	Hereditary disease in which distorted red blood cells carry less oxygen and obstruct blood vessels	Pale skin color, headaches, tiredness, shortness of breath upon exercise, jaundice
RESPIRATORY SYSTEM		
Asthma	Rapid swelling and constriction of airways on exposure to an allergen such as pollen, dust mites and animals' dander	Shortness of breath and wheezing
Bronchitis	Bacterial or viral infection of the larger airways within the lung	Persistent cough with copious sputum production, shortness of breath
Emphysema	Loss of elasticity of lung tissue, commonly caused by long-term smoking	Shortness of breath
ENDOCRINE SYSTEM		
Diabetes mellitus	Insufficient production of insulin by pancreas, leading to excessively high blood sugar; genetic or caused by obesity	Thirst, passing large volumes of urine
Hyperthyroidism	Overproduction of thyroid hormones	Increased metabolic rate, leading to excessive sweating, weight loss and irregular heartbeat
Hypothyroidism	Underproduction of thyroid hormones	Weight gain, hair loss, tiredness, lethargy
URINARY AND DIGESTIVE SYSTEMS		
Coeliac disease	Enzyme deficiency leads to inability to metabolize gluten	May have diarrhea, abdominal pain, bloating, tiredness and anemia
Cystitis	Bacterial infection of the urine within the bladder	Frequent, painful urination, urgency
Gastroenteritis	Bacterial or viral infection, or after ingestion of toxin from contaminated food	Vomiting, diarrhea; may have tiredness, fever
Kidney failure	Can include post-infection, diabetes or injury	Fluid retention, fatigue, nausea, vomiting, shortness of breath
Kidney stones (calculi)	Solid deposits in the kidney, formed from substances in the urine	Severe pain, frequent and painful urination, nausea and vomiting, blood in urine
Peptic ulcer	May be caused by irritants such as smoking, excess alcohol, aspirin or other nonsteroidal anti-inflammatories; may also be associated with *H. pylori* infection	Pain in upper abdomen, loss of appetite, weight loss
BRAIN AND NERVOUS SYSTEM		
Alzheimer's disease	Progressive degeneration of brain tissue	Progressive mental deterioration
Epilepsy	Sudden, uncontrolled discharge from nerve cells in the brain	Episodes of convulsive seizures, periods of loss of consciousness
Motor neuron disease (MND)	Degeneration of nerves in the brain and spinal cord	Wasting and weakness of muscles
Multiple sclerosis (MS)	Progressive, patchy destruction of the myelin sheath that covers and protects nerve fibers	Weakness; loss of sensation; vision, urinary and bowel problems
Parkinson's disease	Degeneration of certain nerve cells in brain	Tremors, rigid posture, loss of spontaneous movements
Stroke	Brain damage caused by rupture or blocking of a blood vessel	Loss of function in the body areas covered by affected brain area

Milestones in understanding the human body

c. 400 BC
The ancient Greek physician Hippocrates is known as "the founder of medicine." Writings attributed to him bring together all medical knowledge of the time. Noting the connection between malaria, locality and climate, Hippocrates paved the way for epidemiology, the study of the causes and distribution of diseases.

1543
Flemish anatomist Andreas Vesalius published the first detailed drawings and descriptions of the human body in his *Seven Books on the Structure of the Human Body*.

1727
English botanist and chemist Stephen Hales measured blood pressure for the first time, by inserting a tube into a blood vessel and allowing the blood to rise.

1796
The vaccination for the deadly disease smallpox was discovered by the English physician Edward Jenner. He inoculated a child with the milder disease cowpox, and then 2 months later with smallpox. The cowpox vaccine prevented smallpox from developing. Within 5 years smallpox vaccination became standard practice.

1842
Ether was first used as an anesthetic by US surgeon Crawford Long, who removed a cyst from a patient's neck after administering the gas.

1844
Rubber condoms began to be mass-produced, thanks to Charles Goodyear's discovery of rubber vulcanization, a process that turns rubber into a strong, elastic material.

1882
The first modern account of cytology, the study of cells, was given by German anatomist Walther Fleming. He established that chromosomes carry genetic information and described the process of cell division, which he named mitosis.

1896
German physicist Wilhelm Röntgen discovered a new form of radiation, which he called X-rays, that could pass through flesh and clothing to produce photographs of the body's bone

1628
English physician William Harvey first described the circulation of the blood in his treatise, *Anatomical Study of the Motion of the Heart and of the Blood in Animals*.

2nd century AD
The Greek physician Galen, through his extensive dissection of animals, discovered that urine was formed in the kidneys (not the bladder) and that arteries carried blood rather than air. He also developed the now standard medical practice of taking the pulse.

1761
Italian physician Giovanni Battista Morgagni wrote *On the Causes of Diseases*. He is considered to be the founder of pathological anatomy, the scientific study of anatomy using cadavers.

1667
In his *Observationes Medicae*, a study on epidemics, English physician Thomas Sydenham set out the principles of medical examination and epidemiology. He was the first to treat anemia with iron and helped popularize the treatment of malaria with quinine.

1816
The stethoscope was invented by René Laënnec, a French physician, who is sometimes called "the founder of thoracic medicine."

1805
The pain-relieving drug morphine was extracted from opium by German chemist Friedrich Sertürner.

1870s–80s
French chemist and microbiologist Louis Pasteur demonstrated that infectious diseases are caused by microorganisms, and subsequently developed vaccines against rabies and anthrax. He also showed how heating foods and beverages to a certain high temperature would destroy disease-carrying organisms, a process now known as pasteurization.

1858
With his publication of *Cellularpathologie*, German pathologist Rudolf Virchow established that only one cell, the fertilised ovum, is the source of all the body's cells.

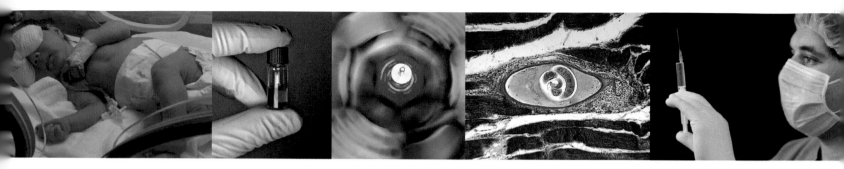

1901
Austrian pathologist Karl Landsteiner discovered the ABO blood-group system and later the Rhesus (RH) system, making it possible to administer safe blood transfusions.

1910
German-born chemist Paul Ehrlich discovered a cure for syphilis— salvarsan, the first synthetic antibacterial drug that could target a specific disease.

1921
The hormone insulin was isolated by Canadian physiologist Frederick Banting and his assistant Charles Best, and was later successfully used to control diabetes.

1929
Electrical activity in the brain was first described by German psychiatrist Hans Berger, who invented the electroencephalograph (EEG) to monitor it.

1952
British geneticist Douglas Bevis invented amniocentesis, a technique to remove and analyze fluid from the womb during pregnancy.

1952
The first polio vaccine was discovered by US doctor Jonas Salk.

1957
The technique of thermography, which records temperature differences in the body, was first applied to the detection of cancer cells and subsequently developed as an additional screening technique for breast cancer.

1973
US biochemist Herbert Boyer and his colleague Stanley Cohen successfully inserted new genes into bacterial cells, a technique that became the foundation of genetic engineering.

1980s
The development of magnetic resonance imaging (MRI) revolutionized diagnostic medicine. Using magnetism and radio waves, clear three-dimensional images of all parts of the human body are produced, from soft tissue as found in the brain to hard tooth enamel.

1987
Positron emission tomography (PET) imaging was developed to show brain abnormalities, such as those caused by strokes. The images are based on emissions from short-life radioactive elements that are administered to the patient.

1897
German chemist Heinrich Dreser developed aspirin from plant extracts.

1909
The first true intrauterine device (IUD) for contraception in women was developed by German doctor Richard Richter.

1916
Social reformer and trained nurse Margaret Sanger started the first US birth-control clinic in New York. She is considered to be the founder of the birth-control movement.

1928
British bacteriologist Alexander Fleming discovered penicillin, the first antibiotic. Australian pathologist Howard Florey and German chemist Ernst Chain later developed it as a commercial drug (1940).

1944
DNA was established to be the carrier of genetic information by Canadian bacteriologist Oswald Avery and his colleagues Maclyn McCarty and Colin MacLeod.

1953
US geneticist James Watson and British biophysicist Francis Crick constructed a molecular model of the double helix structure of DNA.

1950s–1960s
Scottish gynecologist Ian Donald pioneered the development of practical applications for ultrasound in obstetrics and gynecology.

1956
US physiologist Gregory Pincus developed the female contraceptive pill, which was made available in the USA from 1960.

1970s
British physicist Godfrey Hounsfield developed the technique for computer-assisted X-ray imaging, known as CAT scanning, which produces three-dimensional "slices" of the body's interior to aid diagnosis of disease or injury.

1983–1984
French virologist Luc Montagnier and US researcher Robert Gallo independently discovered the virus (later called HIV) responsible for causing AIDS.

2003
The international Human Genome Project consortium announced the completion of the sequence of the human genome, or the genetic blueprint of life, which holds the keys to transforming medicine and understanding disease.

Glossary

ABO blood type A term used to describe a set of markers on red blood cells. Humans may have A markers, B markers, both A and B, or neither—in which case the blood type is O.

Accommodation The ability of the eye to adjust its focal length so that light rays are focused properly on the retina.

Allergy An immune response against a substance that is normally harmless to the body.

Alveoli Tiny air sacs in the lungs where oxygen diffuses from air in the lungs to the blood, and carbon dioxide diffuses from blood to the lungs.

Amino acids Molecules that are the building blocks of proteins.

Amniocentesis Procedure in which a sample of amniotic fluid is removed in early pregnancy and analyzed for evidence that a fetus has genetic defects.

Amnion The protective sac that encloses a developing fetus; it is filled with amniotic fluid.

Antibody Any of the defensive proteins produced by B lymphocytes. An antibody binds to a specific antigen and flags it for destruction by other immune system cells.

Antibody-mediated immunity Immunity conferred by antibodies produced by B cells and that are present in body fluids, especially blood and lymph.

Antigen Any substance that stimulates a response by the immune system.

Antioxidant A substance that can neutralize free radicals before the free radical damages DNA or some other cell part.

Aorta The largest artery in the body, it carries blood from the heart's left ventricle to the rest of the body.

Apoptosis Genetically programmed cell death; a mechanism that helps shape body parts during development and also destroys abnormal body cells.

Appendicular skeleton Pectoral girdle, limbs and pelvic girdle.

Apoxia Lack of oxygen.

Arrhythmia An abnormal heart rhythm.

Arteriole A small blood vessel that branches off an artery and connects to capillaries.

Artery A large blood vessel that carries blood under pressure from the heart to body tissues.

Atrioventricular (AV) node Mass of tissue between the heart's atria and ventricles that receives nerve impulses from the sinoatrial node and transmits them to the ventricles.

Atrioventricular valve One of two heart valves through which blood flows from the atria to the ventricles; prevents return of blood to the atrium.

Atrium Either of the two upper heart chambers. The left atrium receives oxygenated blood from the lungs, the right atrium receives deoxygenated blood from tissues.

Autoimmunity A condition in which the immune system mistakenly attacks the body's own tissues.

Autonomic nerves Nerves associated with the autonomic nervous system, which regulates involuntary or automatic body functions.

Autosome A chromosome that does not carry genes for sexual traits.

Axial skeleton Bones of the head, chest and vertebral column.

Axon A long extension, with branched endings, from the cell body of a neuron. Nerve impulses travel along axons.

Bare nerve endings Pain receptors found in the skin.

Basophil A type of white blood cell that engulfs and destroys microorganisms and also has a role in allergies.

B cell A type of lymphocyte (white blood cell) that arises in bone marrow and can make antibodies against foreign cells or substances during an immune response.

Bilateral symmetry Having a body with two halves that are basically mirror images of each other.

Blastocyst An early embryonic stage that consists of a small cell mass surrounded by a balloon-like ball of cells.

Blood–brain barrier Term for the specialized walls of brain capillaries, which prevent some types of harmful molecules from entering the brain from the blood.

Blood pressure The fluid pressure generated by heart contractions; it keeps blood circulating through the cardiovascular system.

Bone marrow The soft, spongy tissue in the cavities in bones such as the sternum and hip bones; stem cells in bone marrow give rise to red and white blood cells.

Bone tissue Mineralized connective tissue that forms bones.

Bronchus One of the two large airways that branch from the windpipe (trachea) and lead to the lungs.

Capillary The smallest type of blood vessel. Substances in blood move to and from body cells across the walls of capillaries.

Cardiac cycle The complete cycle of one heartbeat.

Cardiac muscle The type of muscle found only in the wall of the heart.

Cardiac pacemaker A cluster of cells in the heart that regulates the heartbeat.

Cartilage Rubbery connective tissue that cushions bones and joints and provides support in areas like the nose and ears.

Cartilaginous joint The type of joint in which cartilage fills the space between the linked bones; a cartilaginous joint permits only slight movement.

Cell-mediated immunity Immunity conferred by the various kinds of activated T cells, which attack infected or abnormal body cells or cells that have been transplanted and also release chemicals that regulate immune responses.

Central nervous system (CNS) The brain and spinal cord.

Cerebellum A region at the rear of the brain where reflexes that maintain posture and adjust limb movements are controlled.

Cerebral cortex The thin, outer layer of the cerebral hemispheres. Some parts receive input from sensory nerves, others coordinate responses.

Cerebrospinal fluid A clear fluid that surrounds and cushions the brain and spinal cord.

Cerebrum The largest and most developed part of the brain, divided into right and left hemispheres. The cerebrum is where cognitive functions such as thought and learning occur.

Chemoreceptor A sensory cell that responds to chemical stimuli.

Chorion One of several membranes that develop around an early embryo; it becomes a major part of the placenta.

Chorionic villus sampling Procedure in which a tissue sample from the placenta is analyzed during early pregnancy to check for possible genetic defects in a fetus.

Chromosome A DNA molecule and its associated proteins.

Chyme The semi-fluid mixture of swallowed food and gastric fluid that passes from the stomach to the small intestine.

Clone An organism or cell that is genetically identical to a single parent organism or cell.

Cochlea Coiled, fluid-filled chamber of the inner ear where pressure waves are translated to signals the brain can interpret as sounds.

Compact bone The dense bone tissue that makes up the outer part of all bones and the shafts of long bones. It contains channels for blood vessels and nerves.

Complement system A set of proteins that help target and destroy pathogens (antigens) during immune responses.

Cones The photoreceptors in the eye that respond to bright light and help provide sharp, color vision.

Connective tissue Tissue, such as bone, cartilage, ligaments and tendons, that supports body organs and other structures.

Cornea The clear, outer layer of the eye. Together with the lens, the cornea focuses light on the retina.

Corpus callosum A band of several hundred million neuron axons that links the right and left cerebral hemispheres.

Corpus luteum A structure that develops after ovulation and secretes hormones (progesterone and estrogen) that prepare the uterus lining for possible pregnancy.

Corticosteroid A steroid hormone secreted by the adrenal glands that reduces inflammation in tissues.

Cytoskeleton A network of protein filaments that provides internal support and structure to a cell and aids in cell movement and division.

Dendrite A short, branched extension from a neuron that receives nerve impulses from other neurons.

Dermis The inner layer of skin, beneath the epidermis. The dermis contains sensory nerve endings, oil glands, hair follicles, and blood and lymph vessels.

Diastole Period during the cardiac cycle when heart chambers relax and fill with blood.

Digestion The physical and chemical breakdown of food into small molecules the body can absorb.

DNA Deoxyribonucleic acid, the genetic material. DNA consists of subunits called nucleotides.

DNA fingerprint The unique sequences of nucleotides in a person's DNA. Except for identical twins, no two individuals have the same DNA sequences.

Dominant gene A version of a gene that masks another, recessive version of the same gene.

Ectoderm The outermost tissue layer that forms in an early human embryo. Its cells give rise to the skin and nerve tissue.

Electrocardiogram A graph of the electrical activity of the heart measured by an electrocardiograph.

Embryo In early development, the term for the stage that begins shortly after conception and lasts until the end of the eighth week of pregnancy.

Endocrine gland A gland that secretes hormones, which usually are carried in the bloodstream to target cells.

Endoderm The innermost tissue layer in an early human embryo. It gives rise to many internal organs.

Endometrium The lining of the uterus.

Endoplasmic reticulum A system of internal cell membranes that are involved in the synthesis, modification and transport of proteins and other molecules.

Endorphin A chemical produced by the brain that acts as a natural pain killer. Enkephalins are a type of endorphin.

Enzyme A protein that speeds ups (catalyzes) chemical reactions.

Eosinophil A type of white blood cell that can digest foreign matter and participates in immune responses to parasites, worms and allergens.

Epidermis The outer layer of the skin.

Epiglottis A small flap of cartilage in the throat that covers the windpipe (trachea) during swallowing so that food does not enter it.

Epithelium A tissue composed of a single sheet of closely packed cells. Epithelium lines internal and external body surfaces.

Erythrocyte A red blood cell.

Eustachian tube A narrow tube connecting the middle ear to the back of the throat.

Extracellular fluid All body fluids that are not inside cells, such as blood plasma.

Fertilization The union of the nuclei of an egg cell and a sperm cell.

Fetus The term for an embryo after it completes 8 weeks of development in the womb.

Fibrous joint A joint in which two bones are connected by fibrous connective tissue.

Fovea A dense area of cone cells in the retina; the part of the eye where vision is sharpest.

Gene A segment of DNA that contains instructions to build a particular protein. Genes are the basic units of heredity.

Gene mutation A change in the nucleotide sequence in a gene, as when a nucleotide is deleted.

Gene sequence The order of nucleotides that make up a person's genes.

Gene therapy Technology in which one or more normal genes are transferred into cells in order to correct a genetic defect.

Genetic code The chemical correspondence between sequences of nucleotides in genes and amino acids in proteins. Each amino acid is encoded by a sequence of three nucleotides.

Genome All the DNA in a full set of chromosomes.

Germ cells The cells in reproductive organs that are specialized to give rise to eggs (in ovaries) and sperm (in testes).

Gland An organ or group of cells that produces and releases one or more substances, such as hormones, digestive juices, sweat and tears.

Glomerulus A cluster of looping blood capillaries in a kidney nephron where water and dissolved substances are filtered from the blood.

Glycogen The chemical form in which the body stores glucose; glycogen is stored mostly in liver and muscle cells.

Golgi apparatus A cell organelle specialized to modify, sort and package proteins for export from the cell.

Gray matter The nerve tissue in the cerebral cortex and spinal cord. It has a gray color because its neurons have grayish, unmyelinated axons.

Growth factor A signaling molecule that can cause cells to divide, leading to growth of tissues and organs.

Gustation The sense of taste.

Hair cell A type of mechanoreceptor that can trigger nerve impulses when it is bent or tilted.

Helper T cell A type of T lymphocyte that facilitates immune system reactions by activating and directing other cells in immune responses.

Hemoglobin The iron-containing protein in red blood cells that binds oxygen.

Hemostasis Stopping blood loss, mainly by blood clotting.

Histamine A chemical released by mast cells that promotes inflammation, as in allergies.

Homeostasis A state of internal balance in the body that is achieved largely by mechanisms that adjust the chemical composition of blood and tissue fluid.

Hormone A communication molecule, produced by an endocrine gland, that travels through the blood to target cells and tissues.

Hypothalamus A region at the base of the brain that regulates various metabolic processes, such as body temperature and appetite.

Immune response The body's physiological response to foreign material (antigens).

Immunodeficiency An abnormal decrease in or total lack of immune responses.

Implantation The process in which an early embryo becomes embedded in the wall of the uterus.

Inflammation A general immune system response to infection or irritation, producing redness, heat, swelling and pain.

Inner ear The innermost portion of the ear; it consists of the cochlea and the vestibular apparatus, the organ vital to balance.

Integument A body covering; in humans, the skin.

Interferons Family of proteins that help regulate immune responses to viruses and some kinds of cancer cells.

Interneuron Neurons in the brain and spinal cord that communicate only with other neurons.

Intervertebral disk Cartilaginous disk between spinal vertebrae.

Iris The colored part of the eye that can dilate or contract the size of the pupil to adjust the amount of light that enters the eye.

Karyotype A complete collection of the chromosomes in a somatic (body) cell. It usually shows all 23 pairs of human chromosomes.

Killer T cell (cytotoxic T cell) A type of lymphocyte that directly destroys cells infected with particular viruses or bacteria.

Larynx The upper part of the windpipe (trachea); it contains the vocal cords.

Lens The thin, transparent structure in the eye that focuses light on the retina.

Ligament A band of strong connective tissue that connects two bones at a joint.

Limbic system Brain regions that collectively control or influence emotions, motivation and memory.

Lipid A greasy or oily compound, used in cells for energy or to build parts such as cell membranes.

Lymph The tissue fluid carried in vessels of the lymphatic system.

Lymph node A small gland that is usually found in clusters in the armpits, groin, neck, chest and abdomen. White blood cells in lymph nodes collect and destroy foreign matter before it enters the blood.

Lymphocyte A type of white blood cell, including T cells and B cells, that participates in immune responses.

Macrophage A type of large white blood cell that engulfs and destroys cellular debris and foreign material such as bacteria.

Mast cell A type of white blood cell in tissues; mast cells release inflammatory chemicals such as histamine.

Mechanoreceptor A sensory receptor that responds to mechanical pressure.

Meiosis The type of cell division that produces gametes (sperm and eggs); only cells in the testes (males) and ovaries (females) divide by meiosis.

Meissner's corpuscles Mechanoreceptors in the skin that are sensitive to low-frequency vibrations and pressure.

Melanin The pigment that gives color to the skin, hair and eyes.

Memory cells A subset of T and B cells that remain in the body after an immune response and that later can mount a stronger, more rapid response to the same antigen.

Meninges Membranes that cover the brain and spinal cord.

Merkel's discs Mechanoreceptors in skin that are sensitive to fine touch and pressure.

Metabolism The sum of chemical reactions in cells by which they obtain and use energy.

Microvillus A hair-like projection from the surface of some epithelial cells, as in the lining of the small intestine; microvilli are specialized to absorb substances.

Middle ear The central cavity of the ear that conducts sound to the inner ear. It contains the three tiny ear bones and the eardrum (tympanic membrane).

Mitochondrion A cell organelle where the cellular fuel ATP is formed.

Mitosis The type of cell division by which body tissues grow and are repaired.

Monoclonal antibodies Antibodies produced in the laboratory; made by cells descended from a single parent cell.

Motor unit In muscle tissue, a neuron and all the muscle fibers it controls.

Mucous membrane A thin, moist membrane containing mucus glands. Mucous membranes line body cavities that open to the outside.

Muscle fiber A single muscle cell, composed of bundled myofibrils.

Muscle tissue Tissue that can contract (shorten) when it is stimulated by nerve impulses. The three types of muscle tissue are cardiac, skeletal and smooth muscle.

Myelin sheath A fatty, insulating wrapping around the axons of many motor and sensory nerves. It is formed by the outer membranes of Schwann cells.

Myocardium Heart muscle tissue.

Myofibril A contractile filament in muscle cells.

Natural killer (NK) cells Lymphocytes that mount a general attack against foreign material in the body.

Negative feedback A mechanism of homeostasis in which a change in some condition triggers a response that counteracts the altered condition.

Nephron A tiny tubule found in the kidney that filters wastes, unneeded substances and fluids from the blood, producing urine.

Nerve A bundle of neuron axons.

Nerve impulse A brief reversal of electrical voltage across the cell membrane of a neuron. Nerve impulses convey information between nerve cells and other parts of the body; also called an action potential.

Neuroendocrine control center In the brain, the parts of the hypothalamus and pituitary gland that interact to control many physiological functions.

Neuroglia Cells that provide physical and metabolic support to neurons; more than half the nerve tissue in the body consists of neuroglia.

Neuromuscular junction The junction where a motor neuron synapses on a muscle fiber.

Neuron A nerve cell.

Neurotransmitter Any of the signaling chemicals released by nerve cells.

Neutrophil A type of white blood cell that ingests and destroys foreign material.

Nociceptor A pain receptor, usually a free nerve ending.

Nucleic acid A long, chain-like molecule composed of nucleotides. DNA and RNA are types of nucleic acids.

Nucleus The cell organelle that contains most of the cell's genetic material, in the form of DNA organized in chromosomes.

Obesity Having excess body fat, typically 20 percent above the recommended weight for a given height and age.

Olfaction The sense of smell.

Organ of Corti A patch of membrane in the inner ear that contains the sensory hair cells involved in hearing.

Organelle In cells, a sac or compartment that is enclosed by a membrane and that has a particular function.

Osmoreceptor A sensory receptor that detects changes in water volume in a body fluid.

Osteon The basic structural unit of compact bone.

Otoliths Calcium carbonate crystals in the inner ear that function in sensing gravity and changes in acceleration.

Ovulation The release of an egg from an ovary during a woman's menstrual cycle.

Pacinian corpuscle A pressure-sensitive mechanoreceptor in skin and some internal organs.

Parasympathetic nerves Nerves involved in physiological processes that maintain normal body functions such as digestion and excretion; part of the autonomic division of the peripheral nervous system.

Pathogen A disease-causing organism.

Pectoral girdle The set of bones that connect and support the upper limbs; includes the shoulder blade (scapula) and collarbone (clavicle).

Pelvic girdle The set of bones that form the pelvis and connect to and help support the lower limbs.

Perception The process of becoming aware of, understanding and interpreting stimuli from sense organs.

Peripheral nervous system The nerves traveling into and out from the spinal cord and brain.

Peristalsis Wave-like muscle contractions that move material through the digestive tract.

pH scale A scale used to measure the relative acidity of blood and other body fluids; it ranges from 0 (most acid) to 14 (most alkaline). A pH of 7 is considered neutral.

Phagocyte A cell that can engulf and destroy matter such as foreign cells, dead body cells and other debris.

Pharynx The throat.

Photoreceptor A sensory receptor that responds to light energy, such as the rods and cones of the retina.

Pituitary gland An endocrine gland in the brain that interacts with the hypothalamus to coordinate and control various physiological functions, including other endocrine glands.

Placenta An organ that delivers nutrients to a developing fetus and carries away wastes. Its structure helps prevent the blood of mother and child from mixing.

Plasma The fluid portion of the blood.

Plasma cell A type of B cell that produces antibodies.

Platelets Cell fragments in blood that function in blood clotting.

Polygenic trait A trait that results from the influence of several genes.

Positive feedback A homeostatic mechanism that intensifies a change from an original condition. Intensifying labor contractions during childbirth are an example.

Prion An infectious particle that consists of only protein.

Prostaglandins Communication chemicals that influence many body functions, for example smooth muscle contraction and blood pressure.

Pulmonary circuit The route of blood circulation leading from the heart to the lungs and back to the heart.

Receptor A protein on or in a cell that is activated by a specific stimulus; sensory receptors are examples.

Recessive gene A version of a gene that can be masked by a different, dominant version of the same gene. A trait governed by a recessive gene usually is seen only when a person inherits the gene from both parents.

Reflex An automatic response to a stimulus.

Retina The light-sensitive nerve tissue in the eye.

RNA Ribonucleic acid; in human cells, genetic instructions in DNA are converted into RNA, which then directly guides the cell's response.

Rods Photoreceptors in the retina that respond mainly to dim light; rods contribute to night vision.

Sclera The white, outer coating of the eyeball.

Secondary sexual characteristic A trait such as beard growth or breast development that is associated with maleness or femaleness but is not directly involved in reproduction.

Self marker Molecules on body cells that identify those cells as a natural part of the body.

Semicircular canals A set of three, fluid-filled canals in the inner ear that provide the sense of balance.

Semilunar valve A half-moon-shaped heart valve that prevents blood pumped by a heart ventricle from flowing backward as it exits.

Sensation The conscious awareness of a stimulus.

Sensory receptor A sensory cell or structure that can detect a particular stimulus, such as light, pressure or a chemical.

Sex chromosome An X or Y chromosome, which carries genes that determine an embryo's gender.

Sinoatrial (SA) node A cluster of cells in the right atrium of the heart that initiates and helps regulate the heartbeat.

Skeletal muscle The muscle tissue that attaches to bones and generally is under voluntary control.

Smooth muscle The muscle tissue in the walls of internal organs; smooth muscle is not usually under conscious control.

Somatic senses The "body" senses of touch, pressure, temperature and pain.

Somatosensory cortex The brain region that processes signals coming from the skin, muscles and joints.

Special senses The senses of smell, taste, vision and hearing.

Sphincter A ring of muscle that can contract and relax to control passage of substances through an opening.

Spleen Organ that filters old blood cells and debris from the blood, stores excess blood cells, and contains infection-fighting white blood cells.

Stem cell An unspecialized cell that can divide repeatedly and give rise to descendants that develop into specialized cells.

Sympathetic nerves Part of the autonomic nervous system: nerves that operate to spur physiological processes such as heart rate in response to stress or excitement.

Synapse The gap between nerve cells across which nerve impulses are transmitted (as by neurotransmitters).

Synovial joint A joint with a fluid-filled cavity between the two linked bones. This type of joint allows for the most movement.

Systemic circuit The cardiovascular circuit running between the heart and body tissues.

Systole In the cardiac cycle, the period in which the heart chambers contract and pump blood out of the heart.

Taste buds Chemoreceptors on the tongue and palate that provide the sense of taste.

T cell A type of lymphocyte (white blood cell) that becomes specialized in the thymus and functions in specific immune responses.

Tendon A strong, fibrous band of connective tissue that attaches skeletal muscle to bone.

Thermoreceptor A sensory receptor that responds to changes in temperature.

Trachea The windpipe.

Triglyceride The most common form of fat in the blood. Triglycerides are the major form of stored fat in the body and are an important source of energy for metabolism.

Urea A waste product of protein metabolism that is excreted by the kidneys in urine.

Vein A blood vessel that carries deoxygenated blood from body tissues back to the heart.

Ventricle Either of the two lower chambers of the heart, both of which pump blood.

Venule A small blood vessel that connects capillaries to larger veins.

Villus Any of the tiny, fingerlike structures on an epithelial surface, such as the mucous membrane lining the intestine. Villi absorb substances such as nutrients.

Visual cortex The brain region that receives signals from the optic nerves.

Visual field The entire area visible to the eye at a given moment, including peripheral vision.

White matter Nerve tissue in spinal cord that is covered with myelin, which gives it a white appearance.

X chromosome A sex chromosome with genes that cause an embryo to develop into a female, if the embryo receives one X from each parent.

Y chromosome A sex chromosome with genes that cause an embryo to develop into a male.

Zygote The first cell of a new individual, produced when a sperm fertilizes an egg.

Index

A

abdominal cavity 56–7
Achilles tendon 34
achondroplasia 270
acromegaly 85
actin (muscle protein filaments) 36–7
activity, physical 42–3
    aging and 260–1
    blood acidity and 226
    body weight and 219
    breathing and 144
    circulation and 167
    endurance sports 149, 159
    energy use and 219
    heart muscle 159
    immunity and 183
    muscles and 39–41
    stretching before 34
    weight-bearing 42, 46
acupuncture 113
adolescents 93, 258–9 see also puberty
    brains wired for risk 79
adrenal gland 90–1
adrenaline (epinephrine) 85, 90–1
aerosol inhalers 145
aging 260–1
    eyes and 132
    muscles and 42
AIDS 188–9
albumin 147
aldosterone 90–1
allergic reactions 97, 176, 178–9, 188
altitude sickness 140–1
alveoli (air sacs) 139–42
amino acids 148, 206
amniocentesis 249
amputees 113
androgens 92–3
anemia 44–5
angioplasty 169
antibiotics 173
antibodies (immunoglobulins) 180, 182–3
    in allergic reactions 178
    in diagnosis 186
antigens 180, 186–7
antioxidants 197
appendices 212–13
appetite 214–15, 218
appetite, hormones and 202
arms 50–1
arrhythmias 159
arteries 155, 163–5
    blocked 168–9

arterioles 164–5
arthroscopes 35
artificial hearts and valves 157
artificial pacemakers 161
asexual reproduction 236–7
asthma 91
atherosclerosis 168
atria 156, 158–61
auditory nerves 108
autoimmune conditions 54, 178, 183, 188–9
autonomic nervous system 72–3
axons 66–71, 74–5, 108

B

B cells 178, 180, 182, 186–8
babies 258–9
    chewing 198
    newborn's skull 46
    premature 254
    senses 104, 115
    vaccination 180
back pain 54
bacteria 172–5, 212
    in defense against other bacteria 174
    defenses against 175, 177
    Pseudomonas 18
    reproduction of 236–7
    toxins, and muscle contractions 38, 41
    tuberculosis 172–3
balance (homeostasis) 73, 194–5, 226–7
balance (movement) 105, 122–3
baldness 269
ballet dancers 32
basophils 150, 176
bats, ultrasonic sound waves 118
bears, winter sleeps of 229
behavior, genes and 270
bicarbonate, in stomach 203
biceps 32, 36–7
bile 206–7, 208
bilirubin 148, 208
biological clock 99
birds 225, 229, 231, 241
birth 30, 195, 256–7
bitter taste 116
bladder 222
blindness 125, 132–3
blood 146–53, 226–7 see also cardiovascular system; red
    blood cells; white blood cells
    acidity 226–7
    bone marrow and 44–5

blood–brain barrier 77
blood clotting (hemostasis) 152–3
blood donations and transfusions 147, 149
blood pressure 158–9, 163–5, 168, 195
blood sugar control 87–9, 206, 214
blood types 149, 270
blood vessels 154–5, 162–9 see also cardiovascular system
    brain 77
    damage to 152–3
    in nerves 68
    skin healing and 26–7
body fat 206, 218–19, 267
    in hypodermis 23
body paint 28–9
body temperature 107, 228–31
    fever 151, 176
body weight 218–19
    exercise and 42–3
bodybuilding 43
bone marrow 44–5, 147, 150–1
bones 44–55 see also joints; skeletons
    fractures of 55
    ligaments and tendons and 34–5
    muscles and 33, 36–7
botulinum toxin injections 38
brain 64–7, 76–7
    blood supply to 162
    breathing and 144
    damage to 81
    disease 195
    gender differences 78–9
    infants 259
    internal bleeding 153
    motor system and 41
    sensory system and 107–11, 114, 126, 130–2
breast cancer 276
breastbone (sternum) 49
breastfeeding 82, 257
breathing (ventilation) 138–45, 254
    controls 144–5
    extreme environments 140–1
    newborn 256
butterfly eggs 241

C

calcitonin 94
calcium 44, 46–7
    electrolyte balance 227
    in muscle contractions 36
    spurs in arthritic bones 54
    thyroid hormones and 94

cancer 276–7 *see also* specific cancers
    diagnosis 277
    killer cells 184–5
    oncogenes 274
    therapy 186–8, 279
capillaries 154, 163, 165–6
carbohydrates 196–7, 201–2
carbon dioxide 138–42, 144–5, 148
carbon monoxide 145
cardiac 156–7 *see also* heart
cardiac cycle 158–9
cardiac (heart) muscle 30–1, 156, 160–1
    genes in 267
cardiac stress tests 159
cardiovascular system 154–69 *see also* blood
    blood circulation 162–3
    disorders of 168–9
carpal tunnel syndrome 35
cartilage 46–7, 52–4
cataracts 132
cell division 264–5
    cancer 276–7
cell membranes 18–19
cells 18–19
    function of 19
    nutrients needed by 196–7
    RNA translation in 266
central nervous system (CNS) 62, 64–5
    natural pain suppressors 112
    sensory system and 107, 110–11
cerebellum 76–7
cerebral cortex (cerebrum) 76–7, 79–81
cerebrospinal fluid 74–6
cerebrovascular accident (stroke) 69, 153, 168
cervical cancer 279
cervical nerves 68
chameleons 201
chemoreceptors 107, 114–17, 144
chemotherapy 188
chest (thoracic) cavity 56–7, 142–3
chewing 198
chick embryos 251
childbirth (labor) 256–7
    uterus muscle action in 30, 195
children 258–9 *see also* adolescents; babies
    bones 45–6
    heart murmurs 158
cholesterol 168–9, 206–7
chromosomes 245, 247, 264–5, 267 *see also* DNA
    mapping the human genome 274–5
    telomeres 261
cilia 19, 138–9
circulation of blood 162–3 *see also* cardiovascular system
cirrhosis 206

clams, mantles of 20
climate, adaptation to 229–31
clones 236, 250
cochlea 119–21
collagen 19
    aging 261
    in bones 44
    in scars 27
    in skin 22–3
    in tendons and ligaments 34, 37
collarbone (clavicle) 49
colon (large intestine) 198, 212–13
color blindness 133
color perception 125, 128
colorectal cancer 212–13
complement proteins 175
connective tissue 19
    nerves 68–9
    tendons and ligaments 34–5
consciousness 78
corals 245
cornea 126, 133
coronary arteries 162, 169
cortisol 86, 90–1
cosmetic surgery 29, 38
cranial cavity 56–7
cranial nerves 62
cystic fibrosis 209, 272, 275
cytoplasm 18, 221

D

defensive strategies (immune system) 174–85 *see also* autoimmune conditions; T cells; white blood cells
    in AIDS 188–9
    lymph system and 170–1
    medical immunology 186–9
    self and non–self 178, 188
    T cells 184–5
defensive strategies (lymphatic system) 170–1
    lymph node cancer 188
dendrites 66–7, 70–1
dermatomes 68
dermis 20, 22–3
development
    embryonic 250–3
    infant 258–9
diabetes 88–9, 188, 222
dialysis 227
diaphragm 142–4
diet 196–7, 212–13
    digestive system 198–201

hormones and 96
    storage and flavoring 116
digestive system 198–203, 208–9
diseases and disorders
    cardiovascular system 168–9
    eyes 132–3
    gene-based 264–5, 268, 270–1, 274–5, 278–9
    inflammation in 176
disorders and diseases
    cardiovascular system 168–9
    eyes 132–3
    gene-based 264–5, 268, 270–1, 274–5, 278–9
    inflammation in 176
diving 141, 144
DNA (deooxyribonucleic acid) 262–7 *see also* chromosomes
    mapping the human genome 274–5
    mutations 272–3, 276
DNA fingerprints 148, 267
dogs, body heat 231
Douglas, Kirk and Michael 270
Down syndrome 247
drinking 224–5
drugs
    affecting neurotransmitters 67
    anti-inflammatory 176
    embryo development 252
    genetic profiling to select 267
    narcotic 112
    performance-enhancing 43
du Pré, Jacqueline 187
dust mites 179
dwarfism 86

E

ear
    balance and 122–3
    hair cells and hearing 118–21
earlobes, attached and detached 271
eating and food 197–203, 208–9, 212–13
    digestive system 198–201
    hormones and 96
    storage and flavoring 116
eggs (oocytes) 240–1, 246–7, 264
ejaculation 244–5
elastin 19, 22, 34
electrolytes (salts) 220, 224–5, 227
elimination 212–13
embryos 248–55
    drugs and infections and 252
emotions 78
emphysema 142

endocrine system (hormones) 82–99
    childbirth 256
    erythropoietin 148
    hunger and satiety and 202, 214–15, 218
    reproductive system 238–40, 242, 248
endometrium 238–9, 248
endorphins 112–13
enkephalins 112
epidermis 20–5
epinephrine 85, 90–1
Escherichia coli 212
esophagus 200
estrogens 92–3, 239–40, 248–9, 256
    hormone replacement therapy 85
exercise 42–3
    aging and 260–1
    blood acidity and 226
    body weight and 219
    breathing and 144
    circulation and 167
    endurance sports 149, 159
    energy use and 219
    heart muscle 159
    immunity and 183
    muscles and 39–41
    stretching before 34
    weight-bearing 42, 46
eyelids 111, 124, 251
eyes 124–33
    babies 258
    color 268
    nervous system control of 73
    problems with 132–3
    rods and cones 128–9

F

facial decoration 28–9
facial muscles 33
facial wrinkles 26–7, 38
family traits 267–9, 274 *see also* genetics
    dominant and recessive genes 268, 270–1
fat (adipose) tissue 206, 218–19, 267
    in hypodermis 23
fats, digestion of 207
feces 212–13
feedback loops 194–5
feelings 78
feet 51
femur (thighbone) 44, 50–1
fertilization 246–7, 264

fetus 252–5
    bones in 48
    childbirth 256–7
fever and pyrogens 151, 176
fiber in diet 212–13
fibrin 152–3
fibrinogen 147, 153
fight-or-flight response 85
    parasympathetic nerves 72–3
fingernails 24–5
fingerprints 269
fingers 251
flies, taste with their feet 117
fontanels 46
food 196–7, 212–13
    digestive system 198–201
    hormones and 96
    storage and flavoring 116
forensic science 50
fractured bones 55
freckles 269
free radicals 197
frogs, breathing 141
fungal infections 172

G

gallbladder 206–9
gas exchange 140–2
gastric bypass surgery 214–15
gastrointestinal tract 198–201
gene-based abnormalities 264–5, 268, 270–1,
    274–5, 278–9
gene isolation 262
gene mutations 272–3, 276
gene therapy 278–9
genetic modification 278–9
    piglets for transplantable organs 179
genetic profiling, for drug targeting 267
genetics 247, 262–79, 266–7 *see also* DNA; family traits
    aging and 260–1
    behavior and 270
    dominant and recessive genes 268, 270–1
    embryo development 250–3
    gene mutations 272–3, 276
    mapping the human genome 274–5
    pregnancy testing 249
    research 279
    single-gene and polygenic traits 268–9
GI tract 198–201
gigantism 86
glucagon 88–9

glucose control 87–9, 206, 214
glycogen 206
goiter 95
goosebumps 23
Graves' disease 95
growth, hormones and 82, 84–6
guano 221
gustation 116–17
Guthrie, Woody 271

H

hair
    derived from skin 24–5
    genetic information from 275
hair cells 120–3
hair follicles 23, 25, 110
hands 51
headache 112
hearing 105, 118–21
heart 156–7 *see also* cardiac; cardiovascular system
    blood supply to 162
    hole in the 254
    hormone production by 96
    murmurs 158
    pacemaker 160–1
    valves 156–7
heart attack 168–9
heartbeat 158–61
heat exhaustion 230
height 268
Helicobacter pylori 172, 203
hemoglobin 148–9
hemophilia 152, 272
hepatitis 206
high blood pressure (hypertension) 158, 168
    kidney failure and 222
hip replacements 55
histamine 150–1, 176–8
HIV (human immunodeficiency virus) 188–9
Hodgkin's disease 188
hole in the heart 254
homeostasis (steady state balance) 73, 194–5, 226–7
homeotherms 228–31
hormone replacement therapy 85
hormones 82–99
    childbirth 256
    erythropoietin 148
    hunger and satiety and 202, 214–15, 218
    reproductive system 238–40, 242, 248
human genome 262, 274–5
hunger 202, 214–15, 218

Huntington's disease 271, 275
hydrochloric acid, in stomach 203
hypertension 158, 168
  kidney failure and 222
hyperventilation 144–5
hypothalamus 86–7, 99, 230
hypothermia 230

I

ileum 210
immune system 174–85
  in AIDS 188–9
  autoimmune conditions 54, 178, 183, 188–9
  lymph system and 170–1
  medical immunology 186–9
  self and non-self 178, 188
  T cells 178, 180, 184–5, 188–9, 278–9
  white blood cells 146–7, 150–1, 170, 174, 176–7, 176–9
immunisation 180–1
immunodeficiency 188–9
immunoglobulins (antibodies) 180, 182–3
  in allergic reactions 178
  in diagnosis 186
in vitro fertilization (IVF) 247
infants (babies) 258–9
  chewing 198
  newborn's skull 46
  premature 254
  senses 104, 115
  vaccination 180
infections 172–3
  embryo development 252
  immune system 174–81, 186–9
  surface barriers to 175
  white blood cells and 151
inflammatory responses 151, 174, 176–7
  sunburn 113
inheritance and genetics 247, 262–79, 266–7 see also DNA;
    family traits
  aging and 260–1
  behavior and 270
  dominant and recessive genes 268, 270–1
  embryo development 250–3
  gene mutations 272–3, 276
  mapping the human genome 274–5
  pregnancy testing 249
  research 279
  single-gene and polygenic traits 268–9
insects 105, 125, 154, 247
insulin 87–9, 214
interferons 184–7

interleukins 184–6
internal environment 194–5, 220–1
interneurons 66, 70–1
intestines 198, 212–13
  blood vessels 163
  cancer 277
iodine, goiters and 95
iris 126–7
islets of Langerhans 82, 88–9

J

jejunum 210
Joan of Arc 93
joints 35, 50, 52–4
  muscles and ligaments and 32, 34, 38
juggling, motor neurons and 63

K

kangaroo rats 225
keratin 22–5
keyboarding 35
kicking 53
kidneys 222–7
killer cells 184–5
knee-jerk reflex 71
knees 34, 52

L

labor 256–7
  uterus muscle action in 30, 195
ladybug sex 247
language 80–1, 143
large intestine (colon) 198, 212–13
legs 44, 50–1
lemurs 50
lens (eye) 126–7, 132
leptin 214–15
ligaments 34–5, 52–4
light 98–9, 107, 124–8
limbic system 78
limbs 48–51, 110–11
lipids 196–7
lips 110
liver 206–7, 267
low blood pressure 158
lumbar nerves 62, 68
lung cancer 277

lungs 139–45, 254
  blood transport to 154, 156, 162–3
  newborn 256
lymphatic system 170–1
  cancer of lymph nodes 188
lymphocytes 150–1, 170, 178, 180
lysosymes 19

M

macrophages 150–1, 176–7, 180, 185
make-up 29
mammography 276
Manson, Marilyn 29
Maori, tattooing 29
Marfan's syndrome 273
mast cells 151, 176–7
mechanoreceptors 107–8, 110–11, 120–1
meiosis 265
melanin 22, 268–9
melanoma 27, 186
melatonin 98–9
membranes 18–19, 56–7
memory 78–9
memory cells 180
meninges 56, 74–5
menstrual cycle 238–9
metabolism
  body temperature and 228–9
  respiratory system and 138
  thyroid hormones and 95
microorganisms 172–4 see also infections
mineralocorticoids 90–1
minerals, in bones 44–7
mitochondria 38, 42, 138
mitosis 265
monoclonal antibodies 186–7
monocytes 150–1
moth hearts 154
motor cortex 77
motor neurons 38–41, 66, 68–71 see also movement
  juggling 63
  sensory system and 109, 111
mountain climbing 140
mountain sheep horns 25
mouth 56, 200–1
  taste 116–17
movement 30–1, 110–11, 122–3
multiple births 240, 251
multiple sclerosis 187
muscle contractions 30–3, 36–41
muscle spindles 110–11

muscle tension 38–40
muscles 30–43
    atrophy 42
    bacterial toxins and 38
    bone movement caused by 36–9
    breathing 142–3
    cardiac 30–1, 156, 160–1, 267
    fibers of 30–1, 38–40
    healthy 42–3
    motor units in 40–1
    pairs of 32
    tendons and 34–5
musical instruments 118
mutations 272–3, 276
myelin 66–7, 69
myofibrils 36–7, 40
myoglobin 38, 42
myopia 133
myosin (muscle protein filaments) 36–7

N
nails 24–5
Neanderthals 228
nearsightedness (myopia) 133
nervous system 62–81
neurons (nerve cells) 18, 62, 65–9, 106–10 see also motor
        neurons
    damage and repair 69, 75
    genes in 267
    nerve tracts 74–5
    neural pathways 70–1
neurotransmitters 67, 70, 85, 112
neutrophils 176–7
nociceptors 106, 112–13
norepinephrine 85, 90
nose 29, 114–15
nuclei, cell 18–19
nucleic acids 196–7, 262–3
nucleotides 262–3, 274–5
nutrients 196–7
    absorption in intestine 210
    transport in blood 154, 163

O
obesity 88, 214–15, 218–19
octopus 20, 125
olfaction (sense of smell) 114–15
oncogenes 274
oocytes (eggs) 240–1, 246–7, 264

optic nerves 108, 126, 128, 130
optical illusions 130–1
organelles 19
organs
    in body cavities 56–7
    embryo development 250–1
    endocrine action on 82
    fetal development 254
    infant development 258
    nervous system control of 72–3
    transplants 179, 184
osmoreceptors 106–7
osteoarthritis 54
osteoporosis 46–7, 54, 85
otoliths 122–3
ovarian cancer 187
ovaries 92, 238–41
overweight 214–15, 218–19
ovulation 240–1
oxygen 138–42, 144–5, 148–9
oxygen debt, in exercise 41, 43
oxytocin 256

P
pacemakers (sinoatrial nodes) 160–1
pain 106, 108, 112–13
painkillers 97
pancreas 82, 88–9, 208–9
paralysis 38, 41
parasites 172
parasympathetic nerves 72–3
parathyroid glands 94–5
pathogens (infections) 172–3
    embryo development 252
    immune system 174–81, 186–9
    surface barriers to 175
    white blood cells and 151
peacock feathers 25
pelvis 50–1, 56–7
penguin guano 221
penis 242–3
perception, defined 106
performance-enhancing drugs 43
peripheral nervous system 68–9, 72–3
    gas-signaling chemicals in 85
peripheral vision 128
peristalsis 202
pesticides 98–9
phagocytes 151
phantom pain 113
pheromones 98

phosphorus, in bones 44, 46
photoreceptors 107, 124–31
physical exercise 42–3
    aging and 260–1
    blood acidity and 226
    body weight and 219
    breathing and 144
    circulation and 167
    endurance sports 149, 159
    energy use and 219
    heart muscle 159
    immunity and 183
    muscles and 39–41
    stretching before 34
    weight–bearing 42, 46
pineal gland 98–9
pituitary gland 82, 84–7
placenta 248–9, 252, 256
plants
    light sensing by 124
    reproduction 236–7
plasma 146–7
platelets 146, 152–3
pollen 178
polydactyly 272
pores 23
position (balance) 105, 122–3
potassium 227
Pré, Jacqueline du 187
pregnancy 93, 248–9
    ectopic 240
    embryo development 250–3
    smoking during 255
    testing 186–7, 249
    ultrasound in 253
premature babies 254
prenatal life 48, 248–55
pressure, sensation of 106–8, 110–11
prions 172
progesterone 92, 239, 248–9
prostaglandins 96–7, 176
prostate gland 242–3
proteins 196–7, 202
    plasma proteins 147
    synthesis 206
Pseudomonas bacterium 18
puberty 99, 238–9, 244, 258 see also adolescents
pulmonary circuit (circulation) 162–3, 166
pulse rate 165
pupils (eye) 126

R

receptors 70, 106–11, 124–31
    ear 120–1
    for hormones 84
    light 111
    olfactory 114–15
    sensory 106–11
    skin 110–11
    taste 116–17
red blood cells (erythrocytes) 44, 146–9, 221
    clotting 152
    sickle cell anemia 148, 272–3
Reeve, Christopher 39
reflexes 70–1
reproductive hormones 92–3, 248
reproductive system 236–7
    female 238–9
    fertilization 246–7
    male 242–5
reproductive technologies 247
reptiles 65, 229
respiratory system 138–43
    exercise and 183
    extreme environments 140–1
    respiratory control centers 144–5
rest and exercise 43
retina 126–30, 132
rheumatoid arthritis 54, 183, 188
rhinoplasty 29
rib cage 49, 142–4
rickets 95
RNA (ribonucleic acid) 188, 266
rubella virus in pregnancy 253

S

sacral nerves 62, 68
saliva 200–1
salivary glands 208
salts (electrolytes) 220, 224–5, 227
sarcomeres 36–7
SARS virus 182
scapula (shoulder blade) 49
scars 29
sciatic nerves 62, 69
sea stars 63, 65
seasonal affective disorder (SAD) 99
secretory glands 56
semen 243
semicircular canals 122–3
senses 62, 104–33
    babies 104

sensory adaptation 108
sensory deprivation 104
sensory hairs 120–3
sensory neurons 66–8, 70–1, 106–10
sex cells
    eggs (oocytes) 240–1, 246–7, 264
    sperm 244–7, 264
sex chromosomes 264, 267
sex hormones 92–3
    DHEA 90
    puberty 258
sexual reproduction 236–7
    why have it? 236
sharks 20, 243
shoulder blade (scapula) 49
sickle cell anemia 148, 272–3
sight (vision) 105, 124–33
    color blindness 133
    color perception 125, 128
    depth perception 130
    interpreting visual signals 130–1
    light rays and 126–7
    problems with 132–3
sign language 80
signaling chemicals (endocrine system) 82–99
    childbirth 256
    erythropoietin 148
    hunger and satiety and 202, 214–15, 218
    reproductive system 238–40, 242, 248
signaling chemicals (neurotransmitters) 67, 70, 85, 112
skeletal muscles 30–43, 110–11
skeletons 48–9
    fetal 255
    forensic science and 50
    injuries or disorders of 54–5
    limbs 48–51
    skin and 20
skin 20–9
    appearance and decoration of 28–9
    body heat and 230–1
    damage and healing 22, 26–7
    hair and nails derived from 24–5
    lower layers of 22–3
    muscles attached to 33
    as a protective barrier 20–1, 174
    receptors in 107, 109–11
    sunburn 26, 113
skin cancer 26–7, 186
skull 46, 48–50
sleeping 98–9, 256
small intestine 210–11
smallpox 181
smell 105, 114–15

smoking 168, 255, 277
sneezing 178
sniffing 115
sodium 224–5, 227
somatic nervous system 72–3, 106, 110–11
sound (hearing) 105, 118–21
speech 80–1, 143
sperm 244–7
    sex chromosomes 264
spiders, waste removal 223
spinal cavity 56–7
spinal cord 64–5, 74–5
    artificial disks 75
    damage and repair 75
    messages to muscles from 41
spinal injuries 39
spinal nerves 62, 68
spine (vertebral column) 48–9, 52–3
    lower back pain 54
spleen 148, 170–1
sponges, digestive systems 198
sprained ligaments 34
steady state (homeostasis) 73, 194–5, 226–7
stem cells 147–8, 150–1
    regeneration of nerve cells from 75
stents 169
steroid hormones 85, 91
stimuli, responses to 106–9
stomach 172, 202–3
strawberry plants 236–7
stress 72–3, 90–1
stretching before exercise 34
stroke (cerebrovascular accident) 69, 153, 168
sucking 255
sugar control 87–9, 206, 214
sun bear claws 25
sunburn 26, 113
surgery 35
survival, senses and 104, 114
swallowing 200–1
sweating 230
sympathetic nerves 72–3, 91
synapses 67, 70–1
synovial joints 52–3
systemic circuit (circulation) 162–3, 166
systolic pressure 158

T

T cells 178, 180, 184–5, 278–9
    in AIDS 188–9
tails 251–2

taste 116–17
  chemoreceptors 107
  odor and 114, 116
taste buds 116–17
tattooing 28–9
tears, as defensive systems 175
teeth 200
  joints and 52–3
temperature 107, 228–31
  fever 151, 176
tendons 34–5, 110–11
testes 92, 242, 244
testosterone 84, 92, 242, 244–5
  baldness and 269
tetanus 41
thermoreceptors 107, 230 see also temperature
thighbone (femur) 44, 49, 50-51
thirst 224
thoracic nerves 62, 68
thrombin 153
thymus 96, 170–1, 184
thyroid gland 94–5
tibia (shinbone) 48, 51
tissues 18–19
tobacco smoking 168, 277
  in pregnancy 255
toenails 24–5
toes 251
  polydactyly 272
tongue 110, 117, 200–1
tonsils 170–1
tooth 200
  joints and 52–3
touch 107, 109–11
  babies and 104
transfusions 149
transplants 179, 184
triceps 32
triglycerides 206
Troyer, Verne 270
tuberculosis bacteria 172–3
tumors (cancer) 276–7 see also specific cancers
  diagnosis 277
  killer cells 184–5
  oncogenes 274
  therapy 186–8, 279
tuna 225, 229
turtles 25, 241
twins and multiple births 240, 251
typing 35

U
ulcers 172, 203
ultrasound in pregnancy 253
ultraviolet radiation, skin and 26
umami (savory taste) 116
umbilical cord 252, 256
urea 221
urethra 222
urinary system 222–3
urine 221, 223–5
uterus 30, 238–9, 256–7

V
vaccination 180–1
  against cancer 186
valves in veins 166–7
varicose veins 167
vascular system 154–69 see also blood
  blood circulation 162–3
  disorders of 168–9
veins 155, 163, 166–7
ventilation (breathing) 138–45, 254
  controls 144–5
  extreme environments 140–1
  newborn 256
ventricles 156, 158–61
venules 166
vertebrae 48–9, 74–5
vertebral column (spine) 48–9, 52–3
  lower back pain 54
Victoria, Queen 272
video technology for blind people 125
villi 210–11
viruses 172–4, 186
  replication 188
vision 105, 124–33
  color blindness 133
  color perception 125, 128
  depth perception 130
  interpreting visual signals 130–1
  light rays and 126–7
  problems with 132–3
vision correction 133
vitamin D 96, 99
vitamin manufacture 212
vocal cords 143
voluntary control over muscles 32
vomit (chyme) 202

W
walking 259
waste removal
  feces 212–13
  urine 221–5
water in body functions 220–1, 224–5
  contaminated 227
weight 218–19
  exercise and 42–3
whales 145
white blood cells (leukocytes) 146–7, 150–1, 174
  immune system 178–9
  inflammatory responses 176–7
  lymphatic system 170
wolves 218
womb (uterus) 30, 238–9, 256–7
work, done by muscles 38–9
worms 173
wounds, healing 26–7, 264–5
wrinkles 26–7, 38
  aging 260–1

Credits and acknowledgments

PHOTOGRAPHS

t=top; l=left; r=right; tl=top left; tcl=top center left; tc=top center; tcr=top center right; tr=top right; cl=center left; c=center; cr=center right; b=bottom; bl=bottom left; bcl=bottom center left; bc=bottom center; bcr=bottom center right; br=bottom right

AAP = Australian Associated Press; ADL = Ad-Libitum; APL = Australian Picture Library; APL/CBT = Australian Picture Library/Corbis ; APL/MP = Australian Picture Library/Minden Pictures; AUS = Auscape International; AUS/BSIP = Auscape International/bsip.com; COR = Corel Corp.; DS = Digital Stock; GI = Getty Images; iS = istockphoto.com; PA = PhotoAlto; PD = Photodisc; PL = photolibrary.com; WA = Wildlife Art Ltd

Front cover bl APL/CBT br, tc GI
Spine GI

1c GI 2c PL 4r GI 6–7c GI 8–9c PL 10c PL l, r GI 11c GI l, r PL 12c PL 14–15c GI 16c PL l, r GI 17c, l PL 18tr PL 19b, c, tr PL 20bl, tr PL 21bc, br, tr PL tl GI 22bl PL r GI 23tl, tr PL 24bl PL br, tr GI 25bl, tr PL br APL/CBT tl Ad-Libitum 26bl GI 27tl PL tr GI 28bl, cr APL/CBT 29bc, tl APL/CBT cr PL 30tr PL 31tl, tr PL 32cr GI cr APL/CBT 33br GI 34bl GI tr APL/CBT 35bc APL/CBT br, tr PL 36tr PL 37tl, tr PL 38bl APL/CBT tr GI 39bc, bl APL/CBT tr GI 40b APL/CBT tl GI tr PL 42tl, tr GI 43b GI tr PL 45bl, tl PL 46bl, br PL 47bc PL 48bl GI c PL 50bl, tr PL 53bl, tl PL 54b GI 55b GI 57bl APL/CBT tr GI 58–59c PL 60c, cr, l PL r GI 61c APL/CBT 63bl, tl iS cl APL/CBT cr APL tr GI 65br PL 67bl PL cr GI 68bl GI 69tl, tr PL 70bc APL/CBT bl GI br AUS 71bc PL 73br, tl, tr GI 74bl PL 75tl PL 76br PL 77l PL 78bl PL tr GI 79bl APL/CBT 80bl iS br, tr PL 81br GI tr PL 82bl, br PL 84bl iS br PL 85tc, tl PL 86bl GI tr APL/CBT 87bl, r, tl iS br APL/MP br PL tr DS 88bl GI 89tl, tr PL 90r iS 91tl GI tr PL 92br GI 93br PL tl, r APL/CBT 94bl PL 95tc APL/CBT tl PL 96b GI 97bc PL br, tl GI 98bl, r PL 99br, tl PL tr APL/CBT 100–101c GI 102cl, l GI r PL 103c PL 104r GI 105bl, cl, tl, tr iS br PL 106bl, br GI tr PL 107br APL/CBT tl PL tr GI 108br GI 109bl, tr GI br, tl iS 110tr GI 111tl, tr GI 112b GI 113bl iS br GI 115bl, br, tr GI tr iS 116br Michael Freeman tr PL 117br GI 118bl PL br iS tr APL/CBT 119tr PL 120tr GI 121cr, tr iS 122r GI 123br GI tl PL 124b APL/CBT tr GI 125bl, tr PL tl GI 127b, tc, tr PL 128bl, cl, tr PL 129b PL tr iS 131b APL/CBT 132b, t PL 133br, tl PL tr APL/CBT 134c PL 136c, l, r PL 137c AAP l PL 138bl, cr PL tr GI 140r PL 141br GI tr iS 142bl, tr PL 144r iS 145br PL tl iS tr DS 146bl GI br APL/CBT tr PL 147tr PL 148bl, tr PL 149br GI tr PL 150bl GI br, tr PL 151bl, br, tr PL 152b PL 153tl, tr PL 154bl iS tr PL 156tr PL 157bl, cr, tr PL 158tr PL 159bl, tr GI 160bl PL 161br, tl PL 162b PL tr APL/CBT 163br iS 165br APL/CBT l PL 166bl PL 167cr PL l GI tr iS 168b GI 169br, tl, tr PL 170bl PL 172bl, br PL 173bl, br, tl PL 174r GI 175br PL tc AAP 176bc, bl PL tr GI 177tr GI 178br PL tc APL/CBT 179b PL tl APL/CBT 180b APL/CBT 181bl, tr APL/CBT 182bl PL tr APL/CBT 183bl GI tr PL 184b PL 185bl, tr PL 186r PL 187bl, br PL tr APL/CBT 188tr GI 190–191c GI 192c PL l, r GI 193c PL l GI 194r PL 195bl, tl PL 196b GI tr iS 197br PL 198bc, tr PL bl iS 200tr GI 201bl GI tl iS 203bl iS br PL tl GI 204tr iS 205br PL tl APL/CBT 207b PL t GI 208br, tr PL 209br, tr PL 210br, tr PL 211r GI tl PL 212r GI 213tl, tr PL 214br AAP tr APL/CBT 215r PL 216tr iS 217br iS t APL/CBT 218br GI l PL 219bl GI r iS 220b iS 221br, tr GI l PL 223br, tl PL tr iS 224bl iS r PL 225cr PL tl COR tr iS 226br, tr PL 227bl AAP tr PL 228r PL tr AUS/BSIP 229br, tc iS tl, tr PL 230br GI tr PL 231bl PA br GI tl PL 232–233c PL 234cl, cr, l PL 235c, l PL 236r GI 237br GI tl iS tr PL 238tr PL 239tr PL 240bl PL tr AAP 241br iS cr, tr APL/MP 242br, tr PL 243tr APL/CBT 244br, tr PL 245r, tl PL 246r PL 247cr PL tr APL/CBT 248tr PL 249cr PL 250b, tr GI tl PL 251br iS tr GI 252r GI 253bl GI br PL 254bl PL 255bc, br, t PL 256bl iS r GI 257tl AUS/BSIP 258tr PL 259r iS tl GI 260r APL/CBT 261br, cl PL t APL/CBT 262bl, tr PL 263c, tl PL 264cr PL 265br PD tr GI 266b GI tr PL 267bl ADL tc, tl PL tr GI 268bl GI 269b GI tc, tl iS tr APL/CBT 270bl AAP br APL/CBT 271bl, tl APL/CBT br, tr GI 272b PL 273b APL/CBT tl PL 274r GI 275br, tc PL 276bl, r PL 277br, t PL 278b PL tr APL/CBT 279br, tl PL 280–281c PL 286bl, cl, tl, tr iS 287bl, br, cl, cr, tl, tr iS 290tcl, tcr, tl, tr iS 291tc, tcl, tcr, tl, tr iS

ILLUSTRATIONS

Leonello Calvetti 258b, 83c
Leonello Calvetti/André Martin (photo) 48cr, 62r, 139c, 154c, 171c, 199c
Andrew Davies/Creative Communication 62bl, 73bc, 89b, 130tr, 131tcl tl tr, 163tr, 195r, 208bl, 225bl, 226bl, 239br, 241bl, 244bl, 274bl
Christer Eriksson 159t
Gino Hasler 36b
Christine Shafner/KE Sweeney Illustration 32bc
Kate Sweeney/KE Sweeney Illustration 30b
Moonrunner Design 33cl
Peter Bull Art Studio front cover tr, 18b, 23b, 27br, 35bl, 40br, 44c, 45tr, 47t, 51br l, 52b, 53r, 55t, 56bl, 57bl, 65bl t, 66b tr, 67tl, 68r, 69br, 70c, 71tr, 72r, 74r, 75br, 76tr, 77br, 79r, 85br, 87bl t, 91br, 92bl, 94tr, 99bc, 108bl, 110bl, 111bc bl, 113tl, 114bl tr, 117bl, 120bl, 123bl tr, 126b, 127tl, 130bl, 133bl, 141bl cl, 142br, 144bl, 149bl, 151tl, 153b, 156bl br, 158b, 160tr, 163bl, 164b, 166r, 170tr, 172tr, 177b, 185tl, 187tl, 188bl, 200bc, 201r, 202br c, 204bl, 206bl, 210bl, 213bl, 214bl, 222r, 223bc, 234tr, 238bl, 242bl, 246bl, 248b, 249br, 253t, 254r, 265bl
photolibrary.com back cover ctl, 95br 189c
John Richards 228bl

ACKNOWLEDGMENTS

The publisher wishes to thank Dr Linda Calabresi for her assistance with the Factfile, and Glenda Browne for the Index.

page 283 Major glands and hormones: Adapted from table 47.1 (Principal endocrine glands and their hormones) in Raven/Johnson/Losos/Singer, *Biology*, 7th edn, McGraw Hill.

page 284–85 Vitamins and minerals: adapted from *Human Biology* (with CD-ROM and Info Trac) 6th edition by STARR/MCMILLAN. © 2005. Reprinted with permission of Brooks/Cole, a division of Thomson Learning: www.thomsonrights.com. Fax 800 730-2215.

page 288–89 Genetic disorders: adapted from *Human Heredity, Principles and Issues* (with Human GeneticsNow/InfoTrac) 7th edition by Cummings. © 2006. Reprinted with permission of Brooks/Cole, a division of Thomson Learning: www.thomsonrights.com. Fax 800 730-2215.

Captions

page 1 A newborn baby presents the world with one of nature's most remarkable feats of biological engineering—the human body.

page 2 An elegant study of the geometry of limb movements, revealing the Italian artist Leonardo da Vinci's life-long fascination with the workings of the human body.

pages 4–5 Great manual dexterity is a hallmark of the human species. A hand's many small articulating bones, with their associated muscles and nerves, underlie this feature.

pages 6–7 Although diverse in their outward features, in terms of basic body functioning all humans are very much alike—the result of having the same basic genetic make-up.

pages 8–9 Sense organs like these taste buds on the tongue supply information to the brain that allows us to perceive and respond to the world in which we live.

pages 12–13 A developing fetus embodies the gene-guided biological journey each of us makes from the moment of conception to ripe old age.

pages 280–81 Human skin at the microscopic level—a landscape that invites us to explore the body's fascinating architecture and the functioning of its parts.

the human body

A VISUAL GUIDE

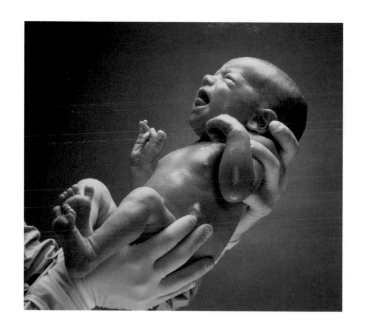

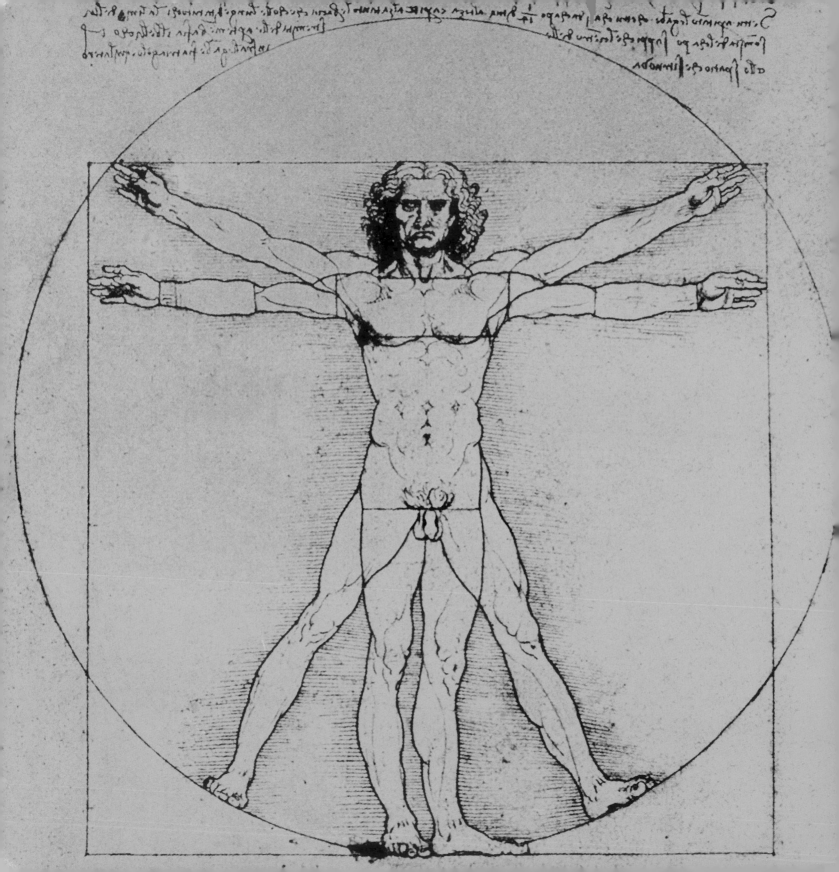